AF796138

Advances in Applied Neurological Sciences 4

Editors
R. J. Joynt, Rochester, USA
A. Weindl, Munich, FRG

Clinical Aspects of Sensory Motor Integration

Edited by
A. Struppler and A. Weindl

With 144 Figures

Springer-Verlag
Berlin Heidelberg New York
London Paris Tokyo

Professor Dr. ALBRECHT STRUPPLER
Privatdozent Dr. ADOLF WEINDL
Neurologische Klinik und Poliklinik
der Technischen Universität München
Möhlstr. 28
D-8000 München 80, FRG

ISBN-13:978-3-642-71542-6 e-ISBN-13:978-3-642-71540-2
DOI: 10.1007/978-3-642-71540-2

Library of Congress Cataloging-in-Publication Data. Clinical aspects of sensory motor integration. (Advances in applied neurological sciences; 4) Includes index. 1. Sensorimotor integration. 2. Movement disorders. I. Struppler, A. (Albrecht), 1919– . II. Weindl, A. III. Series. [DNLM: 1. Nervous System – physiology. 2. Psychomotor Performance – physiology. W1 AD436AH v. 4./ WL 102 C6385]
QP454.C54 1987 612'.743 87-20762
ISBN-13:978-3-642-71542-6

This work is subject to copyright. All rights are reserved, whether the whole or part of the material is concerned, specifically the rights of translation, reprinting, reuse of illustrations, recitation, broadcasting, reproduction on microfilms or in other ways, and storage in data banks. Duplication of this publication or parts thereof is only permitted under the provisions of the German Copyright Law of September 9, 1965, in its version of June 24, 1985, and a copyright fee must always be paid. Violations fall under the prosecution act of the German Copyright Law.

© Springer-Verlag Berlin Heidelberg 1987
Softcover reprint of the hardcover 1st edition 1987

The use of registered names, trademarks, etc. in this publication does not imply, even in the absence of a specific statement, that such names are exempt from the relevant protective laws and regulations and therefore free for general use.

Product Liability: The publisher can give no guarantee for information about drug dosage and application thereof contained in this book. In every individual case the respective user must check its accuracy by consulting other pharmaceutical literature.

2125/3130-543210

Dedicated to the memory of
WOLFGANG PRECHT *and* RUDOLF HASSLER,
eminent pioneers in brain research related
to sensory motor integration

Preface

The ability to use tools skillfully is generally regarded as one of the major achievements in the evolutionary development of the human nervous system. It is possible for controlled movements of muscles to be executed only if sensory information is integrated into complex neural circuits at various hierarchical levels.

The chapters in this volume deal with basic and clinical aspects of integrative processing of sensory and motor activities. New findings emphasize the important influence of somatosensory activity such as tactile, proprioceptive, noxious cutaneous, and articular input on motor output. Furthermore, recordings of evoked potentials as well as unit recordings indicate that sensory and cortical activities are highly interrelated.

Control of muscles by motoneurons is exerted both electrically and chemically. Disturbed muscle-motoneuron interaction is reflected in ultrastructural motoneuron morphology and may be of importance in the pathogenesis of motoneuron disease.

Long loop reflex testing under various pathological conditions provides insight into disturbed sensory motor circuitry in humans.

Electrophysiological recording as well as neurochemical and immunohistochemical studies elucidate the neural circuitry of basal ganglia and their neural connections, thus providing improved therapeutic concepts. The role of the thalamus and thalamocortical connections in sensory motor processing is of particular interest, because motor disturbances such as tremor or dystonia can be effectively relieved by stereotaxic interventions at the subthalamic or thalamic level.

Posture and movement are found to be highly integrated performances of afferent and efferent signals. The considerable potential of modality-specific regeneration of injured peripheral nerves is demonstrated by microneurographic analysis. Transplantation experiments in animals indicate that regeneration in the central nervous system may become a potential therapy of neurodegenerative disorders.

We wish to acknowledge the valuable support by the German Research Council and the International Federation of Societies for Electroencephalography and Clinical Neurophysiology. The assistance of Mrs. A. Kirsch in preparing the subject index is gratefully acknowledged.

The Editors

Contents

I. Somatosensory Activity Relevant for Motor Output

II. Central Motor Actions of Sensory Input

III. The Muscles and Their Neural Control

VI. Motor Functions of Basal Ganglia

VII. Thalamocortical Contributions to Sensory Motor Integration

VIII. Posture and Movement: Interactions and Disturbances

List of Contributors

You will find the addresses at the beginning of the respective contribution

I. Somatosensory Activity Relevant for Motor Output

Tactile Afferent Input Influencing Motor Coordination During Precision Grip

R. S. JOHANSSON [1] and G. WESTLING [1]

The remarkable capacity and versatility of the human hand in precise manipulatory tasks is undoubtedly dependent upon a number of neural factors. One such factor is the tactile sensory innervation of the glabrous skin area, i.e. the hairless skin of the volar aspect of the hand. Indeed, Mott and Sherrington [22], dealing with the motor effects of various patterns of dorsal root sections in *Macacus rhesus,* found that "afferent impulses, both from the skin and from muscles, especially the former, as related to the palm and sole, are necessary for the carrying out of 'highest level' movements." Likewise, Denny-Brown [5] wrote extensively on the capacity of tactile stimuli in eliciting prepatterned integrated hand movements in clinical and experimental material. In patients with frontal lobe lesions, automatic prehensile movements of at least two types could be distinguished: the "grasp reflex" and the more complex "instinctive grasp reaction". In contrast, during parietal lobe lesions, "tactile avoiding reactions" of different complexities were described. The pathological feature of these reactions was considered to be an inability to adequately suppress the first phase, i.e. the reactions appeared to be inappropriately triggered.

Several recent laboratory investigations have dealt with effects of anaesthesia and/or electrical stimulation of human digital nerves on muscle commands mainly influencing the fingers [2–4, 6–10, 19, 20]. A relatively uniform picture is presented: input from digital afferents appears to provide a net facilitatory effect particularly on flexor commands. For instance, Gandevia and McCloskey [7, 8] considered the sensory input arising from both index finger and thumb to assist the motor command signals descending to either the index finger flexors or the thumb flexors. Likewise, Garnett and Stephens [10] demonstrated that continuous electrical stimulation of cutaneous afferents of the index finger promotes the recruitment of high-threshold powerful units in human first dorsal interosseous muscle, whereas the recruitment of low-threshold units may be delayed. It was suggested that the cutaneous input arising from holding an object between forefinger and thumb has an excitatory net effect helping to reinforce the grip (see also [17]). This interpretation agrees with the old observation that vibrations in a grasped object (exciting tactile afferent units) may cause the fingers to adhere to the object, and difficulty may be experienced in attempting to loosen the grip (see [26] and [28]).

Observations of patients with peripheral nerve injuries affecting the sensibility of the fingers may provide a less uniform view. In their "everyday" activities, these patients typically show clumsiness during precise manipulation of small items and difficulties with gripping and holding, i.e. objects are frequently drop-

[1] Department of Physiology, University of Umeå, S-901 87 Umeå, Sweden.

Clinical Aspects of Sensory Motor Integration
Ed. by A. Struppler and A. Weindl
© Springer-Verlag Berlin Heidelberg 1987

ped, etc. (e.g. [21]). Since sufficient voluntary forces can be produced (the muscle innervation is intact), the motor symptoms indicate that the cutaneous input somehow is important for the automatic control of the coordination of the evolving manipulative forces.

The present contribution briefly summarizes some recent studies on the problem of coordination during precision grip between tips of fingers and thumb [14–16, 31, 32]. The nub of these experiments is the automatic adjustments of the grip force to (a) the load forces produced to overcome various forces counteracting the intended manipulation, and to (b) the frictional conditions between the object and the skin. If the grip force is too weak, the object will slip through the grip. Unnecessarily high grip force, on the other hand, would crush a fragile object, and, in the everyday situation, might injure the hand as well as cause unnecessary muscle fatigue. Moreover, it would hamper manipulatory movements further superimposed on the precision grip.

Behavioural Observations

The task of the human subject was to lift off a table a test object equipped with transducers for measuring the vertical lifting force (denoted the *load force*), the *grip force*, and the object's *vertical position*. An *accelerometer* was attached to detect small *slips* between the skin and the object. The touched surfaces were two discs (diameter: 30 mm) symmetrically placed on each side of the object in two parallel planes (30 mm apart). Without changing the visual appearance of the object, its surface structure as well as its weight could be varied between consecutive lifting trials.

During a standard lifting trial, the subject was asked to lift the object, hold it still in the air for ca. 10 s, and then replace it. Figure 1 A shows the sequence of events. At the preload phase (a) the subject grips the object. At the following loading phase (b) the load force and the grip force increase in *parallel* during isometric conditions. Soon the load force overcomes the force of gravity, and the lifting movement (elbow flexion) takes place (c). The grip force shows a peak while the object accelerates, followed by a decay. When the lifting movement is completed the static phase (d) is reached. After the replacement (e), when the object contacts the table, there is a short delay (f), after which the two forces in *parallel* fall to zero (g).

To avoid slips, the ratio grip force/load force must exceed a minimal value, the *slip ratio,* determined by the *coefficient of friction* between the skin and the object. This value is indicated by the horizontal line in Fig. 1 A and B, and measured in Fig. 1 B: starting from the static phase of a lifting trial, the subject was asked to slowly separate his fingers until the slip ratio was reached and the object was dropped. The *safety margin* to prevent slips is the difference between the slip ratio and the grip force/load force ratio automatically coordinated by the subject.

Figure 2A, shows the force coordination with different surface structures: finely textured silk (most slippery), suede (less slippery), and sandpaper (least slip-

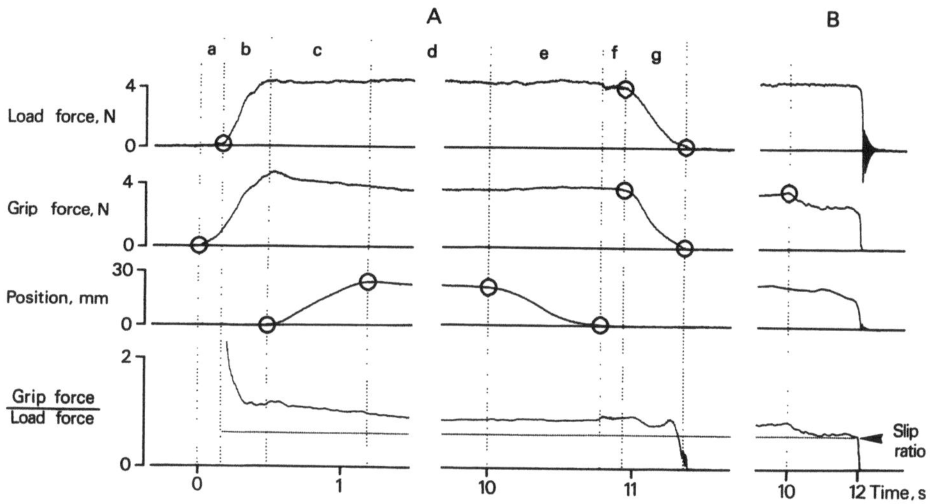

Fig. 1. A Load force, grip force, vertical position, and ratio between grip and load force as a function of time for a sample trial (weight of object = 400 g; surface structure = sandpaper). The periods indicated are: **a** preload phase, **b** loading phase, **c** transitional phase, **d** static phase, **e** replacement phase, **f** delay, **g** unloading phase. Note the interrupted time scale. **B** Slip ratio measurement carried out at the end of a separate trial subsequent to the trial shown in **A** (same surface structure and weight). The *vertical dotted line* indicates the start of the slow voluntary spacing of the fingers. The *horizontal dotted line* in **A** and **B** indicates the obtained slip ratio. (Johansson and Westling [15])

pery). These materials were selected as representative for the frictional range of common materials [14]. The course of the lifting movement and the rate of development of the load force are unaffected by the frictional condition (weight of object constant at 400 g). Only the grip force differs. The more slippery the material, the higher the rate of grip force increase, and the higher the final grip force. Changes of the object's weight, on the other hand, influenced the duration of the phases of parallel force change, but not the balance between the forces, i.e. the heavier the object, the longer the period of parallel isometric force increase before it starts to move.

A comparison between the employed force ratios for the three materials and the corresponding slip ratios (bottom of Fig. 2) indicates that the force coordination was adjusted to the frictional conditions with a fairly small safety margin. (Regarding interindividual differences in safety margin, see [31].)

Manipulation in the "Everyday Situation". Different kinds of experiments were performed to examine whether this adaptation of force coordination to friction may apply in more general contexts during precision grip. In one of these, the subject had to overcome a combined mass and spring load while lifting, and in a second experiment, the subject changed the mass load of the object while holding it in air, by adding or removing weights using the contralateral free hand (see Fig. 6A and B in [15]). Approximately the same force ratio was maintained throughout these manipulations, and it was adequately adapted to the friction.

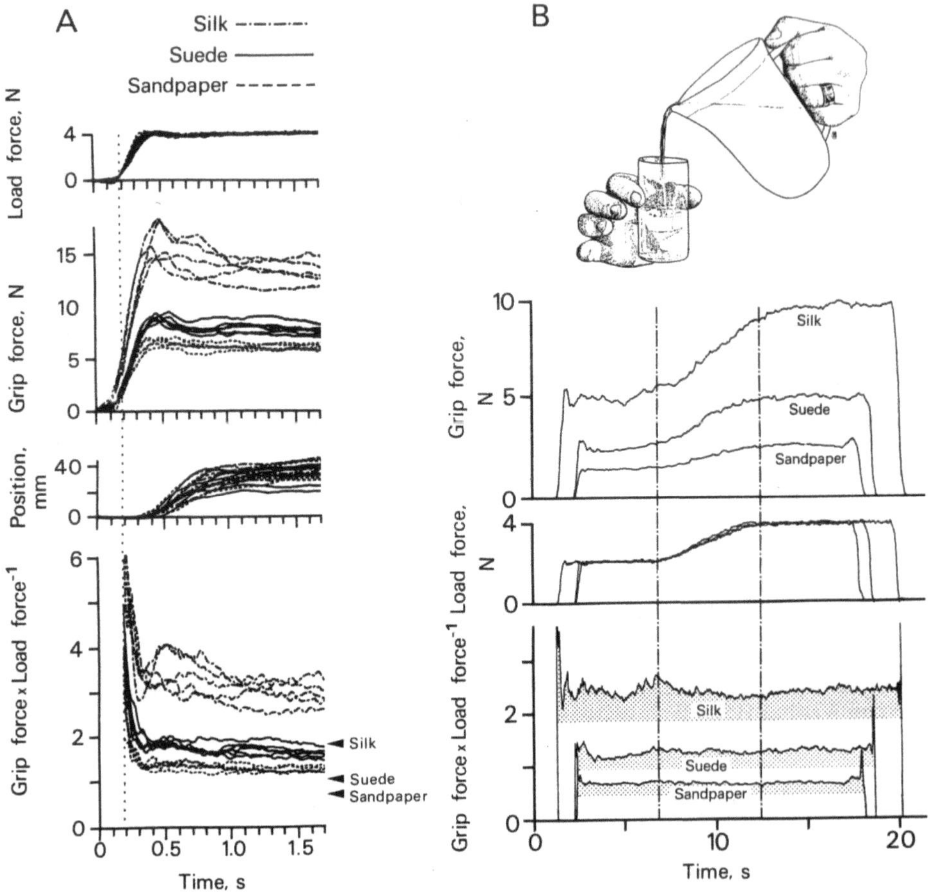

Fig. 2. A Force coordination during initial parts of lifting trials with three different surface structures (16 trials superimposed; weight = 400 g). All trials with the same subject. Trials immediately preceding those shown were carried out using the same surface structure, respectively. *Vertical line* indicates the beginning of the loading phase. *Arrows* indicate mean slip ratios for the three surface structures, respectively. **B** Coordination between grip and load force during a bimanual "everyday" task. An empty glass was attached to the test object which was lifted and held stationary in space (weight = 200 g). The changes of the load force taking place during the period between the two *vertical lines* was accomplished by the subject pouring the glass full with water (ca. 0.2 l) using a jug held by the contralateral hand. Three superimposed trials with three different surface materials in contact with the skin. **A** and **B** Load force, grip force, vertical position (only **A**), and ratio between grip and load force shown as a function of time. Time scale with an arbitrary origin. *Dotted areas* in **B** illustrate safety margins

Another similar experiment – again a bimanual task – is illustrated in Fig. 2 B. An empty glass was placed at the weight carrier of our test object (cf. Fig. 1 in [15]). While the subject was holding this "transducer-equipped-glass" in air, he was asked to take a jug with water with the free hand and fill up the glass. This was done with each of the three surface materials, silk, suede, and sandpaper, respectively. As can be seen, the grip force increases in parallel with the increasing load force while the glass is filled and, again, the force ratio is maintained at a level adequate for the current frictional demands. In further experiments it was shown that the force coordination adequately adapts to the considerable frictional changes occurring after a hand wash with ordinary soap and water [14]. During washing, sweat is removed from the skin and the friction in relation to any given object temporarily decreases. Since the surface structure was constant in these experiments, this finding further indicates that the adaptation is made to the friction per se, rather than on the basis of different texture properties of the touched materials.

Thus, it seems reasonable to conclude that the coordination, or balance, between the motor commands accounting for the grip and load forces is one output parameter critically regulated by the central nervous system during precision manipulation.

The Adaptation to Different Frictions

To study the process of adaptation to a different frictional condition, we analysed lifting trials carried out after unexpected changes in the surface structure (intertrial interval ca. 15 s; weight of object constant at 400 g). Neither were the subjects verbally informed about the new materials, nor could they visually discriminate the structures. Figure 3 A shows the changes in the "safe" direction – from silk to sandpaper – in three consecutive trials. Again, the development of the load force and the trajectory of the lifting movement are unaffected. During the preload phase (period a in Fig. 1 A) the development of the grip force for the first sandpaper trial (solid curves) is the same as in the preceding silk trial (dot-dash curves). At about the moment when the load force starts to rise an adjustment to the new surface (sandpaper) appears, i.e. the rate of grip force increase is slowed down and a lower final value is reached. A comparison with the second sandpaper trial (dashed curves) reveals, however, that this adjustment is not complete, i.e. the safety margin is reduced further. Thus, the motor programme was partially updated initially during the first sandpaper trial, and completely updated initially during the second.

The adjustment to a more slippery material is shown in Fig. 3 B, representing the following three consecutive trials: sandpaper, silk, and silk. Again, the development of the grip force in the trial with the new surface structure (first silk trial: solid curves) and the preceding sandpaper trial (dot-dash curves) is similar during the preload phase, whereas during the following loading phase, the grip force rises more steeply. This initial adjustment was, however, not always sufficient since a "secondary" adjustment of the force balance could occur later during the loading phase (arrow, Fig. 3 B). At this point, the grip force/load force ratio had fallen close to the slip ratio of silk. During the second silk trial (dashed curves) the grip force rose steeply to its proper value, with no secondary adjustments, i.e. it was preprogrammed on the basis of the frictional experiences from the previous silk trial.

Tactile Afferent Input Related to Coordination Adjustments. Impulses in single tactile afferent units (low-threshold mechanoreceptive afferent units) innervating the

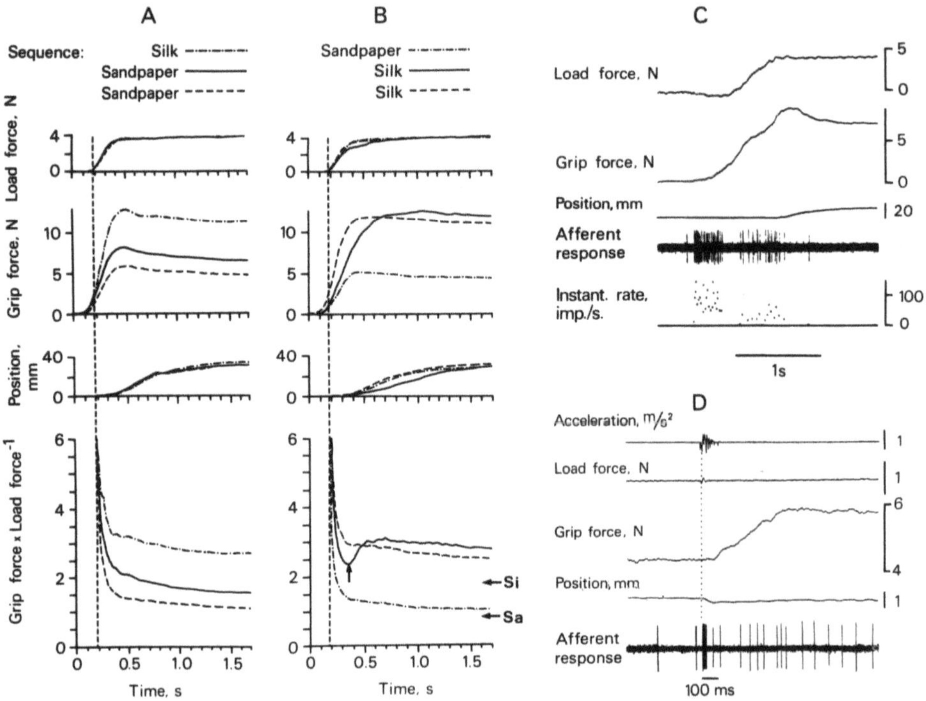

Fig. 3. A and **B** Adjustments to friction between object and skin. Each graph represents a sequence of three consecutive trials selected from lifting series with pseudorandom changes of surface structures. Weight of object was 400 g. Averaged data from a total of 162 trials by 9 subjects. Load force, grip force, vertical position, and ratio between grip and load force as a function of time. *Horizontal arrows* indicate slip ratios for the two surface structures, respectively. *Vertical arrow* in **B** indicates a sudden change of force balance ("secondary" adjustment) accounted for mainly by a decrease in the load force rate. For further details see text and Fig. 2 A. **C** Discharge pattern of a FA I unit during the initial period of a lifting trial with silk. Note the strong response during the preload phase. **D** Late, "secondary", adjustment of the coordination between grip and load forces subsequent to a small slip during the static phase of a trial with silk. "Vertical line" indicates the onset of the slip as revealed by vibrations in the object (acceleration event). *Bottom trace* shows the discharge in a SA-I unit

glabrous skin of the fingers were recorded during similar lifting trials [16, 32]. The microneurographic technique developed by Vallbo and Hagbarth [29] was used, and the median nerve was impaled about 10 cm proximal to the elbow. The findings strongly suggest that tactile signals provide the afferent information accounting for the initial as well as the secondary adjustments of the force coordination [16].

The tactile afferent units in the glabrous skin of the human hand can be divided into four different types mainly on the basis of two pairs of features: adaptation to sustained indentation and structure of the cutaneous receptive field (for refs. see [13, 30]). All four types have strong similarities to well-defined types described in other species [1, 11]. Two types adapt rapidly to maintained skin deformation: the fast-adapting type I (FA I) and the fast-adapting type II (FA II) units, i.e., they only respond to deformation changes. The other two types – the slow-adapting type I (SA I) and the slow-adapting type II (SA II) units – adapt slowly, i.e. in addition to being dynamically sensitive (particularly the SA Is) they exhibit a response related to the strength of maintained skin deformation. The FA Is and the SA Is have small and well-defined receptive fields (typically 10 mm^2). The receptive field properties and the high densities in the skin of these unit types indicate that they account for the detailed spatial resolution of the tactual sense. In contrast, the FA II and SA II units have much larger and less well defined receptive fields, and their densities are lower and more uniform. The FA IIs are particularly susceptible to remote transient mechanical stimulation and vibrations of high frequencies (ca. 50–500 Hz), whereas the SA IIs show an appreciable sensitivity to remote lateral stretching of the skin with a pronounced directional preference. Extrapolations from animal experiments combined with morphological data in man suggest that the FA I units are connected to Meissner corpuscles, the FA II to Pacinian corpuscles and smaller laminated endings (Golgi-Mazzoni bodies), the SA I to Merkel discs, and the SA II to Ruffini endings.

The initial adjustments of the force coordination indicate that afferent information related to friction must exert its influence very early after the object is gripped, i.e. after as little as ca. 0.1 s. Indeed, intense impulse discharges were observed in tactile afferents at the initial skin deformations during the preload phase, i.e. at grip forces below 0.5–1 N. This was true for all FA I units with receptive fields in contact with the object (Fig. 3 C). Most of the SA I units showed similar responses. Interestingly, with some FA I units the strength of this discharge appeared to be related to the surface structure, i.e. the more slippery the material, the higher the discharge rate.

The secondary coordination adjustments occurred in response to small short-lasting slips as revealed by our accelerometer signal. After a latency of 60–80 ms following the onset of the slip, there are changes in the grip- and/or load-force rates resulting in an increase of the grip-force/load-force ratio to a new, higher stable value (cf. arrow in Fig. 3 B), i.e. the force balance parameter is updated and the safety margin increases. Secondary adjustments occur not only during the loading phase, but sometimes also during the static phase as shown in Fig. 3 D. Now, the coordination change is caused by a grip-force increase to a higher stable value. Not surprisingly, the slips elicited brief but intense bursts of action potentials (up to ca. 300 imp./s) in tactile units with high dynamic sensitivity, i.e. the FA I, FA II, and SA I units. These signals most likely triggered the coordination adjustments. The bottom trace in Fig. 3 D represents an SA I response.

Further evidence for this idea was obtained in experiments with cutaneous electrical stimulation delivered through concentric electrodes implanted in the touched surfaces of the object [16]. A single current pulse of 0.2 ms duration, delivered while the object was held in air, could elicit a sustained coordination shift after the same latency as with slips. However, to obtain this motor response, the stimulation intensity has to be very close to the perception threshold: stronger stimuli are recognized as "electrical" and fail to give this response. Likewise, with repeated tests the motor response declines rapidly and usually already disappears on the third to fourth pulse, indicating the operation of an efficient habituation mechanism.

The 60–80 ms latency is about twice that of the most rapid spinal reflex and about half the latency for intended finger movements triggered by exteroceptive cues. The motor response was evidently automatically initiated, which is in agreement with the fact that the coordination adjustments usually proceeded without being noticed by the subject. Still, this latency is compatible with the involvement of supraspinal mechanisms. Interestingly, it is similar to the latency of the late and most pronounced excitatory component of the multiphasic reflex modulation of ongoing EMG in the first human dorsal interosseous muscle, elicited by modest electrical or mechanical stimulation of fingers [9]. Since this muscle contributes significantly to the grip force during pinch grip, it may be assumed that the underlying mechanisms are related. Later studies by the Stephens group suggest that this reflex component is supraspinally mediated, involving transmission of afferent impulses through the dorsal columns, a relay in the sensorimotor cortex and descending transmission to motoneurons via the corticospinal tract [12]. A participation of supraspinal neuromechanisms is in agreement with the multitude of evidence that the motor cortex and the pyramidal tract are of fundamental importance for the performance of precise finger movements [18, 23, 24, 27].

Cutaneous Anaesthesia. During lifting trials by subjects with the tips of the index finger and thumb anaesthetized (local intradermal infiltration of Marcain), the grip and load forces still changed in parallel but the automatic adaptation to the frictional condition was lost. Slipping could now occur when slippery surface materials were used, and the grip force could be unnecessarily high with less slippery materials. For most subjects, the first lifts were successful with suede or sandpaper but not with silk: the fingers slid up the stationary object. Now, the problem with the slip was met by an alternative strategy. The subject consciously attended to the firmness of the grip and to increase it during subsequent lifting attempts until success. A new force balance was then established by voluntary control, which was approximately maintained in an apparently automatic fashion during subsequent trials, including those with less slippery materials.

The CNS programme controlling the anaesthetized hand was also accessible to inputs from the contralateral side. Subjects with fingers on the right hand anaesthetized were asked to lift the object alternately with the right and left intact hand. The surface was always sandpaper while lifting with the numbed hand, whereas the unanaesthetized hand lifted silk, suede, or sandpaper. The more slippery the object on the normal side, the higher the grip-force/load-force ratio on the numbed side. Thus, stored information related to the frictional properties of the object appeared not only to influence the coordination during the entire subsequent trial, but it could be used bilaterally. So if, in everyday life, both hands are informed, the vacant hand stands ready to intervene.

Sensorimotor Memory

The nature of the slip-triggered coordination shifts and the influences on the force coordination of the frictional condition during the previous lifting trials indicate that the coordination of grip and load forces is defined by a memory trace. This memory would be updated on the basis of tactile information received intermittently when inappropriate force coordination is encountered, such as during slips. A reliance on a preprogrammed coordination, adjusted to the current friction, would allow the CNS to simultaneously change the grip and load forces while automatically maintaining an appropriate balance between the two. The intermittent resetting of an open-loop operation would avoid the time lag necessarily involved if the grip force was regulated exclusively on the basis of a continuous "closed feedback loop" involving cutaneous receptors, e.g. a continuous tactile feedback about incipient slip (cf. [25]). Moreover, a preprogrammed force coordination, automatically maintaining an appropriate safety margin preventing slips, disengages a larger fraction of the sensorimotor apparatus for the "higher level" exploratory and/or manipulatory tasks usually superimposed on the basic precision grip.

Summary

The refined force coordination required during precision manipulation of small objects between the tips of the fingers and thumb is heavily disturbed during impairments of the cutaneous sensibility of the fingers. The present contribution deals quantitatively with the regulation of this coordination on the basis of behavioral studies in intact man and during cutaneous anaesthesia, and observations of signals in single tactile afferents from the fingers during precision grip. The most essential feature of this coordination is that the grip force changes in parallel with the load force produced by the subject to overcome various forces counteracting the intended manipulation, i.e., the two forces are generated by coordinated motor commands. The balance, or ratio, between the two forces, is critically adapted to the friction between the skin and the object, providing a relatively small safety margin to prevent slips. This balance appears to be preprogrammed via a sensorimotor memory which is *intermittently* updated by tactile information whenever the frictional conditions are changed. An initial adjustment of the force balance takes place soon after an object is initially touched (ca. 0.1 s). Moreover, brief bursts of action potentials in tactile afferents elicited by small slips can trigger further, "secondary", adjustments, resulting in an increased safety margin to prevent further slips. The latencies between the onset of such slips and the appearance of the adjustments (60–80 ms) indicate that the underlying neural mechanisms operate highly automatically.

Acknowledgments. We wish to thank Prof. Å. Vallbo for valuable comments on the manuscript. This work was supported by grants from the Swedish Medical Research Council (Project No. 3548), the Gunvor och Josef Aner's Stiftelse, and the University of Umeå, which are gratefully acknowledged.

References

1. Burgess PR, Perl ER (1973) Cutaneous mechanoreceptors and nociceptors. In: Iggo A (ed) Somatosensory system. Springer, Berlin Heidelberg New York (Handbook of sensory physiology, vol II, pp 29–78)
2. Caccia MR, McComas AJ, Upton ARM, Blogg T (1973) Cutaneous reflexes in small muscles of the hand. J Neurol Neurosurg Psychiatry 36:960–977
3. Datta AK, Stephens JA (1981) The effect of digital nerve stimulation on the firing of motor units in human first dorsal interosseous muscle. J Physiol (Lond) 318:501–510

4. Day BL, Marsden CD (1982) Accurate repositioning of the human thumb against unpredictable loads is dependent upon peripheral feed-back. J Physiol (Lond) 327:393–407
5. Denny-Brown D (1966) The cerebral control of movement. Liverpool University Press, Liverpool
6. Gandevia SC, McCloskey DI (1977a) Effects of related sensory inputs on motor performance in man studied through changes in perceived heaviness. J Physiol (Lond) 272:653–672
7. Gandevia SC, McCloskey DI (1977b) Changes in motor commands, as shown by changes in perceived heaviness, during partial curarization and peripheral anaesthesia in man. J Physiol (Lond) 272:673–689
8. Gandevia SC, McCloskey DI (1980) Alterations in perceived heaviness during digital anaesthesia. J Physiol (Lond) 306:365–375
9. Garnett R, Stephens JA (1980) The reflex responses of single motor units in human first dorsal interosseous muscle following cutaneous afferent stimulation. J Physiol (Lond) 303:351–364
10. Garnett R, Stephens JA (1981) Changes in the recruitment threshold of motor units produced by cutaneous stimulation in man. J Physiol (Lond) 311:463–473
11. Iggo A (1974) Cutaneous receptors. In: Hubbard JI (ed) Peripheral nervous system. Plenum, New York, pp 347–404
12. Jenner JR, Stephens JA (1982) Cutaneous reflex responses and their central nervous pathways studied in man. J Physiol (Lond) 333:405–419
13. Johansson RS, Vallbo ÅB (1983) Tactile sensory coding in the glabrous skin of the human hand. Trends Neurosci 6:27–31
14. Johansson RS, Westling G (1984a) Influences of cutaneous sensory input on the motor coordination during precision manipulation. In: von Euler C, Franzen O, Lindblom U, Ottoson D (eds) Somatosensory mechanisms. MacMillan, London, pp 249–260
15. Johansson RS, Westling G (1984b) Roles of glabrous skin receptors and sensorimotor memory in automatic control of precision grip when lifting rougher and more slippery objects. Exp Brain Res 56:550–564
16. Johansson RS, Westling G (1987) Signals in tactile afferents from the fingers eliciting adaptive motor responses during precision grip. Exp Brain Res 66:141-154
17. Kanda K, Desmedt JE (1983) Cutaneous facilitation of large motor units and motor control of human fingers in precision grip. In: Desmedt JE (ed) Motor control mechanisms in health and disease. Raven, New York, pp 253–261
18. Lemon RN (1981) Functional properties of monkey motor cortex neurones receiving afferent input from the hand and fingers. J Physiol (Lond) 311:497–519
19. Marsden CD, Merton PA, Morton HB (1977) The sensory mechanism of servoaction in human muscle. J Physiol (Lond) 265:521–535
20. Marsden CD, Rothwell JC, Traub MM (1979) The effects of thumb anaesthesia on weight perception, muscle activity and the stretch reflex in man. J Physiol (Lond) 294:303–317
21. Moberg E (1962) Criticism and study of methods for examining sensibility in the hand. Neurology 12:8–19
22. Mott FW, Sherrington CS (1895) Experiments upon the influence of sensory nerves upon movement and nutrition of the limbs. Proc R Soc B 57:481–488
23. Muir RB, Lemon RN (1983) Corticospinal neurones with a special role in precision grip. Brain Res 261:312–316
24. Phillips CG, Porter R (1977) Corticospinal neurones. Academic Press, London
25. Rack PMH (1981) Limitations of somatosensory feedback in control of posture and movement. In: Brooks VB (ed) Handbook of physiology. The nervous system, vol 2. Am Physiol Soc, Bethesda
26. Rood ON (1860) On contraction of the muscles induced by contact with bodies in vibration. Am J Sci Arts 24:449
27. Smith AM, Frysinger RC, Bourbonnais D (1983) Interactions between motor commands and somatosensory afferents in the control of prehension. In: Desmedt JE (ed) Motor control mechanisms in health and disease. Raven, New York, pp 373–385
28. Torebjörk HE, Hagbarth K-E, Eklund G (1978) Tonic finger flexion reflex induced by vibratory activation of digital mechanoreceptors. In: Gordon G (ed) Active touch. Pergamon, Oxford, pp 197–203

29. Vallbo Å, Hagbarth KE (1968) Activity from skin mechanoreceptors recorded percutaneously in awake human subjects. Exp Neurol 21:270–289
30. Vallbo Å, Johansson RS (1984) Properties of cutaneous mechanoreceptors in the human hand related to touch sensation. Human Neurobiol 3:3–14
31. Westling G, Johansson RS (1984) Factors influencing the force control during precision grip. Exp Brain Res 53:277–284
32. Westling G, Johansson RS (1987) Responses in glabrous skin mechanoreceptors during precision grip in humans. Exp Brain Res 66:128–140

Neurophysiological Mechanisms Underlying Proprioceptive Sensations *

S. C. GANDEVIA [1]

Introduction

The term "cutaneous sensibility" can be defined unambiguously, and the relevant peripheral receptors are easily identified. By contrast, the terms "proprioception" and "kinaesthesia" are now commonly used interchangeably without full regard for the nuances of meaning given them by their originators, Sherrington and Bastian respectively (for discussion see [14, 31]), and there is no agreement on the relevant receptors and neural mechanisms. Proprioception and kinaesthesia are taken to include the *sensations* of joint position and movement along with the *sensations* of force and apparent heaviness which accompany active muscular contraction. Besides the lack of terminological clarity, understanding about this *group* of sensations has been hampered by continual disagreement about the role of specific classes of peripheral receptors which could contribute to it. Much previous debate has been discussed by McCloskey [31].

Sensations of Limb Position and Movement

Until 1972 the clinical term "joint position sense" had appeared aptly named in that there was a body of evidence that joint receptors did, or more particularly that intramuscular receptors did not, contribute to the sensation. (Paradoxically, "joint *position* sense" is usually tested clinically by *movement* of a joint although perceived signals of position and movement can by dissociated – see [6, 15, 30].) However, in 1972, illusions of movement and distortions of position sense produced by transverse vibration of tendons and muscles were described [11, 23]. These illusions were best explained by the perception of discharges in vibration-sensitive, presumably primary muscle spindle, endings. These illusions have also been reported when the vibration was delivered directly (i.e. longitudinally) to the proximal part of a tendon which had been surgically divided [35]. While the occurrence of these vibration-induced illusions has been corroborated (e.g. [8, 44]), the initially proposed explanation has been questioned by Moberg [40]. He argued that a contribution from cutaneous and possibly deep non-muscle receptors had not been excluded and, further, that much of the evidence previously brought forward to support the role of intramuscular receptors could be used to favour

---

* Preparation of this chapter was completed in June, 1985.

[1] Unit of Clinical Neurophysiology, Prince Henry Hospital and School of Medicine, University of New South Wales, P.O. Box 233, Matraville, Sydney 2036, Australia.

Clinical Aspects of Sensory Motor Integration
Ed. by A. Struppler and A. Weindl
© Springer-Verlag Berlin Heidelberg 1987

cutaneous receptors. His view that there was no evidence for a contribution from "musculotendinous" receptors derived from the conceptually simple but difficult experiment of pulling on tendons exposed under local anaesthesia. Detection of any pulls applied to the proximal tendon was believed by him to be due to the deformation of the proximal skin over the muscle belly (see also [22]). McCloskey et al. [34] agreed that detection of tendon pulls occurred but they produced psychophysical evidence to suggest that intramuscular receptors with a sensitivity to longitudinal vibration of low amplitude (less than 100 μm) were responsible (see also [39]).

The present review is restricted, first, to a consideration of recent evidence that reaffirms the previously proposed role for muscle afferents in kinaesthesia, second, to a discussion of the potential role of joint receptors in the sensation of limb movement, and, finally, to the concept that signals related to motor commands are associated with a sensation of force or perceived heaviness. Other reviews with a different scope include those by McCloskey [31], Matthews [38], and Gandevia [14].

Kinaesthetic Illusions During Electrical Stimulation of Muscle Afferents

Electrical stimulation of low-threshold muscle afferents was reported not to produce illusions of movement [23], but the stimuli were delivered to the whole mixed tibial nerve at the knee and evoked cutaneous paraesthesiae which may have obscured the percept. This failure to demonstrate that muscle afferents evoke kinaesthetic illusions has been reassessed using stimulation of the ulnar nerve at the wrist via probe, needle electrodes, and microelectrodes [15]. Illusory movements of the fingers and distortions of position were reported for trains of stimuli at levels which did not evoke cutaneous paraesthesiae or activate any motor axons (Figs. 1 and 2). The direction of illusory movements and their anatomical distribution throughout the hand was such that they could be readily explained by a perceived elongation of one (or more) of the intrinsic muscles of the hand (and not by activity in non-muscle afferents). The commonest illusion was a smooth flexion at the two interphalangeal joints combined with extension at the metacarpophalangeal joint – the movement produced by a lengthening of the lumbrical muscle. The illusions were measured objectively by asking the subjects to track the movements on the unstimulated side. Movements involving only adduction or abduction of the fingers also occurred. For example, adduction of the index finger was produced by stimulation (at below motor threshold) in the motor fascicle innervating the first dorsal interosseous muscle.

The velocity of all components of these illusory movements increased as the frequency of stimulation increased over the range 10–200 Hz (Fig. 1, lower panels). In some subjects illusions were unclear at frequencies below about 10 Hz, while others described fragmentary flick movements even at 1 Hz with "fusion" into a single smooth movement at about 5 Hz. The velocity of the illusory movements also increased as the stimulus strength increased (from, say, 0.5 to 0.9 times motor threshold). Thus, these findings suggest that both the number of afferents activated and their discharge frequency influence the final perceived velocity of

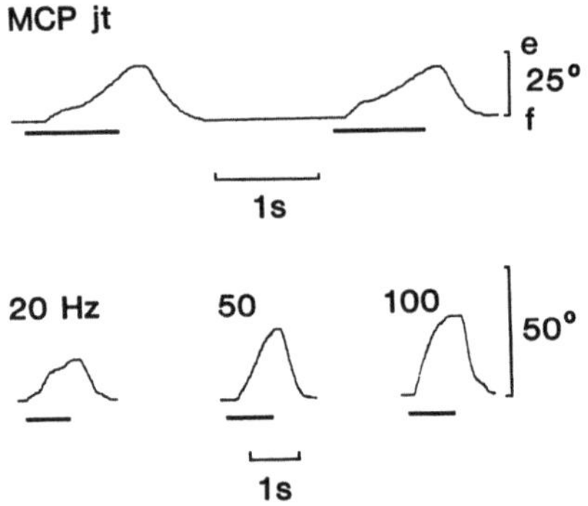

Fig. 1. The subject experienced illusory flexion of the interphalangeal joints and extension of the metacarpophalangeal joint (MCP) of the ring finger during trains of stimuli delivered to the ulnar nerve at the wrist (below motor threshold and without cutaneous paraesthesiae in the digital nerves). The records show the apparent movement at the metacarpophalangeal joint when the subject tracked the illusory movement by moving the metacarpophalangeal joint of the index finger on the contralateral side. The *upper panels* depict two consecutive responses to stimulation at 50 Hz. The *horizontal bars* indicate the periods of stimulation. When the stimulation began, the perceived movement was into extension (*e, upward deflection*) and into flexion when the stimulation ceased (*f, downward deflection*). The *lower panels* indicate the responses to three frequencies of stimulation. The velocity of the illusory movement increased with the frequency of stimulation. In this and Fig. 2 the presence of illusory movements was first established with the probe electrode and then a needle electrode was positioned to obtain the illusions for prolonged periods. (Gandevia [15])

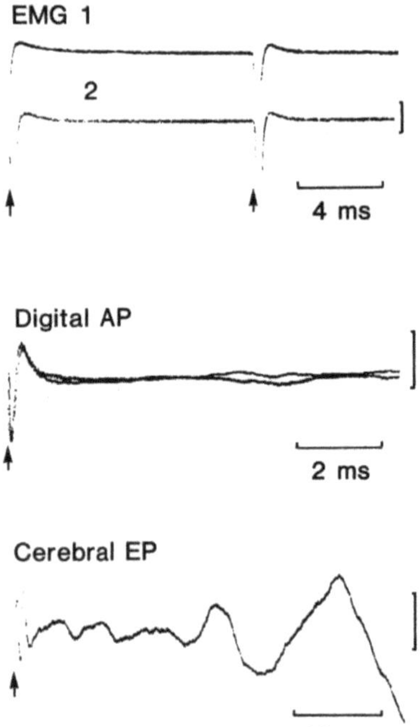

Fig. 2. Recordings made during stimulation of the ulnar nerve at the wrist with a needle electrode which produced illusory extension of the metacarpophalangeal joint and flexion of the interphalangeal joints of the little finger. The *upper traces* are averages at high gain of the electromyographic activity recorded from the hand during the illusion. The *middle traces* are duplicate averages of activity recorded from the digital nerves with ring electrodes over the proximal phalanx. The *lower trace* is the average cerebral potential recorded from the contralateral scalp (earlobe reference) during the illusory movements. This presumably represents the arrival at cortical level of activity from low-threshold muscle afferents. For further details of this projection see Gandevia et al. [19]. The frequency of stimulation for all recordings was 10 Hz. Calibration in the upper traces is 50 μV and 1 μV in the middle and lower trace. 1024 responses were averaged. (Gandevia [15])

the illusory movement. In addition to the "movement" induced by the stimulation there was a distortion of position sense, again in the direction of perceived muscle lengthening. This was demonstrated by asking the subject to indicate the apparent position occupied by the finger on the stimulated side by positioning the appropriate finger on the unstimulated side. These position distortions often carried the finger(s) into positions which were beyond the natural range of joint motion. Thus, these electrically induced illusions are similar in many respects to the illusions produced by tendon vibration (e.g. [7, 8, 23, 44]). Given the low threshold at which these illusions were evoked (0.5–0.9 times motor threshold) and their perceived direction, they are likely to reflect the perception of discharges in afferents of low electrical threshold from muscle stretch receptors, presumably group Ia afferents from primary spindle endings. While a contribution from Golgi tendon organ afferents cannot be ruled out, there is anatomical evidence that the largest diameter fibres within the deep branch of the ulnar nerve (and therefore presumably those of lowest electrical threshold) innervate primary muscle spindle endings [10].

Joint Receptors and Kinaesthesia

A specific role for joint receptors in human kinaesthesia has not been clearly demonstrated. The ability to detect passive movements applied to a finger or toe when the afferents in the digital nerves have been blocked is diminished (e.g. [16, 42]; see also [4]). However, when the velocity of imposed movements increases their occurrence and direction can be detected, and this ability must reflect the minimal contribution of intramuscular receptors – that is, their contribution, without any "facilitation" from the functionally related joint and cutaneous afferents which have been silenced by anaesthesia and without any specific kinaesthetic input from these afferents. In the study by Browne et al. [4], large volumes of anaesthetic were infiltrated around the capsule of the metatarsophalangeal joint of the great toe. It is unclear whether cutaneous afferents were affected by this procedure so that the deficit reported cannot be ascribed unequivocally to anaesthesia of joint afferents. The original studies by Browne et al. [4] and Provins [42] employed relatively low angular velocities of passive movement and thus emphasized the size of the kinaesthetic deficit when local joint and cutaneous receptors were anaesthetized.

By contrast with the reports described above, intracapsular anaesthesia of the human knee joint was without significant effect on "position" sense as tested by matching the perceived position of the knee after passive movements at extremely low angular velocities (less than one degree per min; cf. [6]). This argues against a critical role for joint afferents in the proprioceptive test used. It is difficult to be certain that there was significant joint anaesthesia at the time that the tests were carried out (40 min after injection, including 5–10 min of walking). However, this was suggested on the basis of animal studies. In addition, the slow movements employed were possibly below the velocity threshold at which joint receptors would have been significantly excited. If so, the failure to find a substantial kinaesthetic deficit would not be surprising. Barrack et al. [2] used slow movements (0.5° per s) to determine the threshold for detection of passive motion at the knee. This threshold was unaffected by intracapsular injections of local anaesthetic. Consistent with these findings is the observation that surgical replacement of finger, knee, or hip joints fails to produce a marked kinaesthetic deficit in the detection of passive movement (e.g. [9, 24, 28, 29]) even when comparisons are made with age-matched groups [3].

To reinvestigate the role of joint receptors in kinaesthesia, studies of the kinaesthetic acuity have been carried out at the distal interphalangeal joint of the middle finger [36]. An anatomical peculiarity of the flexor and extensor tendons permits the muscles which operate at the joint to be held at lengths inappropriate for their action of the joint. By positioning the fingers adjacent to the test finger as shown in Fig. 3 (upper panels), it is possible to study kinaesthetic acuity when muscles cannot contribute, or when only the long flexor muscle can operate on the joint, or when there is a full complement of intramuscular receptors in flexor and extensor muscles which can contribute to the detection of applied movements. Results obtained in previous studies are shown in diagrammatic form in Fig. 3 (lower panels). Kinaesthetic acuity is best when all receptors are available, but it deteriorates when the extensor mechanism cannot operate on the joint and deteriorates further when the joint is effectively disengaged from both flexor and extensors [16, 20]. The test of kinaesthetic acuity used in these studies relies on the detection of direction of standard movements applied at different angular velocities.

To determine whether the residual performance (when intramuscular receptors could not contribute) depended on joint receptors, kinaesthetic acuity was studied in the one experimental session under control conditions, following expansion of the intracapsular space with an injection of about 0.1 ml of dextran (a plasma expander) and following injection of a similar volume of local anaesthetic. The consistent response was a significant *enhancement* of performance with injection of the plasma expander and a *deterioration* (below control levels) with injection of the local anaesthetic. This suggests that the deterioration in performance was probably due to paralysis of some joint receptors. It could not be ascribed to anaesthesia of a small area of skin on the dorsal aspect of the joint at the injection site because anaesthesia of this area alone in separate experiments produced no kinaesthetic deficit. The enhancement of performance with expansion of the intracapsular space probably reflects the increased discharge from slowly adapting receptors. Such an increase in the dynamic response to passive movement has been demonstrated for joint receptors at the knee joint of the cat (W. R. Ferrell 1984, personal communication). When the finger joint is in a mid-position, joint receptors seem capable of duplicating the input which arises in intramuscular receptors.

Cutaneous receptors directly over the dorsum of the joint are not required for normal kinaesthetic acuity, at least under the experimental conditions investigated in these studies. Cutaneous anaesthesia around the knee joint also fails to alter proprioceptive performance [6]. Cutaneous receptors may have a general facilitatory role in detection because anaesthesia of the fingers adjacent to the test finger impairs kinaesthetic acuity [16], and cutaneous receptors (probably to-

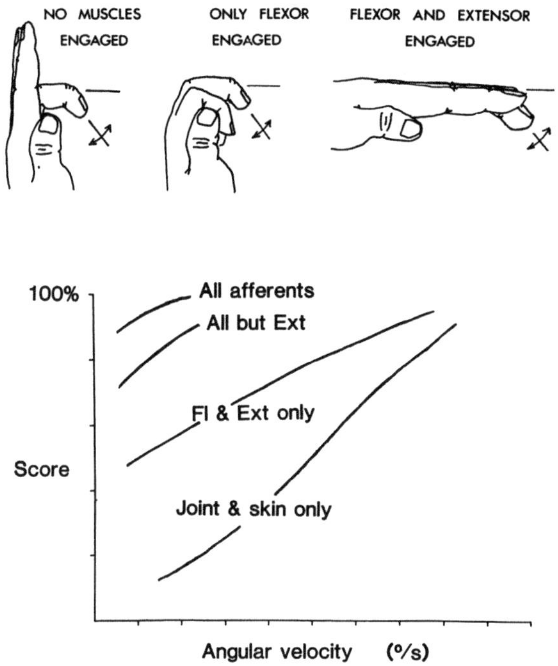

Fig. 3. *Upper panels:* the three hand positions used to assess the contribution of muscle, joint, and cutaneous receptors to kinaesthesia in the hand. Voluntary movement of the distal joint of the middle finger is impossible when the hand is positioned with the middle finger flexed and the other fingers extended (*left*) because the muscles which operate on the joint are held at mechanically inappropriate lengths for action. In this posture applied movements can be detected only by reference to the discharge of local joint and cutaneous receptors. When the adjacent fingers are brought into flexion the long flexor muscle (but not the extensor apparatus) is able to operate on the joint (*middle*), and when all fingers are extended then both active flexion and extension can occur at the joint (*right*). To study the contribution of intramuscular receptors, the hand is positioned as at *right*, and the contribution of joint and cutaneous receptors is eliminated by a local anaesthetic block of the digital nerves at the base of the finger. To study the potential contribution of joint receptors local anaesthetic is injected into the joint from a dorsal approach and kinaesthetic acuity is assessed when the hand is positioned as at left and intramuscular receptors cannot contribute.

Lower panels: the trend of results obtained when kinaesthetic acuity is studied in the three hand positions shown above. Movements of a specific angular extent are applied to the distal joint of the middle finger from an intermediate position at different angular velocities. The subject is required to nominate the direction of the applied movements. Acuity is best when all species of receptor can contribute. It is poor when only joint and cutaneous receptors contribute but improves when first the long flexor is "re-engaged" at the joint and then the extensor mechanism is also able to move the joint. The performance ascribed to the full complement of intramuscular receptors when studied in isolation is intermediate. Further details of the hand positions and the results of this and other kinaesthetic tests are given in Gandevia and McCloskey [16] and Gandevia et al. [20]

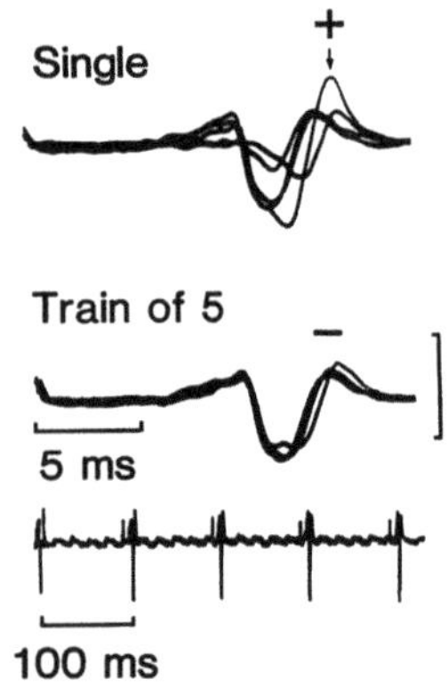

Fig. 4. Detection of a twitch contraction produced by stimulation within a motor fascicle of the tibial nerve in one subject. The *upper trace* presents a series of superimposed compound muscle-action potentials recorded from lateral gastrocnemius during increasing levels of intrafascicular stimulation. Single-twitch contractions remained undetected until the stimulus intensity was sufficient to produce the largest of the compound muscle-action potentials shown. It involved five or more motor units. A train of 5 stimuli at approximately 10 Hz is shown as a series of superimposed muscle action potentials (*middle trace*) and as single responses on a slower time base (*lower trace*). This train of stimuli which produced a sustained contraction of several motor units was also undetected by the subject

gether with joint receptors) may provide the signals used to clarify "ambiguous" signals from receptors in muscles acting over two or more joints (cf. [45]).

Recent studies on human cutaneous sensibility have emphasized the psychophysiological evidence that a single discharge in some single cutaneous mechanoreceptors can evoke specific localized sensations at least for the densely innervated pads of the fingers [41, 47]. Given that synchronization of the discharge of some low-threshold muscle afferents can give rise to sensations of illusory movements of the fingers of the type shown in Fig. 1 (see also [15]), it is interesting to speculate as to whether single muscle afferents have such direct access to consciousness as has been claimed for some cutaneous afferents. Consideration of simple fasciculations may be relevant here because these occasional discharges of a motor unit are often unnoticed, yet they probably evoke discharges in few local muscle spindle receptors (see [37]), and doubtless also in other nearby fascial and cutaneous receptors. In recent experiments the discharge of single motor units has been evoked by intrafascicular stimulation of motor axons innervating hand and leg muscles of normal subjects (S. C. Gandevia and D. Burke, unpublished observations). Unless the twitch contractions of single motor units dimpled the skin, they were not usually perceived until the stimulus intensity recruited 3–5 motor units, as judged by the compound muscle action potential (Fig. 4). Although using unphysiological stimuli these experiments are consistent with the view that some proprioceptive sensations may require activity in a *population* of intramuscular (and/or other) receptors. If so, this would represent a significant difference between the central organization of cutaneous and kinaesthetic sensibility.

Sensations of Muscular Force and Heaviness

Discussion of theoretical central and peripheral mechanisms subserving these sensations has occurred for over a century (for review see [32]). It is clear now that the preferred signal of perceived force is one which is biased by a signal related to the motor command or sense of effort put into the muscular contraction rather than a perceived signal related to the force of the resulting contraction. The clear statement by Holmes [26] that forces exerted by weakened muscles are overestimated has been supported by many studies in which forces have been matched by the same muscle group on the two sides. This has allowed objective quantification of the perception of a particular force under a variety of experimental conditions. Perceived forces are overestimated during "weakness" produced by muscular fatigue [18, 27, 33], neuromuscular blockade (e.g. [17, 43]), changes in the length-tension relationship [5], vibration-induced inhibition of the agonist motoneurone pool [33] and with central motor disorders producing weakness without conventional sensory loss [1, 13, 17, 26]. If judgements about force or heaviness relied only on information directly related to the actual muscular tension generated, then no overestimation should have occurred during weakness. These results have been used to argue for a contribution from a signal related to the size of the motor command required for the contraction (e.g. [13, 33]). However, the specific mechanisms by which such a signal biases the judgement of perceived force and heaviness are not clear. The following points summarize some of the factors which must be considered in an attempt to understand the neurophysiological mechanisms underlying these sensations.

1. Signals of actual achieved tension can be differentiated experimentally from those related to the relevant signal of motor command (or effort). This has been shown for limb muscles during the excitation or inhibition of the motoneurone pools produced by vibration of the agonist or antagonist muscle groups respectively [33] and during muscle fatigue [21, 27]. The two signals are clearly mediated by different neural mechanisms.
2. As would be expected for a centrally generated signal, the earliest time at which the signal of motor command can be perceived precedes the onset of movement [34].
3. Because the perceived force is overestimated when the deficit in the motor pathway lies "upstream" of the motoneurone pool (e.g. following a motor stroke), it is likely that the relevant signal of motor command can be generated above the level of the motoneurone pool.
4. The relevant signal of motor command has access to each cerebral hemisphere despite complete section of the corpus callosum [12].
5. Abolition of the normal perceived motor command which accompanies the attempt to contract paralysed muscles has been described only during the phase of *complete paralysis* of pure motor hemiplegia ([13]; C. M. Fisher 1985, personal communication). This has been described for lesions of the internal capsule and motor cortex.
6. The relevant signal of motor command can be dissociated from the central command signals which are thought to contribute to the usual cardiovascular responses (increases in blood pressure and heart rate) during static muscular contraction [25].
7. Interpretation of specific signals of motor command as particular forces requires reference to afferent inputs which point to the success of the motor command in developing muscular force or lifting the required object. The relevant peripheral indicator for calibration of these motor commands may be fairly crude. It does not need to be directly proportional to the force of the achieved contraction for the signal of motor command to be calibrated accurately [18].

In addition, visual and other nonproprioceptive inputs may alter the interpretation of signals of motor command as they do for intramuscular signals of joint position (e.g. [46]).

While there is evidence both for a peripherally originating percept of muscular tension and for a centrally originating percept of motor command, each of which can influence judgements of force, understanding of the underlying neural mechanisms is limited. The relevant signal of voluntary motor command is related to the size of some central motor drive which requires the motor cortex for its expression. In pathological conditions involving the motor system (ranging from the muscle weakness of myopathy to the akinesia of Parkinson's disease) it intrudes into consciousness and is perceived as a sensation of "fatigue", "heaviness", or "increased effort".

Conclusion

Kinaesthesia encompasses a group of sensations including those of limb position and movement and those of force and heaviness. No single mechanism should be proposed to explain them all. While electrophysiological techniques are available to study specifically the peripheral and central conduction of kinaesthetic afferents, psychophysiological techniques can be used to quantify a kinaesthetic deficit and to determine whether it involves muscle, joint, or cutaneous afferents.

Summary

Proprioceptive and kinaesthetic sensations include the sensations of limb position and movement and the sensations of muscle force and tension. While signals from cutaneous, joint and intramuscular receptors together with signals of motor command (or "effort") can theoretically contribute to this group of sensations, the actual role of each signal is not well defined.

Evidence for a contribution from muscle spindle afferents in the sensation of limb position and movement is presented. This includes the illusions of movement and distortions of position produced by trains of weak electrical stimuli delivered at below motor threshold to muscle fascicles in the ulnar nerve. Kinaesthetic acuity at the distal interphalangeal joint deteriorates when the muscles acting at the joint are effectively disengaged with further deterioration occurring when the capsule of the joint is anaesthetized as well. This suggests that, under some circumstances, inputs from joint receptors may duplicate those provided by intramuscular receptors. While there may be a specific role for cutaneous afferents in kinaesthesia, no definite evidence for this has emerged. However, there is evidence that they may play a facilitatory role. The preferred signal of muscular force (or perceived heaviness) appears to be biased by a signal related to the size of the outgoing motor command rather than by one related to the actual force achieved. The properties of the former command-related signal are reviewed.

Acknowledgment. Much of the work described in this chapter has been supported by the National Health and Medical Research Council of Australia.

References

1. Angel R (1980) Barognosis in an patient with hemiataxia. Ann Neurol 7:73–77
2. Barrack RL, Skinner HB, Brunet ME, Haddad RJ (1983 a) Functional performance of the knee after intraarticular anesthesia. Am J Sports Med 11:258–261
3. Barrack RL, Skinner HB, Cook SD, Haddad RJ (1983 b) Effect of articular disease and total knee arthroplasty on knee joint-position sense. J Neurophysiol 50:684–687
4. Browne K, Lee J, Ring PA (1954) The sensation of passive movement at the metatarso-phalangeal joint. J Physiol (Lond) 126:448–458
5. Cafarelli E, Bigland-Ritchie B (1979) Sensation of static force in muscles of different length. Exp Neurol 65:511–525
6. Clark FJ, Horch KW, Bach SM, Larson GF (1979 a) Contribution of cutaneous and joint receptors to static knee-position sense in man. J Neurophysiol 42:878–888
7. Clark FJ, Matthews PBC, Muir RB (1979 b) Effect of the amplitude of muscle vibration on the subjectively experienced illusion of movement. J Physiol (Lond) 296:14–15P
8. Craske B (1977) Perception of impossible limb positions induced by tendon vibration. Science 196:71–73
9. Cross MJ, McCloskey DI (1973) Position sense following surgical removal of joints in man. Brain Res 55:443–445
10. Devanandan MS, Ghosh S, Simoes EAF (1980) The myelinated fibers of the deep branch of the ulnar nerve at the wrist in bonnet monkeys (*Macaca radiata*) and some of its branches to the hand. Anat Rec 197:387–396
11. Eklund G (1972) Position sense and state of contraction: the effects of vibration. J Neurol Neurosurg Psychiatry 35:606–611
12. Gandevia SC (1978) The sensation of heaviness after surgical disconnection of the cerebral hemispheres in man. Brain 101:295–305
13. Gandevia SC (1982) The perception of motor commands or effort during muscular paralysis. Brain 105:151–159
14. Gandevia SC (1983) Psychophysical and electrophysiological evidence for the cortical projection of muscle afferents in man. Proc Aust Physiol Pharmac Soc 14:241–259
15. Gandevia SC (1985) Illusory movements produced by stimulation of low-threshold muscle afferents from the hand. Brain 108
16. Gandevia SC, McCloskey DI (1976) Joint sense, muscle sense, and their combination as position sense, measured at the distal interphalangeal joint of the middle finger. J Physiol (Lond) 260:387–407
17. Gandevia SC, McCloskey DI (1977) Sensations of heaviness. Brain 100:345–354
18. Gandevia SC, McCloskey DI (1978) Interpretation of perceived motor commands by reference to afferent signals. J Physiol (Lond) 283:493–499
19. Gandevia SC, Burke D, McKeon B (1984) The projection of muscle afferents from the hand to cerebral cortex in man. Brain 107:1–13
20. Gandevia SC, Hall LA, McCloskey DI, Potter EK (1983) Proprioceptive sensation at the terminal joint of the middle finger. J Physiol (Lond) 335:507–517
21. Gandevia SC, Killian KJ, Campbell EJM (1981) The effect of respiratory muscle fatigue on respiratory sensations. Clin Sci 60:463–466
22. Gelfan S, Carter S (1967) Muscle sense in man. Expl Neurol 18:448–473
23. Goodwin GM, McCloskey DI, Matthews PBC (1972) The contribution of muscle afferents to kinaesthesia shown by vibration-induced illusions of movement and by the effects of paralysing joint afferents. Brain 95:705–748
24. Grigg P, Finerman GA, Riley LH (1973) Joint position sense after total hip replacement. J Bone Joint Surg 55A:1016–1025
25. Hobbs SF, Gandevia SC (1985) Cardiovascular responses and the sense of effort during attempted muscular contractions: role of the spinal cord. Neurosci Lett 57:83–90

26. Holmes G (1922) The Croonian Lectures on the clinical symptoms of cerebellar disease and their interpretation. Lecture IV. Lancet II:111–115
27. Jones LA (1983) Role of central and peripheral signals in force sensation during fatigue. Exp Neurol 81:497–503
28. Karanjia PN, Ferguson JH (1983) Passive joint position sense after total hip replacement surgery. Ann Neurol 13:654–657
29. Kelso JAS, Holt KG (1980) The role of proprioception in the perception and control of human movement: toward a theoretical assessment. Percept Psychophys 28:45–52
30. McCloskey DI (1973) Differences between the sense of movement and position shown by the effects of loading and vibration of muscles in man. Brain Res 61:119–131
31. McCloskey DI (1978) Kinesthetic sensibility. Physiol Rev 58:763–820
32. McCloskey DI (1981) Corollary discharges: motor commands and perception. In: Brooks VB (ed) Handbook of physiology. I. The nervous system. American Physiological Society, Bethesda, pp 1415–1448
33. McCloskey DI, Ebeling P, Goodwin GM (1974) Estimation of weights and tensions and apparent involvement of a "sense of effort". Exp Neurol 42:220–232
34. McCloskey DI, Colebatch JG, Potter EK, Burke D (1983 a) Judgements about onset of rapid voluntary movements in man. J Neurophysiol 49:851–863
35. McCloskey DI, Cross MJ, Honner R, Potter EK (1983 b) Sensory effects of pulling or vibrating exposed tendons in man. Brain 106:21–37
36. McCloskey DI, Ferrell WR, Gandevia SC (1985) Contribution from joint receptors to detection of movements imposed on the finger joint while muscle afferents cannot contribute. Proc Aust Physiol Pharmac Soc 16:34P
37. McKeon B, Burke D (1983) Muscle spindle discharge in response to contraction of single motor units. J Neurophysiol 49:291–302
38. Matthews PBC (1982) Where does Sherrington's "muscular sense" originate? Muscles, joints, corollary discharges. Ann Rev Neurosci 5:189–219
39. Matthews PBC, Simmonds A (1974) Sensations of finger movement elicited by pulling upon flexor tendons in man. J Physiol (Lond) 239:27–28P
40. Moberg E (1983) The role of cutaneous afferents in position sense, kinaesthesia, and motor function of the hand. Brain 106:1–19
41. Ochoa J, Torebjörk E (1983) Sensations evoked by intraneural microstimulation of single mechanoreceptive units innervating the human hand. J Physiol (Lond) 342:633–654
42. Provins KA (1958) The effect of peripheral nerve block on the appreciation and execution of finger movements. J Physiol (Lond) 143:55–67
43. Roland PE, Ladegaard-Pedersen H (1977) A quantitative analysis of sensations of tension and kinaesthesia in man. Brain 100:671–692
44. Roll JP, Vedel JP (1982) Kinaesthetic role of muscle afferents in man, studied by tendon vibration and microneurography. Exp Brain Res 47:177–190
45. Simon J, Wei JY, Randic M, Burgess PR (1984) Signaling of ankle joint position by receptors in different muscles. Somatosens Res 2:127–147
46. Tardy-Gervet MF, Gilhodes JC, Roll JP (1982) Demonstration of an illusory limb movement and associated motor activities induced by a moving visual stimulus in man. A descriptive study. Neurosci Lett 28:187–192
47. Vallbo ÅB, Olsson KÅ, Westberg K-G, Clark FJ (1984) Microstimulation of single tactile afferents from the human hand. Sensory attributes related to unit type and properties of receptive fields. Brain 107:727–749

Noxious Cutaneous Input and the Tactile Exploratory Function of the Skin of the Hand

J. Gybels [2], H. Adriaensen [1], H. O. Handwerker [3], and J. Van Hees [2]

Introduction

In daily life, during skillful motor behaviour such as violin playing, or during tactile exploratory behaviour, the skin of the hand is subjected to possible noxious agents without the subject's experiencing a pain sensation. In other words, under certain conditions, the skin's nociception function is subordinated to the explorative role of the skin.

Since 1972, work in Leuven has accumulated many data on nociceptors and conscious experience. In the present account, which is based mainly on the publications of the Leuven group, findings will be highlighted which may shed some light on this subordination of nociception to mechanoception.

To do this, we will examine the conditions in which there is congruence and mismatch between nociceptor activity and sensation, during both thermal and mechanical stimulation of the skin. We will then briefly indicate what might be the physiological significance of these findings and what they might mean for the neurological clinic.

Congruence Between Nociceptor Activity and Sensation

Figure 1 summarizes the data of a series of experiments in which the subject was asked to judge thermal stimuli with graded intensities as painful or non-painful. Stimuli from 40° to 55 °C (radiant heat) were given in a random order in 8 experiments on 8 different polymodal C-nociceptors. In addition to the signalling of the first pricking or burning sensation the subjects were asked to interrupt the stimulus by saying "stop" when pain became strong. It can be seen that C-fibre activity of more than 0.4 spikes/s is usually (86%) accompanied by pain reports; stimuli provoking discharges of less than 0.2 spikes/s were usually (86%) judged as non-painful. Statistical analysis (point-biserial correlation coefficient) led to the conclusion that the subjective judgment correlated both with the C-fibre response and with the skin temperature [10].

In a second series of experiments the discharge pattern in C-nociceptors was computed during controlled heating of the receptive field with the estimate of the sensory magnitude of identical stimuli. Sensory magnitude estimation on a six-

[1] Departments of Anesthesiology, [2] Neurology and Neurosurgery, University of Leuven (K.U.L.), B-3000 Leuven, Belgium.
[3] II. Physiologisches Institut, University of Heidelberg, D-6900 Heidelberg, FRG.

Clinical Aspects of Sensory Motor Integration
Ed. by A. Struppler and A. Weindl
© Springer-Verlag Berlin Heidelberg 1987

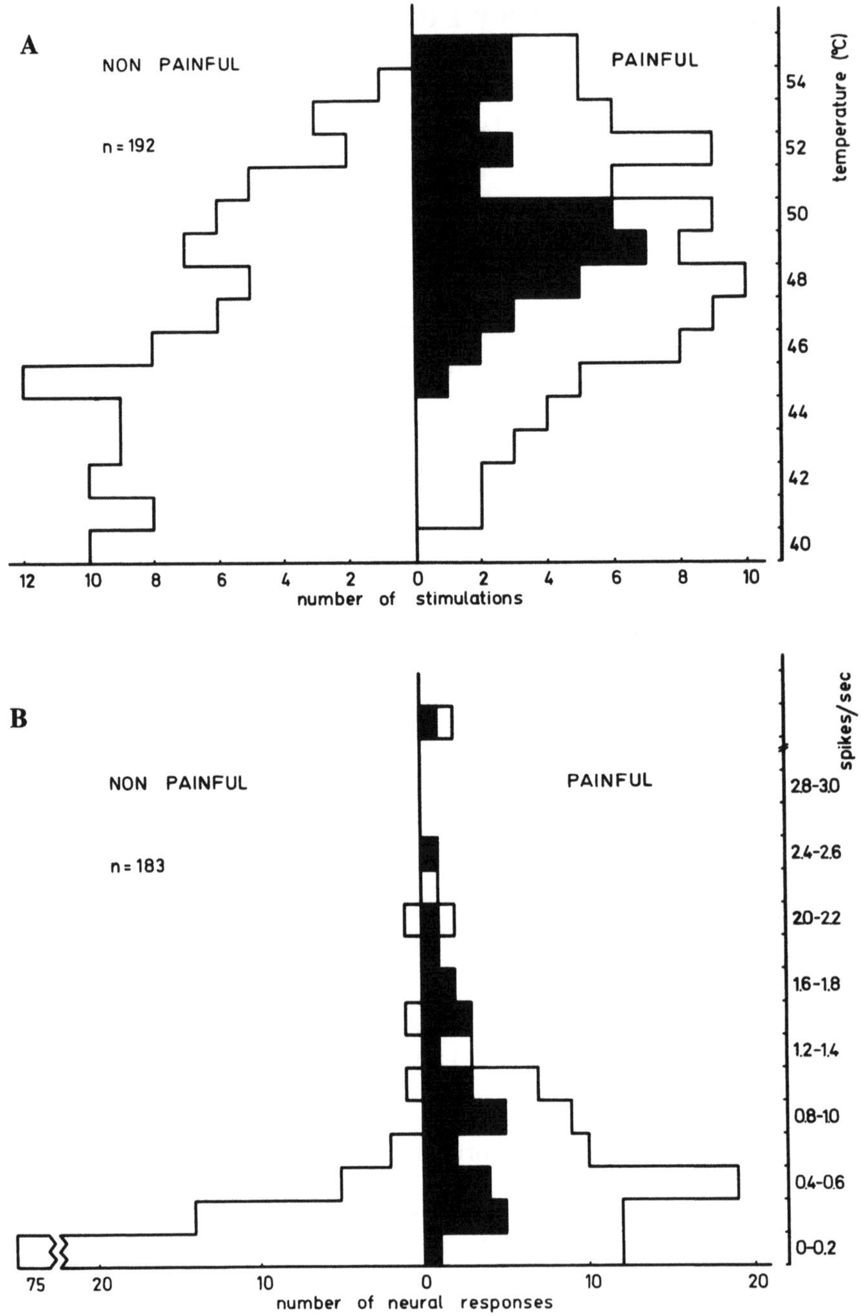

Fig. 1. A Distribution of 192 heat stimuli (from 8 experiments) in the classes of judgment in function of the skin temperature. *Black* represents strong pain. **B** Distribution of 183 C-fibre responses (from the same 8 experiments) according to the subject's judgments. C-fibre activity higher than 0.4 spikes/s is mostly (86%) accompanied by report of pain (Van Hees and Gybels [10])

Table 1. Six-point scale of sensory magnitude estimation (Gybels et al. [4])

Rating	Definition		
	Dutch	German	English
1	Juist voelbaar	Eben merklich	Just noticeable
2	Lichte warmte	Etwas wärmer	Slightly warmer
3	Duidelijke warmte	Deutlich wärmer	Clearly warmer
4	Heet, licht prikkend	Heiß, leicht stechend	Hot, slightly stinging
5	Heet, duidelijk prikkend	Heiß, deutlich stechend	Hot, clearly stinging
6	Pijnlijke warmte	Schmerzhaft heiß	Very hot, painful

point scale was used; Table 1 specifies this scale. The three lower points of the scale were designed to specify degrees of sense of warmth; the ratings four or higher had to be selected when the sensation had changed quantitatively and was of a stinging character.

Figure 2 shows the relationship between number of spikes/stimulus in 4 C-nociceptors and the ratings of the subjects.

These results suggest that the C-fibres recorded in these 4 experiments are better discriminators of stimulus levels producing stinging heat sensation than levels producing the sense of warmth. A comparison of the discriminative power of spike discharge and ratings indicated that both systems could about equally discriminate between the three stimulus levels used (2°, 2.5°, and 3 °C above a 43 °C or 43.5 °C base temperature). From this analysis it could be concluded that C-fibre input provides the necessary information for subjective estimation of painful heat stimuli [4]. Is the well-known suppression of C-fibre activity, i.e. a diminished

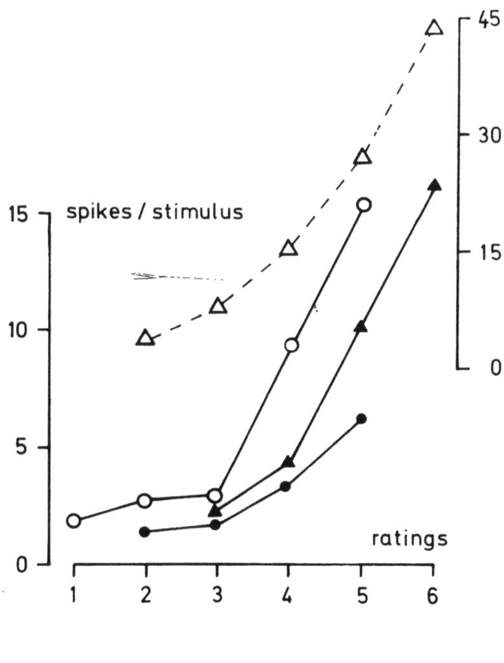

Fig. 2. Relationship between number of spikes/stimulus in 4 °C units and the ratings of the subjects. Discharges of unit △ are calibrated in the *right* ordinate, the others in the *left* one. The table below shows the statistical significance of the difference between the discharges related to each two adjoining ratings (Mann-Whitney *U*-test). (Gybels et al. [4])

○	n.s.	n.s.	$P<0.01$	$P<0.01$	–
▲	–	–	$P<0.05$	$P<0.01$	$P<0.01$
●	–	n.s.	$P<0.01$	$P<0.01$	–
△	–	$P<0.01$	$P<0.01$	$P<0.01$	$P<0.01$

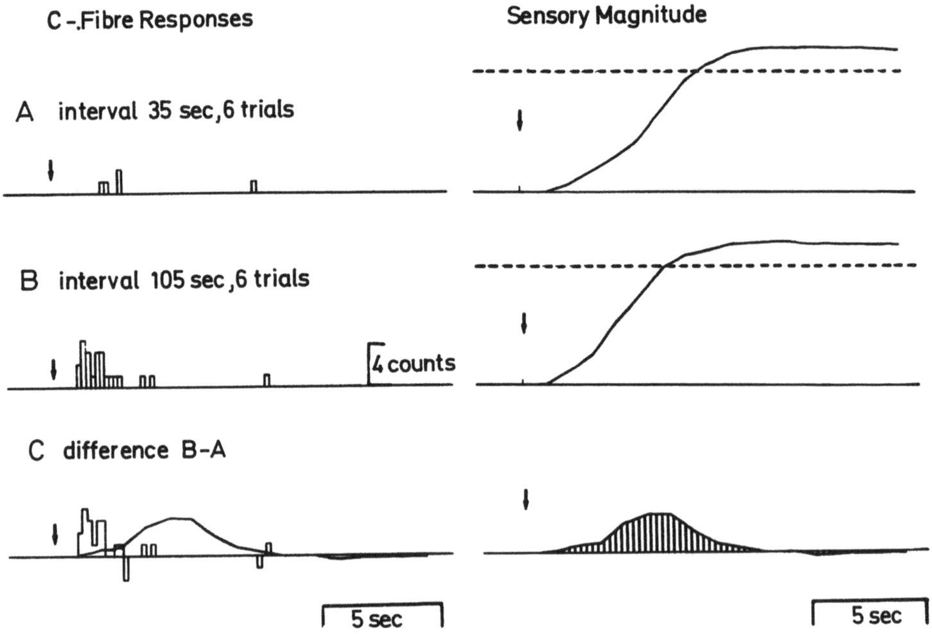

Fig. 3. Comparison of interval influences on spike discharges and on subjective sensory ratings. **A** and **B** represent frequency histograms of spike activity (*left*) and sensory magnitude ratings (*right*) after a 35-s and a 105-s interval, respectively. Data are from a single experiment (adapting temperature 35 °C; stimulus temperature 41 °C), and the reported mean values are averaged from 6 trials each. The subtraction of **A** from **B** is shown in **C**. The difference in sensory rating is superposed on the respective differences in spike discharges in the left diagram. The *horizontal dotted line* in the right diagram indicates the pain threshold (20 LEDs). The *arrow* in each diagram indicates the start of the stimulus. Latency to electrical stimulation of the receptive field was 165 ms in this fibre (cv 1.1 m/s). No correction for conduction delays was made. (Adriaensen et al. [2])

response to the following stimulus, reflected in a reduction of pain sensation? Figure 3, in which a comparison is shown of interval influences on spike discharges and on subjective sensory ratings, demonstrates that this is the case for the unit under study. The same conclusion can be drawn when a C-fibre population is analysed [2].

From these three experiments it can be concluded that nociceptive input provoked by thermal stimuli correlates well with pain sensation.

Mismatch Between Nociceptor Activity and Sensation

Figure 4 shows responses of a single C-nociceptor recorded from the superficial radial nerve to mechanical and warmth stimuli.

It is apparent that C-fibre activity up to 10 spikes/s evoked by von Frey hairs can be accompanied by a sensation of pressure without any pain. Bearing in mind,

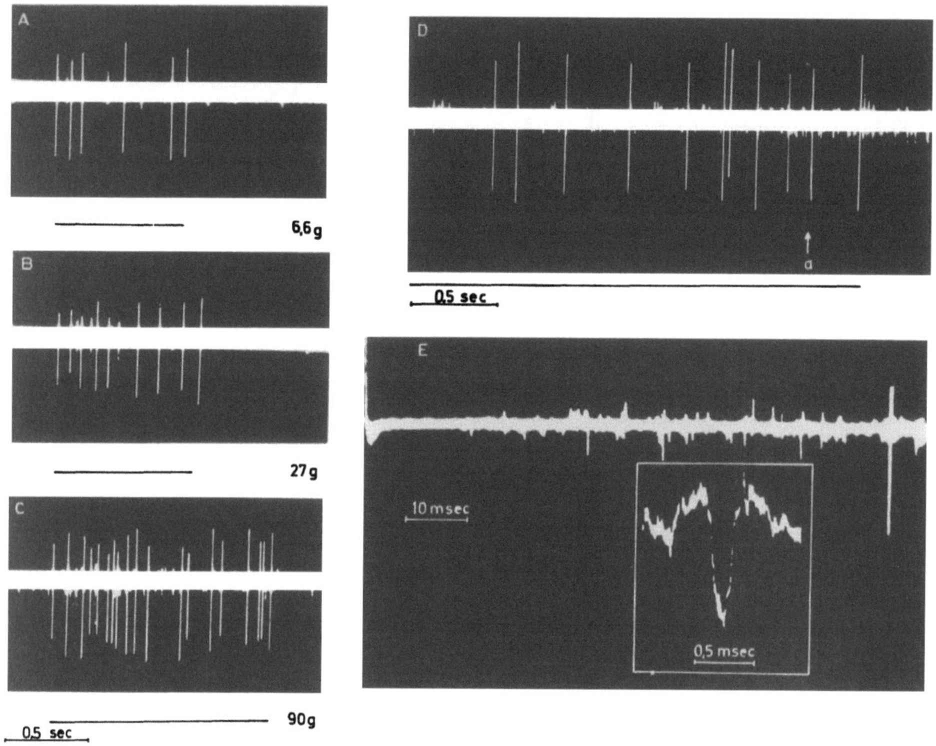

Fig. 4. Example of afferent C-fibre unit activity in human radial nerve, in response to skin stimulation with von Frey hairs of different intensity in **A, B,** and **C,** and to irradiation warmth in **D.** The *horizontal lines* indicate the duration of the stimulus. In **A, B,** and **C** the subject reports a sensation of pressure. At *arrow a* (**D**) the subject reports the onset of pain. **E** Response to electrical stimulation of receptive skin area; latency time, 90 ms; distance between stimulating and recording point, 8.5 cm. Noise amplitude is reduced by amplitude discriminator. (Van Hees [9])

as we have shown before, that strong heat pain was only rarely accompanied by C-fibre discharge above 2 spikes/s (see Fig. 1), it is evident that we are actually dealing with a strong activation of a peripheral nociceptive system without evoking pain sensation. It could be argued that von Frey hairs activate a far smaller number of C-nociceptors than the used heat stimulus (which in our experiments irradiated a skin surface of about 1.75 cm^2), and that the pain sensation evoked by heat could in part be the result of a greater degree of spatial summation. However, stimulating with several von Frey hairs at the same time at neighbouring points does not change the sensation of pressure into a sensation of pain, although it has been estimated [9] that in this condition as many C-fibres can be excited as by the heat stimulus used.

In a second series of experiments we examined quantitatively the relationship between single nerve discharges and pain sensations induced by mechanical stimuli. To do this, skinfolds were squeezed with a feedback-controlled forceps, whereby a defined force was exerted for 120 s between 2 surfaces of 16 mm^2. The subjects were instructed to rate the magnitude of the stimulus-induced sensations

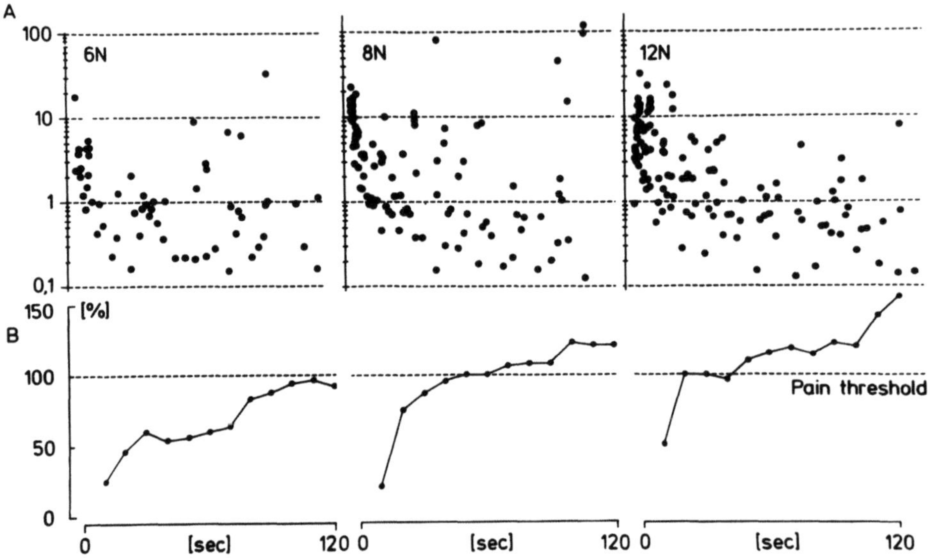

Fig. 5 A, B. Response of a human C-fibre polymodal nociceptor to squeezing of a skin fold for 120 s with different forces and concomitant pain ratings of the subjects. **A** Instantaneous firing frequencies of the nociceptive unit computed from the interspike intervals are plotted on a logarithmic scale (ordinate) versus time (abscissa). The three diagrams are from stimulation with 6, 8, and 12 N respectively. **B** The pain ratings are represented on the ordinate in percentage of the pain threshold. (Adriaensen et al. [3])

by moving a lever that controlled a chain of LEDs representing a visual analogue scale. Figure 5 shows data from an experiment in which discharges of a C-fibre nociceptor were recorded (Fig. 5 A) together with the sensory reports of the subject (Fig. 5 B). The latter shows an increase of painfulness during the course of a stimulation period. In contrast, the nociceptive unit shows discharges of a relatively high frequency at the beginning of each stimulus, i.e. a dynamic response followed by slowly adapting tonic discharges. Thus polymodal C-fibre nociceptors do not seem to provide the information leading to increased pain responses in the time course of a sustained mechanical stimulus [3].

We have postulated, when observing the lack of pain induced by stimulation with von Frey hairs exciting polymodal C-fibre nociceptors, that coactivation of A-beta mechanoreceptors by these stimuli has an inhibitory influence on the nociceptive C-fibre input somewhere in the central nervous system. To test this hypothesis, we have computed the mean activity of C-fibre polymodal nociceptors and compared it with the respective activity of SA I and SA II units during conditions of prolonged suprathreshold mechanical stimulation, and computed the ratio of C and SA fibre activity second by second (Fig. 6). When only the SA I fibre population is used for this computation a slightly increasing ratio throughout the 120-s stimulation period is found. This increase is more evident when the ratio between the C-fibre and the total SA fibre activity is calculated. The assumption that C-fibre input may be suppressed by concomitant A-fibre in-

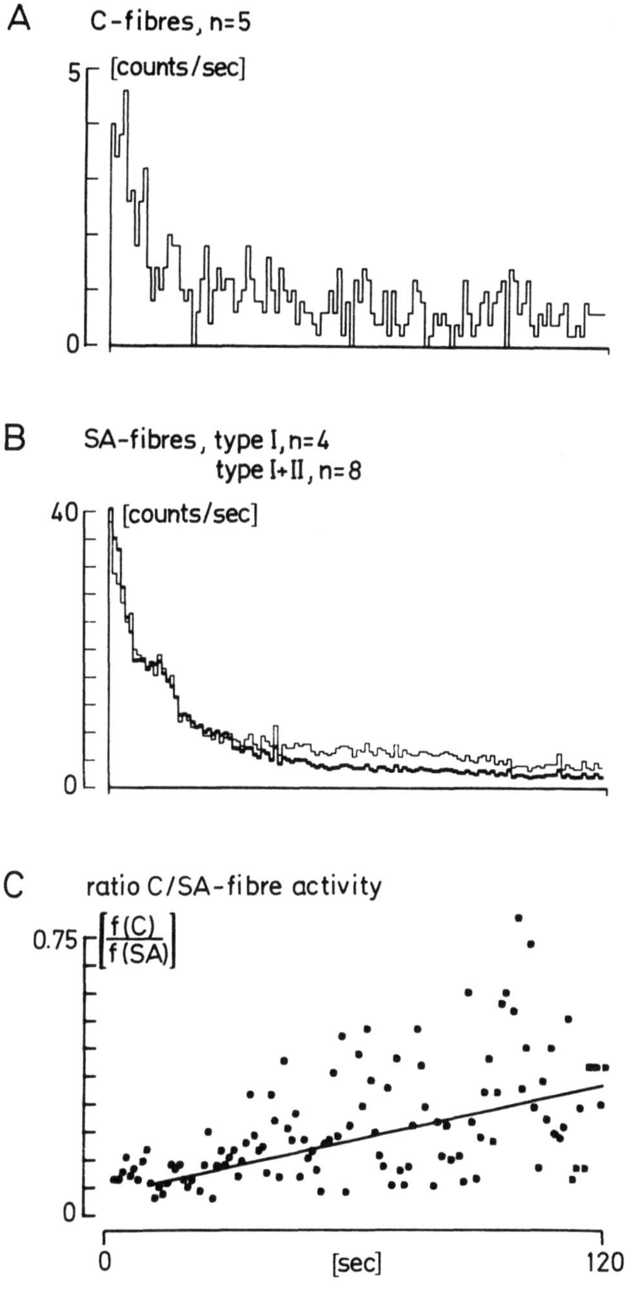

Fig. 6. Population responses of **A** polymodal C-nociceptors and **B** SA mechanoreceptors to squeeze stimuli of 12 N. Histograms in **A** and **B** represent arithmetic means of the responses obtained, with a bin width of 1 s. The *thin line* in B shows the histogram obtained from a small population of SA I fibres, the *thick line* shows the respective response of a population of SA I and SA II mechanoreceptors. **C** Ratio of C to total SA fibre activity computed for each bin. The *straight line* represents the regression line computed for data obtained 10–120 s after the stimulus onset. The correlation coefficient was 0.51. (Adriaensen et al. [3])

put from mechanoreceptors during the first seconds of a mechanical stimulus seems even more justified when consideration is given not only to the slowly adapting but also to the rapidly adapting mechanoreceptors, which fire vigorously but only for some seconds at the beginning of a stimulus.

Physiological and Clinical Implications

Several hypotheses may explain the mismatch between nociceptor discharge and sensory reports. We have postulated that coactivation of A-beta mechanoception has an inhibitory influence somewhere in the central nervous system on the nociceptive input, a hypothesis which was put forward by Noordenbos [7] in his "pattern theory" and by Melzack and Wall [6] in their "gate-control theory". Afferent impulses in nociceptive fibres apparently only evoke pain sensations if other peripheral conditions are satisfied. Our experience regarding the mechanical excitability of C-nociceptors has forced us to realise that considerable activation of cutaneous nociceptors takes place in many conditions of daily life that are not experienced as painful, such as combing the hair, using scissors, eating with knife and fork, etc. This inhibitory influence of beta input on nociceptive C input could be interpreted as a way of subordinating the function of nociception to the explorative role of the skin. Cutaneous exploration is served in the first place by mechanoception. It would be impaired if every mechanical stimulus which threatens the skin gave rise to pain sensation. Of course, nature could have raised the mechanical threshold for cutaneous nociceptors, but by keeping the threshold to the strictly noxious level and providing at the same time a pain inhibitory system, another function of the C-nociceptors remained possible, namely, a contribution to the repair process by means of the axon reflex-induced vasodilatation [5].

In this context, it is important to consider the response properties of A-delta fibres. In a study concerning 140 human A-delta fibres, we have shown that A-delta fibres with high mechanical threshold show a higher receptor specificity than C polymodal nociceptors, and that the firing frequency on noxious stimulation is often higher in A-delta fibres than in C-fibres, suggesting that individual A-delta receptors may contribute more information on stimulus quality than individual C-fibres. Our data do not allow, however, a quantitative study of whether the discharge patterns of individual A-delta fibres or the ratings of the subject discriminate better between different stimulus levels [1]. We still do not know why we need a double pain-triggering system although some intriguing suggestions have been formulated [11].

The receptor characteristics of polymodal C-nociceptors and their relation to sensations also have clinical implications, of which we will mention only two examples:

- When in the neurological clinic "epicritic" sensation is being tested, such as two-point discrimination or graphesthesia, we are not only examining the so-called "lemniscal system", but C-polymodal nociceptors are strongly activated by the mechanical stimulus.
- In certain pathological conditions, in which A-beta fibre input is diminished or absent, e.g. during nerve compression or ischemia, or in certain forms of polyneuropathy, a mechanical stimulus which normaly does not provoke pain will be felt as painful; in such conditions a severe burning pain sensation is experienced when the skin is slightly pinched or a fingernail is stroked over it.

Much work remains to be done in this field. Nonetheless, it has already been demonstrated that microneurography in patients [8] is a powerful tool for the investigation of clinical problems.

Summary

This paper examines the conditions in which there is congruence and mismatch in the human between nociceptor activity as recorded by microneurography and sensation, during thermal and mechanical stimulation of the skin.

In the experimental conditions used, congruence was observed when the noxious stimulus was a thermal stimulus, while mismatch occured when the noxious stimulus was a mechanical one.

It is postulated that co-activation of A-beta mechanoception has an inhibitory influence somewhere in the nervous system on the nociceptive input, thereby providing a mechanism by which the function of nociception is subordinated to the explorative role of the skin.

Acknowledgments. This work was supported by the F.G.W.O. of Belgium, Grant 3.0053.83, and by the DFG, Grant Ha 831/8. The informed consent of all subjects was obtained according to the Declaration of Helsinki [British Medical Journal (1964) 2, 1977].

References

1. Adriaensen H, Gybels J, Handwerker HO, Van Hees J (1983) Response properties of thin myelinated (A-delta) fibers in human skin nerves. J Neurophysiol 49:111–122
2. Adriaensen H, Gybels J, Handwerker HO, Van Hees J (1984a) Suppression of C-fibre discharges upon repeated heat stimulation may explain characteristics of concomitant pain sensations. Brain Res 302:203–211
3. Adriaensen H, Gybels J, Handwerker HO, Van Hees J (1984b) Nociceptor discharges and sensations due to prolonged noxious mechanical stimulation – a paradox. Human Neurobiol 3:53–58
4. Gybels J, Handwerker HO, Van Hees J (1979) A comparison between the discharges of human nociceptive nerve fibres and the subject's ratings of his sensations. J Physiol (Lond) 292:193–206
5. Lewis T (1942 – Facsimile edition 1981) Pain. Macmillan, London
6. Melzack R, Wall PD (1965) Pain mechanisms: a new theory. Science 150:971–979
7. Noordenbos W (1959) Pain. Elsevier/North-Holland, Amsterdam
8. Nordin M, Nyström B, Wallin U, Hagbarth KE (1984) Ectopic sensory discharges and paresthesiae in patients with disorders of peripheral nerves, dorsal roots and dorsal columns. Pain 20:231–245
9. Van Hees J (1979) De C-nociceptor bij de mens en zijn rol in pijnsensatie. Thesis. Leuven
10. Van Hees J, Gybels J (1981) C nociceptor activity in human nerve during painful and nonpainful skin stimulation. J Neurol Neurosurg Psychiatry 44:600–607
11. Wall PD (1984) Introduction. In: Wall PD, Melzack R (eds) Textbook of pain. Churchill-Livingstone, Edinburgh, pp 1–16

New Aspects of the Role of Articular Receptors in Motor Control

H.-G. Schaible [1], R. F. Schmidt [1], and W. D. Willis [2]

Historical Perspective

Our present knowledge of the effects of impulses in articular afferent units on the motor system is characterized by variable and in part conflicting results and opinions. In his comprehensive review on joint receptors Skoglund [24] summarized the evidence available at that time. He pointed out that volleys in joint afferents are able to elicit polysynaptic ventral root discharges [4, 10] and that most authors seem to have obtained facilitation of flexor and inhibition of extensor motoneurons [4, 10, 23]. The fact that opposite effects were sometimes observed and that the reflex effects tended to be highly variable in different animals [10, 23] was taken to indicate that the experimental methods either favored one or the other component of the joint afferent input or that the central excitability state of the animal was different in the various experimental settings.

More recently Lundberg et al. [14] found that electrical stimulation of the posterior articular nerve (PAN) facilitated transmission in disynaptic and trisynaptic inhibitory and excitatory reflex pathways from Ib afferents. They suggested that impulses from joint receptors can influence the regulation of muscle tension from Golgi tendon organs. Since joint receptors seem mainly to be activated at the extremes of joint position (cf. [5]), their activity may contribute to a purposeful decrease of muscle tension in the terminal phase of a movement.

When joint afferents were excited using small extension movements of the knee, Grigg et al. [11] observed that monosynaptic reflexes of knee extensors (vasti) were facilitated and those of knee flexors (posterior biceps, semitendinosus) were inhibited. The effects disappeared when the PAN was cut, regardless of whether or not the medial articular nerve (MAN) was intact. These results of a positive feedback induced by joint extension are difficult to reconcile with the findings of Lundberg et al. [14], which point to a negative feedback action of articular afferent impulses.

Finally, Baxendale and Ferrell [2], in agreement with Lundberg et al. [14], reported that the excitability of flexion withdrawal and crossed extensor reflexes was modulated by knee joint position. Flexion withdrawal reflexes were most easily elicited when the knee was extended, and crossed extensor reflexes were most easily elicited when the knee was flexed. These reflexes should act to decrease the probability of the knee being hyperflexed or hyperextended by movements.

[1] Physiologisches Institut der Julius-Maximilians-Universität, Röntgenring 9, D-8700 Würzburg, FRG.
[2] Marine Biomedical Institute, University of Texas Medical Branch, 200 University Boulevard, Galveston, TX 77550-2772, USA.

Clinical Aspects of Sensory Motor Integration
Ed. by A. Struppler and A. Weindl
© Springer-Verlag Berlin Heidelberg 1987

Present Status

In all the studies just reviewed it was more or less taken for granted that joint receptors with large myelinated afferents (group II afferents) have low to medium thresholds to movements and that the fine myelinated (group III) as well as the unmyelinated afferents (group IV) do not discharge to movements in the physiological, innocuous working range of the joint. Accordingly, low-strength electrical stimulation of either PAN or MAN was supposed to elicit volleys in nonnociceptive afferent units, whereas high-strength stimulation should almost exclusively activate nociceptive units.

As will be briefly outlined below, these assumptions probably have all to be abandoned in view of the results of our recent studies on the receptive properties of fine afferent units in the knee joint of the cat. It will be shown that these fiber groups contain afferent units which in their sensitivity to movement range from very low threshold ones to those which in normal joints are not activated by any movement, even a very noxious one, and that joint inflammation (or other sensitizing processes) is required to "wake up" such "sleeping nociceptors". As a consequence of these findings the role of articular units in motor control has to be reconsidered, and new experiments which take this new situation into account will have to be designed and executed. Some preliminary results will be briefly discussed.

Composition of Joint Nerves

Articular nerves are small compared to the overall somatosensory input into the spinal cord and brain stem. For instance, in the cat (and, similarly, in other mammals, including humans) the knee joint is innervated mainly by two small nerves of about equal size, namely, the medial and posterior articular nerves, MAN and PAN, respectively. Together they contain just about 450 medullated and 800 unmedullated afferent nerve fibers [13], whereas the combined nerve to the lateral gastrocnemius and soleus muscle alone contains some 480 medullated and about 1300 unmedullated afferent nerve fibers (cf. [16]). Thus it may not be unexpected that articular afferent inflow may have much more subtle actions on the motor system than either muscle or cutaneous afferent input.

The myelinated fiber group contains large myelinated fibers (group II afferents) with corpuscular endings in the joint tissue (cf. [9]) and thin myelinated fibers (group III afferents), which, just as the unmyelinated afferents (group IV), terminate in so-called "free nerve endings". The exact percentages of group II versus group III afferents are not yet known, but from our own physiological as well as histological results it appears that at least 50% of the myelinated afferents belong to group III. Thus the knee joint appears to be innervated by about 200 large medullated afferents with corpuscular endings and more than 1000 fine afferents with unencapsulated nerve endings (cf. [1]), i.e., the vast majority of articular afferents (at least in MAN and PAN but presumably also in other joint nerves) are fine afferents of groups III and IV. This distribution of coarse versus fine articular afferent units has not generally been appreciated in the past, and in regard to the motor action of these units it has also not been fully taken into account that many of the fine units may have other than nociceptive function (see below).

Mechanosensibility of Fine Articular Afferent Units

Identification of Fine Joint Units

We recently carried out an intensive investigation of the properties of fine articular afferent units in the MAN [17–19] and in the PAN [12]. In these studies single units were isolated by microdissection and classified by determining their conduction velocities. Units conducting at less than 2.5 m/s were classified as group IV fibers, and those with conduction velocities from 2.5 to 20 m/s as group III units. The majority of group III units have conduction velocities between 3 and 12 m/s, whereas most group IV units have conduction velocities of less than 1.5 m/s.

Afferent units were only accepted as "articular" if they could be excited by electrical stimulation of either the MAN or PAN and if they clearly had one or several receptive fields in the joint tissue as revealed by probing with glass rods and a set of von Frey hairs. It was one of the surprises of this study that the receptive fields of mechanosensitive fine afferent units were nearly always small, regardless of their location on the fibrous capsule and the ligaments. In units which had more than one receptive field these were always clearly separated and sometimes far apart from each other.

Resting Activity and Responses to Local Mechanical Stimulation

Resting activity, defined as the occurence of discharges with a frequency of more than one impulse per minute when the joint is in mid position (normal resting position of the awake as well as sleeping cat), is found in about 30% of all group III and IV units with local mechanosensitive receptive fields [17]. The discharges are irregular, and their average frequency rarely exceeds 40 impulses per minute. Often it is less than ten impulses per minute.

Testing the local thresholds with von Frey hairs had the unexpected result that many of the mechanosensitive fine afferent units have very low thresholds – well outside the nociceptive range – to such stimuli (Fig. 1). In the group III range about one-third of the units have thresholds below 1 g and at least another third below 6 g. Thus about 70% of the group III units have low to moderate von Frey thresholds. With the group IV units the distribution of thresholds clearly shifts to higher values. The cumulative percentages for thresholds up to 1 g, 2 g, and 6 g were 3%, 33%, and 50%, respectively. Even if it is taken into account that the von Frey thresholds were tested after the removal of the overlying skin, these findings imply that half of the mechanosensitive units with myelinated fibers probably respond to everyday local mechanical stimuli of moderate intensity.

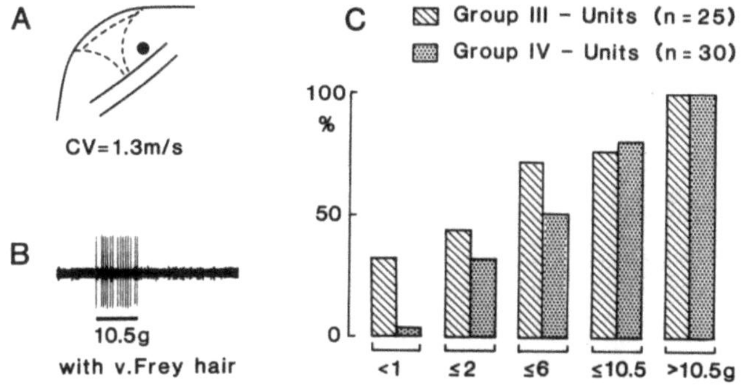

Fig. 1 A–C. Thresholds of fine articular afferent units in the medial articular nerve to local mechanical stimulation of their receptive fields. **A** Receptive field (*dot*) of a group IV unit with a conduction velocity (CV) of 1.3 m/s. **B** Tonic response of this unit to local mechanical stimulation within its receptive field. The stimulus strength of the von Frey hair was 10.5 g (threshold strength). In regard to its movement responses the unit belonged to category 3. **C** Diagram showing the cumulative distribution of local thresholds of 25 group III (*hatched columns*) and 30 group IV (*stippled columns*) articular afferent units. Note that 72% of the group III and 50% of the group IV units had thresholds ≦6 g. (Data in **C** from [19])

There is, however, a continuum of von Frey thresholds rather than a clear-cut separation of the low-threshold from the high-threshold ones.

Responses to Passive Joint Movements

With passive innocuous and noxious movements of the knee joint (flexion, extension, rotatory movements in the long axis of the tibia), four distinct categories of units were found in both fine fiber groups [18, 19]. An example of each category is illustrated in Fig. 2, and the proportion of each category in our sample from normal joints is shown in the upper graphs of Fig. 3 B.

The first category, comprising about 20% of all fine units, responds to innocuous joint movements, particularly rotations, in the normal working range of the joint (Fig. 2 A). Usually the units respond to more than one type of movement. Units of this type are more frequent in the group III than in the group IV range (cf. Fig. 3 B). Practically all of them increase their rate of discharge when the rotations are extended to the noxious range or when extreme flexions or extensions are applied.

The second type of unit shows very distinct differences in response rates to innocuous and noxious movements (Fig. 2 B). Such units were only weakly or occasionally activated by innocuous movements, whereas noxious movements led to pronounced discharges. Usually flexions and extensions were ineffective, and outward rotations were more effective than inward ones. About 16% of all fine articular units are in this category.

The third category is made up of units which do not respond to any innocuous joint movement but fire clearly and consistently to noxious ones (Fig. 2 C). All in

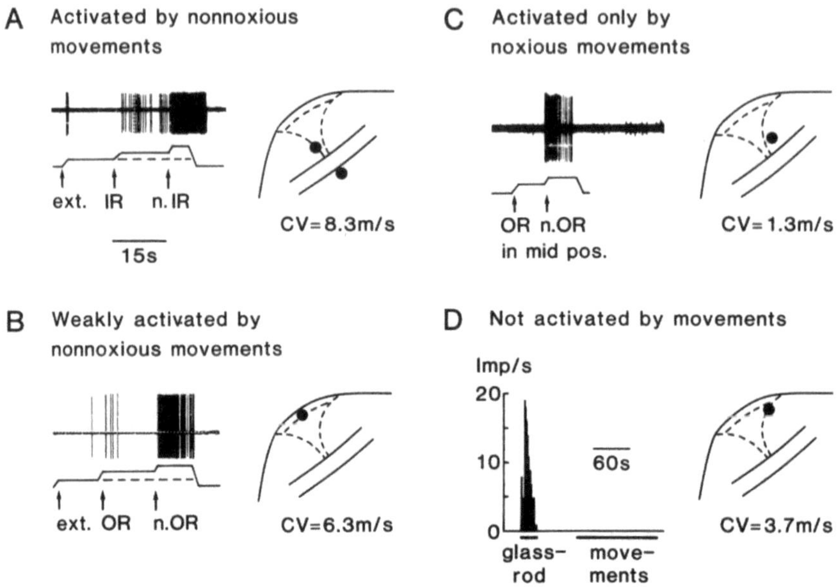

Fig. 2 A–D. Categories of fine articular afferent units in the medial articular nerve as revealed by their response behavior to passive movements of the knee joint. **A** Response behavior of a group III unit of category 1 to innocuous extension (ext.), innocuous inward rotation (IR), and noxious inward rotation (n.IR). The latter two movements started from the extended position. The unit had 2 receptive fields and a conduction velocity of CV = 8.3 m/s. **B** Response behavior of a group III unit of category 2 to innocuous extension (ext.), innocuous outward rotation (OR), and noxious outward rotation (n.OR). The latter two movements started from the extended position. The unit had one receptive field and a conduction velocity of CV = 6.3 m/s. **C** Response behavior of a group IV unit of category 3 to innocuous (OR) and noxious (n.OR) outward rotation in mid-position. The unit had one receptive field and a conduction velocity of CV = 1.3 m/s. **D** Response behavior of a group III unit of category 4 to local mechanical stimulation of its receptive field with a glass rod. The unit did not respond to any innocuous or noxious joint movements. It had one receptive field and a conduction velocity of CV = 3.7 m/s

all about 33% of fine units are in this category. Some of them respond to a number of outward and inward noxious rotations starting from various joint positions; others were only sensitive to outward rotations from one position.

The fourth and last type of unit has distinct and typical receptive fields, but they do not respond to any joint movement whatsoever ("sleeping nociceptor"; Fig. 2D). This finding is particularly surprising for units which respond well to high-strength local stimulation and which upon intense local stimulation exhibit distinct and long-lasting afterdischarges. All together about 31% of all fine units are of this category (Fig. 3B, upper graphs, stippled columns).

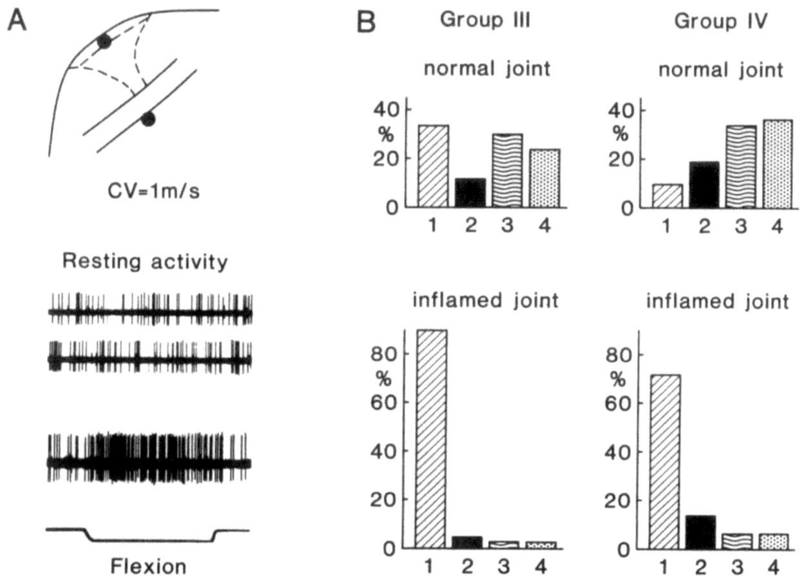

Fig. 3 A, B. Categories of fine articular afferent units in the medial articular nerve according to their response behavior to passive movements of normal (*upper columns* in **B**) and inflamed knee joints (*lower columns*). **A** Specimen records of the resting and evoked activity (Flexion) of a group IV unit of category 1 from an inflamed joint. The unit had two receptive fields and a conduction velocity of CV = 1 m/s. **B** The control population consisted of 33 group III and 41 group IV units, the inflamed population of 38 group III and 29 group IV units. (Data in **B** from [19, 20])

Modification of the Response Properties by an Acute Arthritis

Several rather impressive changes in the resting and response properties of fine articular afferent units can be observed in the course of an acute experimental arthritis [8, 20]. First of all, three-quarters of all units develop resting activity in mid-position of the joint which in frequency and pattern differs from that observed in normal joints: the frequency is much higher than in normal joints, and quite often considerable burst discharges are superimposed on the elevated background activity (Fig. 3 A).

The second major finding is that many units in inflamed joints respond to movements much more readily than in normal joints (compare the lower graphs in Fig. 3 B with the upper ones). In particular, pronounced responses to flexions and extensions are rare to absent in units from normal joints but frequent in those from inflamed ones.

Third, many units change their responses to local mechanical stimulation in the course of inflammation. The number of their receptive fields increases, and these fields are often surrounded by areas from which the unit can be excited by high-strength local stimulation, a feature which is practically never seen in normal joints.

Finally, it appears that the receptive properties of all four categories of fine articular units can be modified in their receptive properties by an acute experimental arthritis, although it has to be pointed out that not all units in each category may be similarly affected. In both fine fiber groups, units with several of the above-mentioned "signs of inflammation" were seen, whereas others displayed only one or two or none at all. In this respect an increased sensitivity to movement certainly is the best single indicator of inflammation.

Spinal Actions of Joint Afferent Volleys

Spinal Cord Dorsum and Field Potentials

A recent study of the population responses of spinal cord neurons to electrically induced volleys in the PAN revealed that such volleys evoke activity at levels of the spinal cord from at least L4 to S2 [21]. The population responses were examined by recordings of cord dorsum potentials (Fig. 4 A) and of the field potentials detected by a microelectrode inserted into the substance of the spinal cord (Fig. 4 B).

The cord dorsum responses consist of a series of negative waves: NI, NII, and NIII. The initial wave is due to the largest afferent fibers in the nerve (having con-

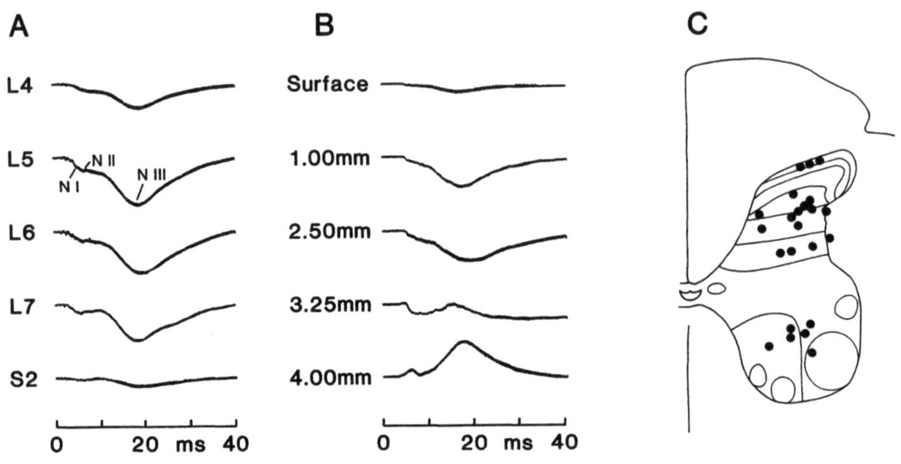

Fig. 4 A–C. Spinal cord field potentials evoked by stimulation of the posterior articular nerve (**A**, **B**), and localization of single spinal cord units responding to articular afferent volleys (**C**). **A** Relationship between the recording site along the length of the spinal cord and the amplitudes of the cord dorsum potentials. The potentials were signal averaged at the indicated sites along the lumbosacral enlargement. **B** Relationship between the depth of recording within the spinal cord and the amplitudes and signs of the field potentials. The signal-averaged records were made at the indicated depths from the dorsal surface of the L5 segment. **C** Localization of spinal cord neurons which were responsive to volleys evoked by electrical stimulation of the posterior articular nerve. Data from 10 cats spinalized at the lower thoracic level. Note that such units could not only be found in the dorsal but also in the ventral horn of the spinal cord

duction velocities of 43–87 m/s), the second wave is due to the action of the medium-sized afferent fibers (conduction velocity 33–53 m/s), and the NIII wave is produced by the smallest myelinated fiber group in the PAN with conduction velocities of 27 m/s or less. The relatively large size of this wave may be related to the large number of group III fibers in the nerve (see above: for a comparison of these waves with those induced by volleys in cutaneous and muscle afferents, see [21]).

No additional negative wave was noted when the stimulus intensity was elevated to include unmyelinated fibers in the afferent volley. This negative observation presumably reflects the asynchronous arrival of nerve impulses in the unmyelinated fibers from the joint due to wide range of slow conduction velocities in this fiber group.

The recording of field potentials in the depth of the spinal cord revealed that the potentials were negative when the recording electrode was in the dorsal horn, but reversed to become positive in the ventral horn (Fig. 4 B). Somewhat different generators for the different N waves are suggested by the observation that the different N waves had different reversal points along the same electrode track. However, considerable overlap of the generators is indicated by the finding that the maximum N waves were generally found at the same segmental level and at the same depth within the spinal cord gray matter.

Responses of Single Cord Neurons to Input from Normal and Inflamed Knee Joints

Recordings from interneurons revealed that afferent volleys in the PAN could excite neurons in segments L5 to S1 and at depths from less than 1–4 mm below the dorsal surface of the cord [21]. The largest number of units activated by articular afferent input were in L5 and L6, the same segments that generally showed the largest population response. Most marked recording sites were in laminae I, IV–VI, or VIII (Fig. 4 C).

Many units were activated just by the group II + III (A-fiber) afferent volley, but others could also be excited by the group IV (C-fiber) component. None were excited by group IV volleys alone. All of the units had a convergent input from nerves in addition to the PAN; i.e., none of the cells was excited exclusively by articular afferents.

Mechanical stimulation of the knee joint showed that many spinal cord neurons have receptive fields within joint structures and can be activated by joint movements [21]. On the basis of careful local exploration we can be confident that the receptive fields are in fact articular, particularly when localized probing or pressure was applied directly to the joint capsule or if pressure was applied through the skin when stimulation of the skin itself was ineffective. On the other hand, it was sometimes difficult to decide whether the effective local stimuli were better characterized as innocuous or noxious. With joint movements the intensity of the stimulus is easier to judge [18, 19]. However, when recording from spinal or other central neurons the effects of joint movements cannot be exclusively assigned to an action of knee joint afferents, since mechanical coupling to distant tissue could, to some extent, have activated receptors in other parts of the limb.

In limbs denervated except for the joint nerves, the largest proportion of joint-activated units could be excited by innocuous movements of the knee joint. These neurons could generally be excited still more vigorously by noxious movements. Thus most of these cells may have a convergent input from joint mechanoreceptors and nociceptors. They resemble in this property the "multiconvergent" or "wide dynamic range" neurons of the dorsal horn that respond to innocuous and noxious mechanical stimulation of the skin [15, 25]. Another group of units responded only to noxious intensities of joint movements. These cells resemble in this respect the "high-threshold" or "nociceptive-specific" neurons of the dorsal horn [6, 7].

In the course of the development of an acute arthritis in the knee joint spinal cord neurons with a weak excitatory input from the normal joint become extremely excited by even small movements in the physiological working range of the joint. As is exemplified in Fig. 5, continuous single-unit recording from lumbar neurons with axons projecting at least to the thoracic spinal cord (most of them presumably spinoreticular neurons) revealed that both the resting and the evoked activity of these neurons increases in parallel to the sensitization of the fine primary afferents by the experimental arthritis. It appears likely that in par-

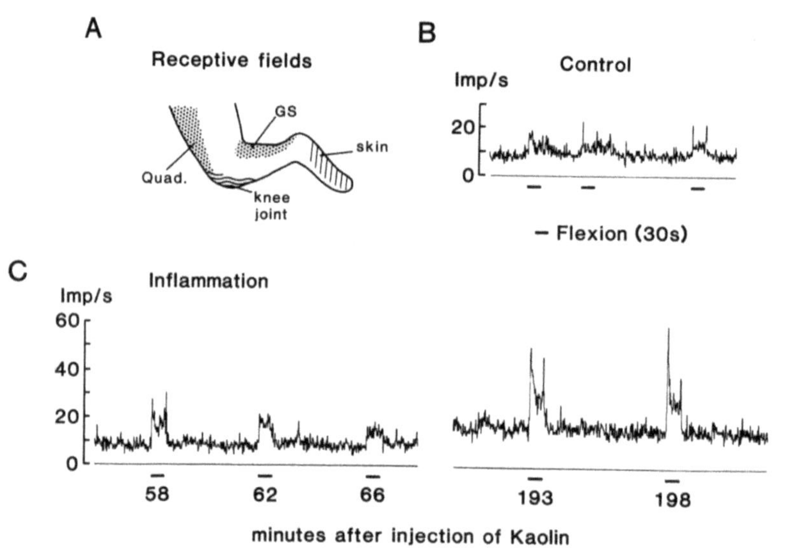

Fig. 5 A–C. Changes in the response pattern to passive joint flexions of a spinal cord neuron during the development of joint inflammation. The unit could be antidromically excited from the lower thoracic level. Spinalized preparation. **A** Receptive fields of the unit as revealed by careful manipulation of the leg. Such receptive fields could be detected on the skin of the foot, inside the quadriceps and gastrocnemius-soleus muscle bellies, and in the joint tissue (not in the skin overlying the muscles and the joint). **B** Changes in background activity in the course of three flexion movements of 30 s duration (indicated by *bars*). **C** Responses to the same movements at the indicated times (in min) after an acute experimental arthritis had been induced in the knee joint

ticular the "waking up" of the "sleeping nociceptors" (category 4, see above) results in a powerful additional excitatory input from the joint to these spinal neurons.

Motor Actions of Fine Articular Afferents

Several of the findings in the new studies on the properties and spinal actions of fine articular afferents briefly summarized above may be of immediate relevance for the participation of articular receptors in motor reflexes. Some of them will be pointed out here, and a proposal on the contribution of these receptors to motor control of joint movements will be made.

First, it has to be appreciated that the afferent innervation of joints is dominated by fine afferent fibers which outnumber the thick afferents with corpuscular endings severalfold. Up to now the motor actions of these fine afferents have more or less been neglected. At best, they were considered to constitute nothing but the noxious input from joints.

Second, in the normal joint, the local mechanical thresholds as well as those to movements of the joint range in fine afferent units from very low to very high in a nearly continuous way (see description of categories 1 to 3 above). Correspondingly, it is to be expected that the afferent inflow from joints will increase steadily as local stimuli increase in strength or – even more importantly – as the joint moves to the extremes of its working range and even starts to leave it.

Figure 6 illustrates how such a population of fine articular afferents with graded thresholds may be hooked up in a topographically and functionally organized way to spinal cord neurons in motor pathways. The arrangement shown provides rapidly increasing inhibition to motoneurons supplying those muscles which threaten to move the joint outside its working range. At the same time, it gives excitatory input to the respective antagonists. This excitatory input will also become effective during passive joint movements. Thus, probably long before noxious signals reach consciousness, such a network would induce reflexes counteracting excess active or passive movement to prevent joint damage.

The model has still to be put to rigorous experimental testing. At present support in regard to the inhibitory actions of fine joint afferents comes from the findings of Lundberg et al. [14] and Baxendale and Ferrell [2]. Their findings have been discussed above (see "Historical Perspective"). As to its functional role, it may be pointed out that patients with congenital insensitivity to pain suffer profound damage and destruction of their joints very early in their lives [3, 22]. It may well be that this is due mainly to a loss of the protective reflex circuits as pictured in Fig. 6.

The third point is that in inflamed joints activity over the various inputs as depicted in Fig. 6 will increase and the additional input line from the "sleeping nociceptors" will be opened (dashed line in Fig. 6). Both events will dramatically increase the negative feedback of the circuit. As a consequence, the spinal motor apparatus will strongly resist any movement of the inflamed joint – as indeed can be observed in many clinical situations involving arthritic afflictions of joints.

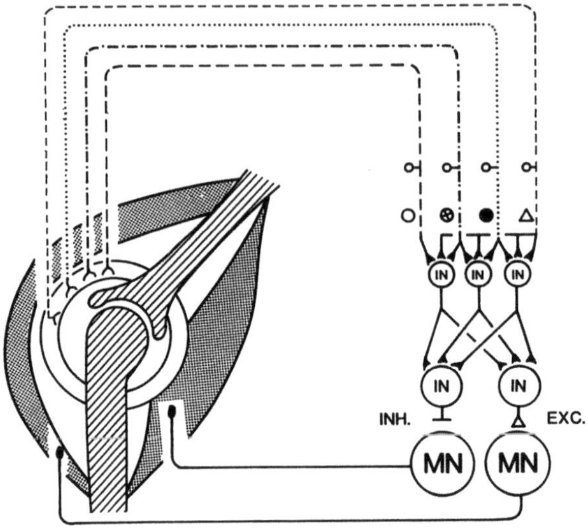

Fig. 6. Hypothetical circuit diagram of the connections of the various categories of fine articular afferent units to the neuronal pools of muscles acting on a joint. The diagram takes into account that the afferent inflow from such joint units depends on the direction as well as on the extent of a movement. Such a network will induce reflex actions counteracting excess movement to prevent joint damage. Inflammation will "awake" the "sleeping" nociceptors, and in this way increase the negative feedback gain of the circuit. Further explanation and discussion in the text

Finally, it may be worth mentioning that during joint inflammation the augmented articular afferent input reaches not only the spinal cord but also the supraspinal motor centers. This input may be several times its normal value (see the increase of activity in projecting spinal neurons described in the preceding section and depicted in Fig. 6 C). Under these conditions the central motor influence of the joint afferent input may well equal or exceed that of the muscle and cutaneous input from the same body region.

Summary

The vast majority of sensory endings in joint nerves stem from fine myelinated and unmyelinated nerve fibers. Contrary to previous assumptions on the functional role of these fine afferents, recent investigations in our laboratory have shown that these fiber groups contain afferent units which in their sensitivity to movement range from ones with very low thresholds to those which in normal joints are not activated by any movement, even a very noxious one. Joint inflammation or other sensitizing processes "wake up such sleeping nociceptors". The fine articular afferents also have specific actions on spinal cord neurons, particularly in the course of an acute arthritis. Evidence is presented that the various types of fine articular afferents participate in spinal motor control in a topographically and functionally organized fashion to prevent the joint from operating outside its physiological working range.

Acknowledgment. This work was supported by grants from the Deutsche Forschungsgemeinschaft.

References

1. Andres KH, Düring M v, Schmidt RF (1985) Sensory innervation of the achilles tendon by groups III and IV afferent fibers. Anat Embryol 172:145–156
2. Baxendale RH, Ferrell WR (1981) The effect of knee joint afferent discharge on transmission in flexion reflex pathways in decerebrate cats. J Physiol (Lond) 315:231–242
3. Baxter DW, Olszewski J (1960) Congenital universal insensitivity to pain. Brain 83:381–393
4. Beswick FB, Glockey NJ, Evanson JM (1955) Some effects of the stimulation of articular nerves. J Physiol (Lond) 128:83–84
5. Burgess PR, Clark FJ (1969) Characteristics of knee joint receptors in the cat. J Physiol (Lond) 203:317–335
6. Cervero F, Iggo A, Ogawa H (1976) Nociceptor-driven dorsal horn neurones in the lumbar spinal cord of the cat. Pain 2:5–24
7. Christensen BN, Perl ER (1970) Spinal neurons specifically excited by noxious or thermal stimuli: marginal zone of the dorsal horn. J Neurophysiol (Lond) 33:293–307
8. Coggeshall RE, Hong KAP, Langford LA, Schaible H-G, Schmidt RF (1983) Discharge characteristics of fine medial articular afferents at rest and during passive movements of inflamed knee joints. Brain Res 272:185–188
9. Freeman MAR, Wyke B (1967) The innervation of the knee joint. An anatomical and histological study in the cat. J Anat 101:502–532
10. Gardner E (1950) Reflex muscular responses to stimulation of articular nerves in the cat. Am J Physiol 161:133–141
11. Grigg P, Harrigan EP, Fogarty E (1978) Segmental reflexes mediated by joint afferent neurons in cat knee. J Neurophysiol 41:9–14
12. Grigg P, Schaible H-G, Schmidt RF (1986) Mechanical sensitivity of group III and IV afferents from posterior articular nerve in normal and inflamed cat knee. J Neurophysiol 55:635–643
13. Langford LA, Schmidt RF (1983) Afferent and efferent axons in the medial and posterior articular nerves of the cat. Anat Rec 206:71–78
14. Lundberg A, Malmgren K, Schomburg ED (1978) Role of joint afferents in motor control exemplified by effects on reflex pathways from Ib afferents. J Physiol (Lond) 284:327–343
15. Mendell LM (1966) Physiological properties of unmyelinated fiber projection to the spinal cord. Exp Neurol 16:316–332
16. Mitchell JH, Schmidt RF (1983) Cardiovascular reflex control by afferent fibers from skeletal muscle receptors. In: Shepherd JT, Abboud FM (eds) Handbook of physiology, sect 2. The cardiovascular system, vol 3. Peripheral circulation and organ blood flow, part 2. American Physiological Society, pp 623–658
17. Schaible H-G, Schmidt RF (1983a) Activation of groups III and IV sensory units in medial articular nerve by local mechanical stimulation of knee joint. J Neurophysiol (Lond) 49:35–44
18. Schaible H-G, Schmidt RF (1983b) Responses of fine medial articular nerve afferents to passive movements of knee joint. J Neurophysiol (Lond) 49:1118–1126
19. Schaible H-G, Schmidt RF (1984) Mechanosensibility of joint receptors with fine afferent fibers. In: Creutzfeldt O, Schmidt RF, Willis WD (eds) Sensory-motor integration in the nervous system. Springer, Berlin Heidelberg New York, pp 284–297
20. Schaible H-G, Schmidt RF (1986) Effects of an experimental arthritis on the sensory properties of fine articular afferent units. J Neurophysiol 54:1109–1122
21. Schaible H-G, Schmidt RF, Willis WD (1986) Responses of spinal cord interneurons to stimulation of articular afferent fibres in the cat. J Physiol (Lond) 372:575–593
22. Schmidt RF, Struppler A (1983) Der Schmerz. 2. Aufl, (Serie Piper, Bd 241). Piper, München, pp 1–304
23. Skoglund S (1956) Anatomical and physiological studies of knee joint innervation in the cat. Acta Physiol Scand [Suppl 124] 36:1–101
24. Skoglund S (1973) Joint receptor and kinaesthesis. In: Iggo A (ed) Handbook in sensory physiology, vol II. Somatosensory system. Springer, Berlin Heidelberg New York, pp 111–136
25. Wall PD (1967) The laminar organization of dorsal horn and effects of descending impulses. J Physiol (Lond) 188:403–423

II. Central Motor Actions of Sensory Input

Exteroceptive Input to the Motor Cortex in Man

J. E. Desmedt [1]

Exteroceptive Control of Motor Commands

Whereas the proprioceptive control of voluntary motor commands is well established (see, e.g. [13, 14, 31]), the role of exteroceptive feedback is less clearly documented, even though this would appear to be essential in conjunction with the precision handling of objects by hands and fingers.

Napier [30] distinguished two categories of prehensile movements in the primate hand: the power grip, in which prehensile strength is the primary objective; and the precision grip, which involves a fine control of independent finger movements to achieve adroit manipulation of light objects.

On the motor side, the latter must involve the corticospinal tract from the motor area 4, which exerts monosynaptic control on motoneurons of wrist and finger muscles and thereby can achieve differentially graded muscle force [23, 24, 31, 33]. On the sensory side, one must emphasize the rich endowment of the hand and fingers with mechanoreceptors readily activated by passive or active touch [19].

Clinical observations on patients with disturbance of finger sensitivity from peripheral nerve lesions as well as experiments involving local anesthesia of digital nerves suggest that exteroceptive input is important for the patterning of motor control in precision finger movements [16, 26, 27, 32].

In experiments on man, mild electrical stimulation of digital nerves of the index finger has been shown to influence the recruitment of single motor units in the first dorsal interosseous muscle [15]. The predominant effect is increased probability of firing of high-threshold motor units and a reduction in firing of the lower-threshold motor units.

In experimental studies of palpation of objects as in the precision grip, Kanda and Desmedt [22] recorded a clear drop of the recruitment threshold of the higher-threshold motor units of the first dorsal interosseous muscle. This recruitment of large motor units in active touch is considered to assist and reinforce the precision grip in skilled manipulatory activities of the hand [22]. Lack of such facilitatory effects in patients with neuropathies or nerve lesions that deafferent the fingers prevents reliable prehension and manipulation, with the patient unintentionally dropping small objects held by the affected hand.

There is reason to believe that exteroceptive inputs exert such actions, not only at segmental level, but also through long loops involving the rolandic motor cortex. Jenner and Stephens [18] observed both a short-latency and a long-latency [17 ms later) excitation of the first dorsal interosseous muscle after an electric

[1] Brain Research Unit, University of Brussels 115, boulevard de Waterloo, B-1000 Brussels, Belgium.

Clinical Aspects of Sensory Motor Integration
Ed. by A. Struppler and A. Weindl
© Springer-Verlag Berlin Heidelberg 1987

stimulus to the index finger in normal man. The long-latency response is reduced or absent in patients with dorsal column or perirolandic lesions, which suggests that it involves a long loop to and from the cortex. The question then arises as to whether and how such exteroceptive impulses are conveyed to the brain and influence the patterning of motor commands in motor cortex.

Somatosensory Evoked Response of the Motor Cortex of Man

Most studies of somatosensory evoked potentials (SEP) have emphasized the contralateral response that is recorded from the parietal scalp, usually with a frontal reference electrode. For stimulation by brief electrical pulses delivered to the median nerve or to fingers, this response includes a negative component with a peak at about 20 ms, hence its label N20 (see [11]), followed by positive P27 and P45 components.

However, using a frontal scalp electrode as reference obviously confounds any response that would be generated in the anterior half of the scalp. In our studies searching for any distinct prerolandic response to exteroceptive input, we used recording montages with either an earlobe or a noncephalic reference whereby distinct prerolandic SEP responses were disclosed, namely the positive P22 and the negative N30 [6, 8, 9]. These papers can also be consulted for description of methods used in SEP studies (see also [5]).

Figure 1 illustrates typical SEPs recorded in normal male of 23 years. The parietal and frontal electrodes were referred to a noncephalic reference electrode that was on the dorsum of the hand on the unstimulated side. The electric stimuli are delivered to fingers II and III of the left hand.

With these noncephalic montages, the scalp traces disclose first a series of brief positive farfield potentials of widespread distribution that are labeled P9 (reflecting the afferent volley in the brachial plexus) and P14 (reflecting the afferent volley in the medial lemniscus) (see [7]). The P11 farfield (reflecting the afferent volley in the dorsal column) is inconstant and is not seen in this example.

All scalp traces then show a prolonged negativity called N18, which is indicated by the hatching in trace 1 of Fig. 1 B [9]. Both P14 and N18 persist in patients with a thalamic vascular lesion that eliminates all cortical SEP responses, which implies that they are generated below the thalamus, in the brain stem [17, 29].

The cortical SEP components indeed appear as deflexions superimposed on the N18 baseline in the case of noncephalic reference recording. N18 is recorded virtually undistorted at the parietal electrode that is ipsilateral to the stimulated hand (trace 1). The focal parietal N20 and P27 components appear at the contralateral parietal electrode (trace 2). At the contralateral prerolandic electrode (trace 4) the P22 and N30 components are also clearly identified. Notice that, in this example, P22 is not seen at the ipsilateral prerolandic (trace 3), which shows however a clear frontal negativity N30. Superimposition of the 4 traces (Fig. 1 C) emphasizes the distinct latencies of N20 and P22 and the fact that these responses are not mere mirror images.

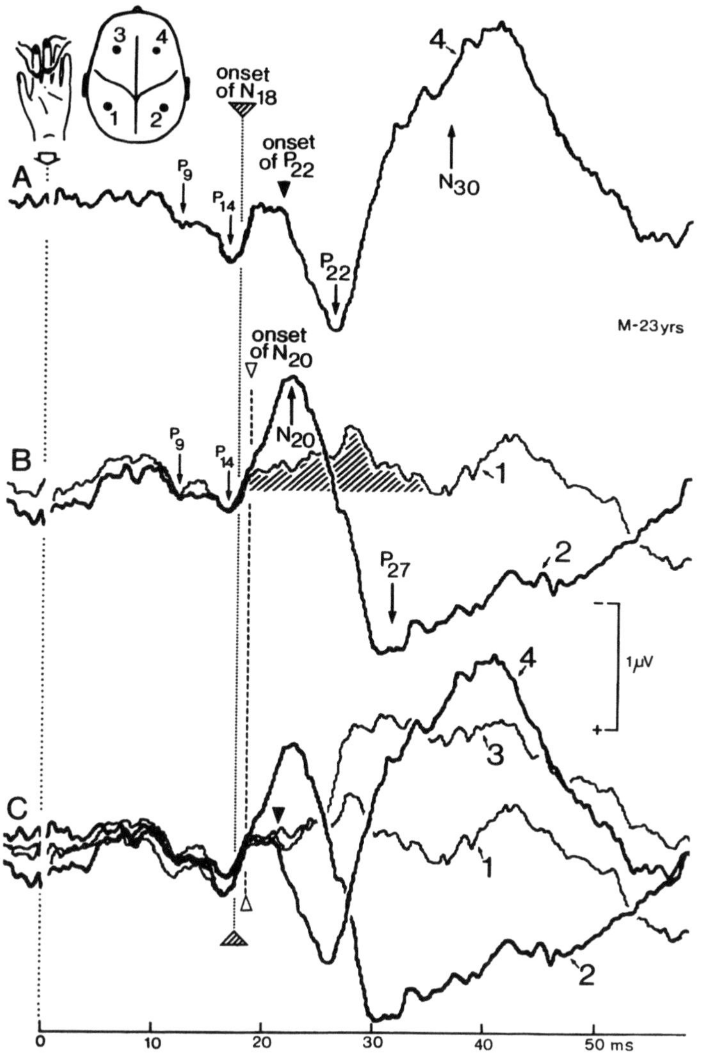

Fig. 1 A–C. Somatosensory evoked potentials recorded from 4 scalp sites (*sketch at upper right*) with a noncephalic reference electrode on the dorsum of the right hand. Electric stimulus to left fingers II and III. Normal male of 23 years. **A** Contralateral prerolandic electrode. **B** Superimposed traces from contralateral and ipsilateral parietal sites. N18 at the ipsilateral parietal site is indicated by oblique hatching. **C** Superimposed traces recorded from the 4 sites. The prerolandic P22 diverges from the N18 "baseline" later than N20. The P9 and P14 farfields are present in all traces. (Desmedt and Cheron [9])

Bit-Mapped Color Imaging of SEP Fields over the Scalp

The interpretation of single SEP traces may be unintendedly biased by the selected electrode sites used. Therefore the current data base can be extended by detailed spatial delineation of SEP fields to clarify the functional organization of the somatosensory system. Using many channels raises problems of data management and analysis unless appropriate electronic imaging procedures are used [10, 12]. The time history of bit-mapped SEP fields can be visualized by animation effects through displaying series of frozen maps at, say, 1-ms intervals.

Sixteen electrodes were arranged over both sides of the scalp and the common reference was a silver clip on the right earlobe. The electrodes were spaced by 20% of the nasion-inion distance. The head, assumed to be spherical, was projected onto a two-dimensional circular outline such that the arc length from each electrode to Cz was proportional to its radial distance from center. Bit-mapped images were drawn on paper by the microprocessor-based 4695 Tektronix Color Graphics Copier, each map being created by 1074 pixels. In map computation the pixel values were computed through a four-point linear interpolation algorithm using the 4 nearest electrodes as identified by standard analytic geometry. Potential values were each given a weight that was in inverse linear proportion to the distance from the pixel to each of the 4 electrodes considered. Voltage values were then fitted to a scale of equal discrete levels imaged by different hues. Details of method can be found in Desmedt and Nguyen [10] and Desmedt and Bourguet [6].

Figure 2 displays scalp potential fields as a series of frozen maps at selected latencies in a normal subject of 23 years. Brief electric pulses are delivered to fingers I–IV stimulation. Similar SEP field patterns are found for median nerve stimulation. The scalp being initially at about zero potential (grey), the widespread P14 farfield (blue) appears at 14–16 ms.

The next event is a focal negativity (red) over the right (contralateral) parietal scalp at 19 ms, which is associated with a widespread positivity "P20" (blue) over the anterior scalp. This pattern is accentuated at 20 ms with a deepening of the colors (see scale). However, at 21 ms, the frontal positivity further develops with a clear prerolandic focus on the right side while the parietal negativity begins to fade away. This pattern of prerolandic P22 is seen in virtual isolation at 22 and 23 ms when the parietal N20 has disappeared. P22 is seen as a clear focus of positivity that is radially oriented with respect to the cortex [6]. This contrasts with the early stage at 19 and 20 ms when the N20-"P20" pattern rather suggested a tangential dipole generator with its negative pole at the back and its positive pole in front. At 27 ms, a large bilateral N30 develops frontally while a focal P27 appears at the right parietal region. No significant N20-P27 appear ipsilaterally [4, 8, 9]. From about 42 ms, P45 develops over the central scalp.

In our present hypothesis, slightly modified from Desmedt and Bourguet [6], the tangential dipole in the posterior wall of the Rolando fissure (suggested by Broughton [3] and Allison et al. [1]) is indeed seen in the virtually concomitant appearance of the parietal N20 and frontal "P20" potential fields at about 19 ms. We take exception, however, to their view that this is the whole story for early

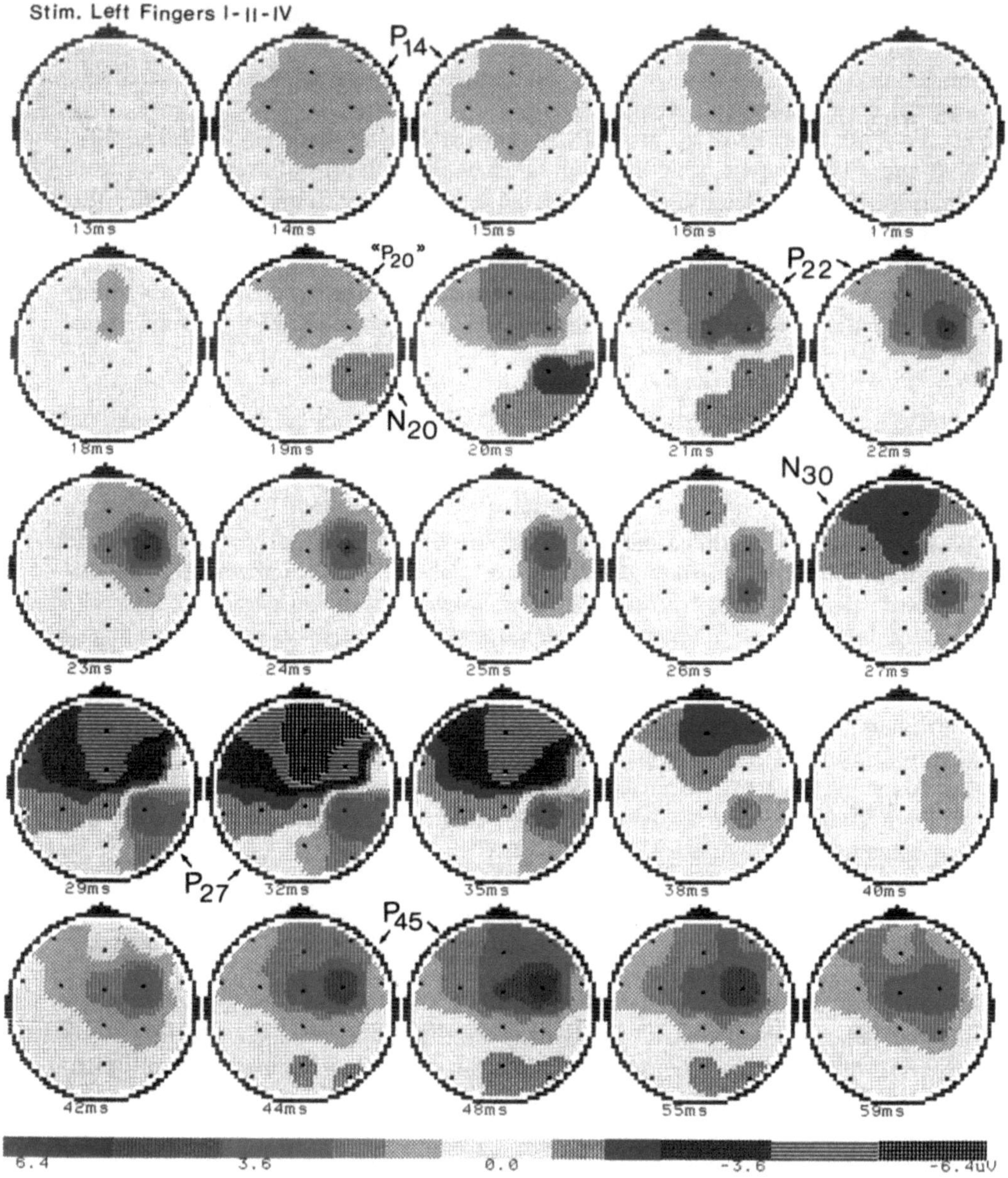

Fig. 2. Bit-mapped color imaging of scalp potential fields to stimulation of left fingers I-II-IV in a normal male of 23 years. Electrode sites over the entire scalp indicated by black pixels. Reference electrode at right earlobe. Latency of each "frozen" map indicated in ms at lower left. The zero baseline (*grey*) has been set for each recording site by the mean potential level during 5 ms before and after the stimulus. Voltage increments are in red hues for negative and blue for positive (calibration scale). (Modified from Desmedt and Bourguet [6])

SEP responses, emphasizing that there is in addition a genuine prerolandic P22 that develops with a distinct time scale than the parietal N20. P22 indeed is imaged as a positive focus which is centered over the contralateral motor cortex and reflects a radial generator [6].

We think that the prerolandic P22 is generated independently of the parietal N20 and P27 responses. P22 represents cortical activation in area 4 by a direct thalamocortical projection from VPLo-VLc, as documented in the monkey [20, 35]. Units in VPLo can discharge at very short latencies to electrical stimulation of peripheral nerves [2, 25]. Neurons in VPLo in turn have fast direct projections to area 4 [35]. Thus the short latency of exteroceptive input to motor area 4 might involve a direct lemniscal projection to VPLo which directly projects to area 4 [34].

Evidence from Patients with Unilateral Focal Brain Lesions

Patients with a unilateral vascular lesion destroying the parietal cortex on one side present a contralateral hemianesthesia, and their SEPs recorded with earlobe reference disclose a loss of N20 and P27 while preserving prerolandic P22 and N30 components; because the parietal cortex destruction results in the retrograde degeneration of the thalamocortical neurons in VPLc, the preserved short-latency P22 must involve the separate thalamocortical projection from VPLo-VLc to motor area 4 [28]. On the other hand, patients with a prerolandic lesion and central hemiplegia, but without hemianesthesia, have SEPs with parietal N20 and P27, but they lose the P22 prerolandic response [28]. Such double dissociations through focal lesions imply distinct neural generators and thalamocortical projection systems for N20 and P22, respectively.

Various types of prefrontal lesions located rostral to the anterior horn of the lateral ventricle do not affect N30 nor P22 [28]. Hence the P22 must be generated in front of the central sulcus, but behind the prefrontal cortical areas. The persistence of P22 after extensive prefrontal lesions indicates that it must be generated primarily in the motor area 4, which indeed is the recipient of the short-latency direct thalamic projections. Furthermore, the Supplementary Motor Area (SMA), which receives projections from area 4, may add some contribution to P22 and N30. As SMA is close to the midline, this would account for the extension of their potential fields over the ipsilateral scalp [6].

Our emphasis on direct projections from thalamus to motor cortex does not exclude a role for the corticocortical connexions from parietal area 2 to motor area 4, and from areas 1, 2, and 5 to SMA [20, 21], but the latter intervention must be of longer latency.

Conclusion

Exteroceptive input from the skin and joints of fingers and hand are significant for motor control in precision grip and adroit manipulation of objects. Natural mechanical activation of fingers during active touch behavior results in a drop of recruitment threshold of the larger motor units in the active muscle, thereby assisting the precision grip. Stimulation of fingers also elicits long-latency responses that appear to involve long loops through the cortex. Studies in man have now identified a short-latency response called P22 (mean peak latency 22 ms) recorded from the prerolandic scalp. Bit-mapped color imaging of potential fields indicates that P22 is distinct from the parietal N20 or P27 potentials. Moreover, double dissociations between the two sets of concomitant responses to somatosensory inputs are documented in patients with focal unilateral vascular brain lesions.

In conclusion, the P22 prerolandic response reflects a radial generator in motor area 4 that is activated by short-latency somatosensory inputs from the contralateral hand in man. This P22 response is independent of the postrolandic N20 response that reflects the posterior pole of a tangential neural generator in the parietal receiving cortex in the posterior wall of Rolando fissure.

Acknowledgment. This work has been supported by grants from the Fonds de la Recherche Scientifique Médicale.

References

1. Allison T, Goff WR, Williamson PD, Van Gilder JC (1980) On the neural origin of early components of the somatosensory evoked potential. In: Desmedt JE (ed) Clinical uses of cerebral, brainstem and spinal somatosensory evoked potentials. Karger, Basel (Progress in clinical neurophysiology, vol 7, pp 51–68)
2. Asanuma H, Larsen KD, Yumiya H (1979) Receptive fields of thalamic neurons projecting to the motor cortex in the cat. Brain Res 172:217–228
3. Broughton RJ (1969) Discussion. In: Donchin E, Lindsley DB (eds) Average evoked potentials. US Government Printing Office, Washington DC, NASA SP-191:79–84
4. Desmedt JE (1971) Cerebral somatosensory evoked potentials in man. In: Rémond A (ed) Handbook of EEG and clinical neurophysiology, vol 9. Elsevier, Amsterdam, pp 55–82
5. Desmedt JE (1977) Some observations on the methodology of cerebral evoked potentials in man. In: Desmedt JE (Ed) Attention, voluntary contraction, and event-related cerebral potentials. Karger, Basel (Progress in clinical neurophysiology, vol 1, pp 12–29
6. Desmedt JE, Bourguet M (1985) Color imaging of scalp topography of parietal and frontal components of somatosensory evoked potentials to stimulation of median or posterior tibial nerve in man. Electroencephalogr Clin Neurophysiol 62:1–17
7. Desmedt JE, Cheron G (1980a) Central somatosensory conduction in man: neural generators and interpeak latencies of the far-field components recorded from neck and right or left scalp and earlobes. Electroencephalogr Clin Neurophysiol 50:382–403
8. Desmedt JE, Cheron G (1980b) Somatosensory evoked potentials to finger stimulation in healthy octogenarians and in young adults: waveforms, scalp topography and transit times of parietal and frontal components. Electroencephalogr Clin Neurophysiol 50:404–425
9. Desmedt JE, Cheron G (1981) Non-cephalic reference recording of early somatosensory potentials to finger stimulation in adult or aging man: differentiation of widespread N18 and contralateral N20 from the prerolandic P22 and N30 components. Electroencephalogr Clin Neurophysiol 52:553–570

10. Desmedt JE, Nguyen TH (1984) Bit-mapped color imaging of the potential fields of propagated and segmental subcortical components of somatosensory evoked potentials in man. Electroencephalogr Clin Neurophysiol 58:481–497
11. Donchin E, Callaway E, Cooper R, Desmedt JE, Goff WR, Hillyard SA, Sutton S (1977) Publication criteria for studies of evoked potentials in man (report of a committee). In: Desmedt JE (ed) Attention, voluntary contraction and event-related cerebral potentials. Karger, Basel (Progress in clinical neurophysiology, vol 1, pp 1–11
12. Duffy F, Burchfiel JL, Lombroso CT (1978) Brain electrical activity mapping: a method for extending the clinical utility of EEG and evoked potentials data. Ann Neurol 5:309–321
13. Evarts EV (1973) Motor cortex reflexes associated with learned movement. Science 179:501–503
14. Evarts EV, Fromm C (1978) The pyramidal tract neuron as summing point in a closed-loop control system in the monkey. In: Desmedt JE (ed) Cerebral motor control in man: long loop mechanisms. Karger, Basel (Progress in clinical neurophysiology, vol 4, pp 56–59
15. Garnett R, Stephens JA (1981) Changes in the recruitment threshold of motor units produced by cutaneous stimulation in man. J Physiol (Lond) 311:463–473
16. Goodwin GM, McCloskey DI, Matthews PBC (1972) The contribution of muscle afferents to kinaesthesia shown by vibration-induced illusions of movement and by the effects of paralysing joint afferents. Brain 95:705–748
17. Hashimoto I (1984) Somatosensory evoked potentials from the human brainstem: origins of short-latency potentials. Electroencephalogr Clin Neurophysiol 57:221–227
18. Jenner JR, Stephens JA (1982) Cutaneous reflex responses and their central nervous pathways studied in man. J Physiol (Lond) 333:405–419
19. Johansson RS, Vallbo AB (1979) Spatial properties of the population of mechanoreceptive units in the glabrous skin of the human hand. Brain 184:353–366
20. Jones EG (1983) The nature of the afferent pathways conveying short-latency inputs to the primate motor cortex. In: Desmedt JE (ed) Motor control in health and disease. Raven, New York (Advances in neurology, vol 39, pp 263–285)
21. Jones EG, Coulter JD, Hendry SHC (1978) Intracortical connectivity of architectonic fields in the somatic sensory, motor and parietal cortex of monkeys. J Comp Neurol 181:291–348
22. Kanda K, Desmedt JE (1983) Cutaneous facilitation of large motor units and motor control of human fingers in precision grip. In: Desmedt JE (ed) Motor control mechanisms in health and disease. Raven, New York (Advances in neurology, vol 39, pp 253–262)
23. Kuypers H (1973) The anatomical organization of the descending pathways and their contributions to motor control especially in primates. In: Desmedt JE (ed) New developments in electromyography and clinical neurophysiology. Karger, Basel, pp 38–68
24. Landgren S, Phillips CG, Porter R (1962) Cortical fields of origin of the monosynaptic pyramidal pathways to some alpha motoneurones of the baboon's hand and forearm. J Physiol (Lond) 161:112–125
25. Lemon RN, van der Burg J (1979) Short-latency peripheral inputs to thalamic neurons projecting to the motor cortex in the monkey. Exp Brain Res 36:445–462
26. McCloskey DI (1978) Kinaesthetic sensibility. Physiol Rev 58:763–820
27. McCloskey DI, Gandevia S, Potter EK, Colebatch JG (1983) Muscle sense and effort: motor commands and judgments about muscular contractions. In: Desmedt JE (ed) Motor control mechanisms in health and disease. Raven, New York (Advances in neurology, vol 39, pp 151–168)
28. Mauguière F, Desmedt JE, Courjon J (1983a) Astereognosis and dissociated loss of frontal or parietal components of somatosensory evoked potentials in hemispheric lesions: detailed correlations with clinical signs and computerized tomography scanning. Brain 106:271–311
29. Mauguière F, Desmedt JE, Courjon J (1983b) Neural generators of N18 and P14 far field somatosensory evoked potentials: patients with lesion of thalamus or thalamocortical radiations. Electroencephalogr Clin Neurophysiol 56:283–292
30. Napier JR (1956) The prehensile movements of the human hand. J Bone Joint Surg 38b:902–913
31. Phillips CG, Porter R (1977) Corticospinal neurones. Academic, London

32. Sanes JN, Mauritz KH, Evarts EV, Dalakas MC, Chu A (1984) Motor deficits in patients with large-fiber sensory neuropathy. Proc Natl Acad Sci USA 81:979–982
33. Smith AM, Frysinger RC, Bourbonnais D (1983) Interaction between motor commands and somatosensory afferents in the control of prehension. In: Desmedt JE (ed) Motor control mechanisms in health and disease. Raven, New York (Advances in neurology, vol 39, pp 373–386)
34. Tamai Y, Waters RS, Asanuma H (1984) Caudal cuneate nucleus projection to the direct thalamic relay to the motor cortex in cat: an electrophysiological and anatomical study. Brain Res 323:360–364
35. Tracey DJ, Asanuma C, Jones EG, Porter R (1980) Thalamic relay to motor cortex: afferent pathways from brainstem, cerebellum and spinal cord in monkeys. J Neurophysiol 44:532–554

Reorganization of Projection from the Sensory Cortex to the Motor Cortex Following Deprivation of Thalamocortical Projection

H. Asanuma [1]

Recovery of function following partial destruction of the central nervous system and its neuronal mechanism has been an intriguing question among neurologists and neurophysiologists. It has long been wondered whether the functional recovery is accomplished by anatomical reorganization in the central nervous system or by compensation of the function by the remaining systems. Ramon y Cajal [8] transected the spinal cord of young cats and dogs and reported that a large number of the severed intraspinal fibers sprouted new processes with cones of growth similar to those observed in the peripheral nerves. However, after several weeks these sprouting fibers disappeared, and he concluded that the processes of regeneration were followed by atrophy and absorption. Recent progress in anatomical and physiological techniques, however, has made it possible to reexamine the genesis of the functional recovery. Raisman and Field [7], using the electron microscope, demonstrated that elimination of one input to the septal nucleus is substituted by new synapses which were formed by the sprouting of the remaining fibers in the rat. Tsukahara et al. [9], using an intracellular recording technique, demonstrated that the elimination of cerebellar input to the red nucleus resulted in the formation of new synapses by fibers from the cerebral peduncle in the cat. Thus, it seems that at least a part, and probably a major part, of the functional recovery is due to the formation of new synapses by the remaining fibers.

We have recently shown that elimination of sensory input to the motor cortex produces severe motor deficits in the monkey [1]. In these experiments, the deprivation of sensory input was accomplished by section of the dorsal column and suction of the sensory cortex. Section of the dorsal column by itself produced some motor deficits, but the function was recovered within two weeks as long as the sensory cortex was intact. The results strongly suggested that the recovery of the function was accomplished by reorganization of the projection from the sensory cortex to the motor cortex.

To examine what kind of reorganization takes place during the recovery of motor function, we have examined characteristics of the projection from the sensory cortex to the motor cortex in normal [6] and rostral thalamus (VL and rostral VPL) lesioned cats [4]. Thalamic lesion was used instead of dorsal column section because in the monkey, destruction of rostral thalamus could also produce motor deficit; the function, however, was recovered as long as the sensory cortex was intact (Bornschlegl and Asanuma, unpublished observation). The experiments were carried out under Nembutal anesthesia (initially 35 mg/kg, followed by 5 mg/h). Several tungsten microelectrodes (3–5) were implanted into the depth of anterior bank of the ansate sulcus (area 2), and intracortical microstimulation (ICMS) was

[1] The Rockefeller University, New York, NY 10021, USA.

Clinical Aspects of Sensory Motor Integration
Ed. by A. Struppler and A. Weindl
© Springer-Verlag Berlin Heidelberg 1987

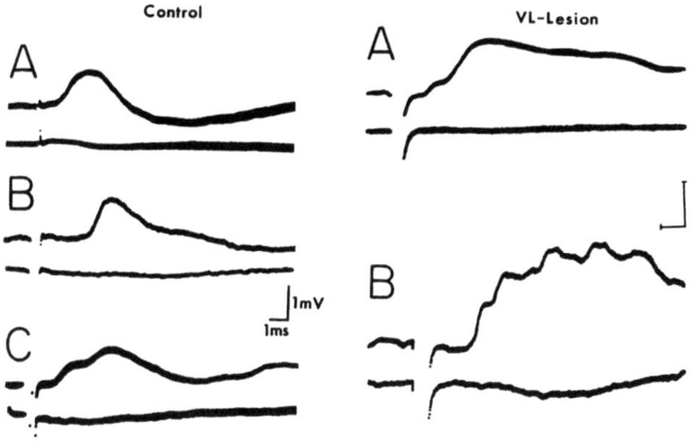

Fig. 1 A–C. Examples of monosynaptic EPSPs in motor cortical neurons elicited by stimulation of area 2 of the sensory cortex. Upper lines are intracellular potentials and lower lines are extracellular field potentials. *Left column* Examples of EPSPs recorded in normal cats. **A** Short-latency slow-rising EPSP. **B** Long-latency fast-rising EPSP. **C** Double-peaked EPSP. *Right column* Examples of EPSPs recorded in cats in which rostral thalamus was lesioned one month previously. Note the multiple peaks

delivered through each electrode (0.2 ms duration; 30 μA intensity). A glass pipette microelectrode filled with 2.0 mol potassium acetate was inserted into the lateral cruciate gyrus (area 4) and intracellular recordings were made. Whenever short-latency postsynaptic potentials (PSP) were recorded, they were averaged and photographed. Then the pipette electrode was withdrawn, a tungsten microelectrode was inserted into the same position, and a lesion was made by passing a negative current of 10 μA for 10 s. By repeating this procedure, we were able to locate the sites of recordings and types of PSPs. Then the same procedure was repeated in the cats in which the rostral thalamus had been lesioned a month earlier by injecting 1.0 μl kainic acid.

Figure 1 shows typical examples of the PSPs in normal and thalamus-lesioned cats. In normal cats, two-thirds of the potentials showed smooth, monophasic shape and only one-third had two or three humps as shown in the left column. In the chronic cats, three-fourths of the PSPs showed multiple humps and some of these had several notches as shown in the right column. In both normal and chronic cats, monosynaptic inhibitory postsynaptic potential (IPSP) was never recorded. Utilizing these averaged excitatory postsynaptic potentials (EPSPs), we measured the latencies, the amplitudes, and the time from the start to the peak of the potentials for the earliest EPSPs. A total of 51 EPSPs in normal cats and 49 EPSPs in chronic cats were analyzed. There were significant differences between these two groups of EPSPs in amplitude and location. In the control group, the average amplitude of the EPSPs was 1.0 mV whereas in the chronic cats, the average was 1.6 mV, suggesting that there was a significant reinforcement of monosynaptic cortico-cortical connections following the elimination of thalamic input. In normal animals, monosynaptic EPSPs could be recorded only in the superficial layers (II and III) or at the border area between layers III and V. In the chronic animals, monosynaptic EPSPs could also be recorded in layer V, suggesting either the formation of new synapses or the activation of synapses which were inactive in normal animals.

To identify the neurons receiving monosynaptic input, we examined terminal sites of cortico-cortical fibers in 3 normal [5] and 3 chronic (Ichikawa et al., in preparation) cats in which rostral thalamus had been lesioned 2 months earlier. Under Nembutal anesthesia (35 mg/kg) the anterior bank of ansate sulcus (area 2) was removed by suction, and 5 days later the animals were reanesthetized and perfused. The motor cortex was removed and stained with the rapid Golgi method. The tissue was then embedded in celloidin, cut serially at 80–100 μm thickness, and mounted on slides. Well-impregnated neurons were selected from the motor cortex, and the tissue containing the cell was trimmed into a small piece, refixed with OsO_4, and embedded in Epon 812. The Golgi image of the selected neurons was drawn by means of a camera lucida, and the tissue was cut into series of a semi-thin sections (0.5–1.0 μm) followed by ultra-thin sections (60–100 nm). Using the semi-thin sections as a guide, we were able to reconstruct almost all parts of the neurons with an electron microscope. In the control animals, the degenerating terminals of cortico-cortical fibers were examined in 23 identified neurons, of which 11 were pyramidal and 12 were stellate. Stellate cells in layer III received many degenerating terminals (average dendrite length: 8.4/1.0 mm), and 95% of these were found on the proximal dendrites or on the cell bodies. On the other hand, the pyramidal cells in both layers III and V received fewer degenerating terminals (2.1/1.0 mm), and these were located on more distal dendritic shafts or on dendritic spines.

To compare this pattern of distribution to that of thalamocortical projection, we made a lesion in the rostral thalamus by injecting 1.0 μl of kainic acid and examined degenerating terminals in the same way. The pattern of distribution was different from that of cortico-cortical fibers. Pyramidal neurons received many terminals (6.0/1.0 mm) and these were located on the basal and the apical dendrites and on dendritic spines. Stellate neurons received fewer terminals (4.2/1.0 mm) and these were located primarily on proximal dendritic shafts. Thus, there was a substantial difference between these two projection systems.

The pattern of projection in the chronic cats was examined using 8 stellate cells and 11 pyramidal cells. The density of degenerating terminals on stellate cells was not much different from that in the control animals, but terminals on the pyramidal cells increased to 4.7/1.0 mm from 2.1 in the controls. The results clearly demonstrated that elimination of thalamocortical input resulted in a substantial change in the projection from the sensory cortex to the motor cortex.

Discussion and Conclusions

What is the functional significance of the changes following the elimination of thalamocortical projection? In the cats examined, injection of kainic acid produced only minor behavioral changes. These consisted of disturbances in walking, such as circling in one direction or inability to stand straight. These symptoms disappeared rapidly, and within a few days the animals appeared normal. In monkeys, however, injection of kainic acid produced marked motor disturbances de-

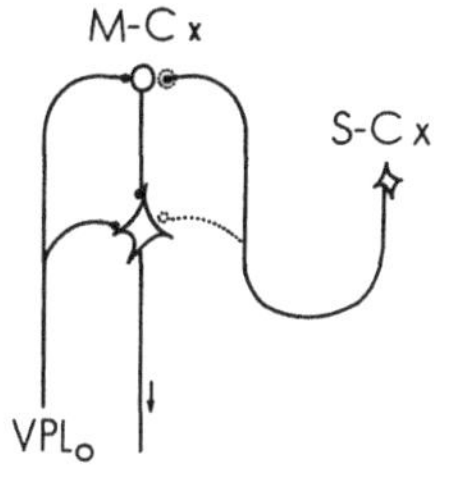

Fig. 2. Simplified diagram of thalamocortical and cortico-cortical connections in the motor cortex In normal cats, projection fibers from the thalamus synapse on both interneurons and corticofugal neurons whereas projection fibers from the sensory cortex synapse primarily on interneurons. When thalamocortical projection was removed, cortico-cortical fibers make new synapses on corticofugal neurons as shown by *dotted line*, probably taking over the function of thalamocortical projection. (*M-Cx*, motor cortex; *S-Cx*, sensory cortex; *VPLo*, n. ventralis posterolateralis pars oralis of thalamus)

pending on the site of injection. These were mainly ataxia and loss of orientation in hand movement in addition to the loss of hand skills. These symptoms disappeared within 1–2 weeks, and the animal appeared normal except for a small loss of hand skills. Removal of the sensory cortex after recovery had a severely disabling effect on the motor skill of the monkeys, as was the case when dorsal-column section and sensory-cortex removal were combined. From these observations, it is likely that recovery of function was accomplished by the reorganization of the projection from the sensory cortex to the motor cortex in the monkey. We have performed the present series of experiments on the assumption that the basic changes following thalamic lesion are the same in both the cat and monkey. Figure 2 shows a simplified summary of the results obtained in this series of experiments. The thalamocortical projection in normal cats innervates both cortical interneurons in the superficial layers and corticofugal neurons in the deep layer. On the other hand, projection from the sensory cortex to the motor cortex synapses primarily on cortical interneurons, although not exclusively. These patterns of projections (Fig. 2, solid lines) suggest that thalamic input plays a more important role in regulating the movement than the cortico-cortical input, but also that the gain of the thalamic input is modulated by the latter input. In fact, it has been shown that microstimulation in area 3a of the sensory cortex did not produce excitation of motor cortical neurons but facilitated the activities evoked by peripheral stimulation [2]. When the thalamic input was eliminated, the cortico-cortical projection changed its pattern and came close to that of the thalamocortical projection, as shown by the dotted lines in Fig. 2. If this change of pattern is the same in the cat and the monkey, then it might be said that it is the basis for functional recovery following deprivation of the input from the thalamus to the motor cortex. Although we do not know what input was deprived by the thalamic lesion, it is certain that the direct sensory input to the motor cortex [3], which is known to arise from the n. ventralis posterolateralis pars oralis (CVPLo) of the thalamus, is also deprived. As already stated, elimination of this input produces some motor deficit, but the function recovered as long as the sensory cortex is intact. Since the deficit produced by the thalamic lesion somewhat resembles that produced by dorsal-column section, it is possible that this reorganization of projection from the sensory cortex to the motor cortex is the basis for the recovery of function following deprivation of the direct sensory input from the thalamus to the motor cortex.

There are still many questions as to the role of the projection from the sensory cortex to the motor cortex. If this controls the gain of the thalamic input, how does this control fit into the purposeful movements? Is it a transient change of the gain depending on the input to the corresponding part of the sensory cortex, or is it a long-lasting modulation of the gain such as those represented by the "motor memory", which is semipermanent? Is this long-lasting gain change achieved by the long-term potentiation of the neurons controlling the gain, or by continuous reorganization of connections within the motor cortex by the sprouting of the existing fibers? The answers to these questions still await further elucidation.

Summary

Anatomical and physiological changes of the projection from the sensory cortex to the motor cortex following chronic lesion of the VL in the thalamus were examined in cats. It was found that monosynaptic EPSPs in motor cortical neurons elicited by stimulation of the sensory cortex became much greater after the lesion and that these EPSPs could be recorded in deep-layer neurons which in normal cats could rarely be recorded. Anatomical studies revealed that in normal cats, most association fibers from the sensory cortex terminated on dendritic spines of neurons in all cortical layers, but some terminated on dendritic shafts of stellate neurons in the superficial layers. In the chronic cats, many terminals were found on the dendritic shafts of deep pyramidal neurons. It is suggested that these physiological and anatomical changes contribute to the recovery of function following lesion in the thalamus.

References

1. Asanuma H, Arissian K (1984) Experiments on functional role of peripheral input to motor cortex during voluntary movements in the monkey. J Neurophysiol 52:212–227
2. Asanuma H, Waters RS, Yumiya H (1982) Physiological properties of neurons projecting from area 3a to area 4 of feline cerebral cortex. J Neurophysiol 48:1048–1057
3. Asanuma H, Larsen KD, Yumiya H (1979) Direct sensory pathways to the motor cortex in the monkey: a basis of cortical reflexes. In: Asanuma H, Wilson VJ (eds) Integration in the nervous system. Igaku-shoin, Tokyo, pp 223–238
4. Asanuma H, Kosar E, Tsukahara N, Robinson H (1985) Modification of the projection from the sensory cortex to the motor cortex following the elimination of thalamic projections to the motor cortex in cats. Brain Res 345:79–86
5. Ichikawa M, Arissian K, Asanuma H (1985) Distribution of cortico-cortical and thalamocortical synapses on identified motor cortical neurons in the cat: Golgi, electron microscopic and degeneration study. Brain Res 345:87–101
6. Kosar E, Waters RS, Tsukahara N, Asanuma H (1985) Anatomical and physiological properties of the projection from the sensory cortex to the motor cortex in normal cats: the difference between cortico-cortical and thalamocortical projections. Brain Res 345:68–78
7. Raisman G, Field PM (1973) A quantitative investigation of the development of collateral reinnervation after partial deafferentation of the septal nuclei. Brain Res 50:241–264
8. Ramon y Cajal S (1928) Degeneration and regeneration of the nervous system. Oxford University Press, London New York
9. Tsukahara N, Hultborn H, Murakami F, Fujito Y (1975) Electrophysiological study of formation of new synapses and collateral sprouting in red nucleus neurons after partial denervation. J Neurophysiol 38:1359–1372

III. The Muscles and Their Neural Control

Properties of Motoneurones and Motor Units in Relation to Problems of Sensorimotor Integration

D. KERNELL [1]

Introduction

The present contribution concerns the properties and the organization of single muscles and their motoneurone pools. Thus, it will not directly deal with any sensory functions, and it will only discuss a most elementary and fragmentary kind of motor action: the contraction of an individual muscle. There are several reasons, however, why the properties of single muscles and single motoneurone pools actually are of importance also for the understanding of problems concerning the integration of motor behaviour. Two of the various possible reasons are:

1. During motor behaviour, important "integrative" functions take place also at the level of single muscles and motoneurone pools. This concerns, for instance, the choice of the right type of motor unit for a given task, or the matching between motoneuronal discharge rates and muscular response properties. Many such integrative functions, if we may call them that, take place without any direct sensory control.
2. The quantitative input-output properties of the motoneurone-muscle system are not invariable, but they may be changed in complex ways by preceding activity. For the adequate performance of motor functions, such "usage-dependent" alterations in force-output would have to be measured and corrected for. Thus, the usage-dependent variability in the properties of the motoneurone-muscle system contributes to the need for sensory control of motor execution.

In this chapter a brief but fairly general survey of functional properties of the motoneurone-muscle system will be given. First, it will be described how the components of such a system tend to be used by the central nervous system for the gradation of muscle force. Second, some of the ways will be described in which the gradation properties of the system may be altered by preceding activity. The discussion will be centered around experimental work from our own laboratory concerning the properties of motoneurones and hindlimb muscles of cats.

Gradation of Force and the Properties of Motoneurones and Motor Units

It is well known that there are two ways in which the central nervous system may cause an increase of the contractile force of a skeletal muscle: (a) by increasing the number of active motoneurones (recruitment modulation), and (b) by increas-

[1] Department of Neurophysiology, University of Amsterdam, Academisch Medisch Centrum, Meibergdreef 15, NL-1105 AZ Amsterdam, The Netherlands.

Clinical Aspects of Sensory Motor Integration
Ed. by A. Struppler and A. Weindl
© Springer-Verlag Berlin Heidelberg 1987

ing the discharge rate of already active motoneurones (rate modulation). The use of either kind of strategy is complicated, however, by the fact that the various motor units of a muscle differ so markedly from each other in contractile properties.

Within a given muscle, the *contractile properties of the motor units* generally co-vary in the manner illustrated by Fig. 1 (for review and references, see [3]). Fatigue resistance has been plotted versus a measure of contractile speed, the twitch contraction time. Furthermore, the diameter of the symbols is proportional to the maximum force of the various groups of units. At one end of the distribution we find units that are strong, fast-twitch and fatigue-sensitive (FF units, according to terminology of [3, 4]), and at the other extreme the units tend to be weak, slow-twitch and fatigue-resistant (S units). In between there are categories of units with a relatively high twitch-speed, an intermediate force and an intermediate [F(int)] or high (FR) fatigue resistance. Figure 1 shows the average properties for these various groups of units in the cat's m. peroneus longus, as categorized according to Burke's criteria (see [28]). Such a categorization is of practical value for analytical and descriptive purposes. It should be stressed, however, that the properties in question commonly display a continuous variation across the motor unit population of a given muscle.

The evidence available indicates that the *recruitment modulation* of muscle force usually occurs, statistically speaking, by recruiting the units in the general order schematically indicated by the arrows in Fig. 1 (for reviews and references, see [3, 14]). Thus, the weak and fatigue-resistant S units tend to be more easily recruited than faster and stronger units. The S units also tend to have thinner axons than those of the faster and stronger units. Thus, to an important extent, the recruitment order indicated in this graph occurs according to the "size principle" as formulated by Henneman and colleagues [2, 14, 15]. The "ascending-size order" (or "ascending-force order") of recruitment is of great functional significance. Maintained postural contractions will, for instance, primarily have to be executed by the most easily recruited units. Hence, it is very appropriate that these units are slow and fatigue-resistant.

The typical ascending-size order of recruitment depends on the distribution of activated synapses within the motoneurone pool as well as on the distribution of neuronal excitability among the respective motoneurones. Recent studies have

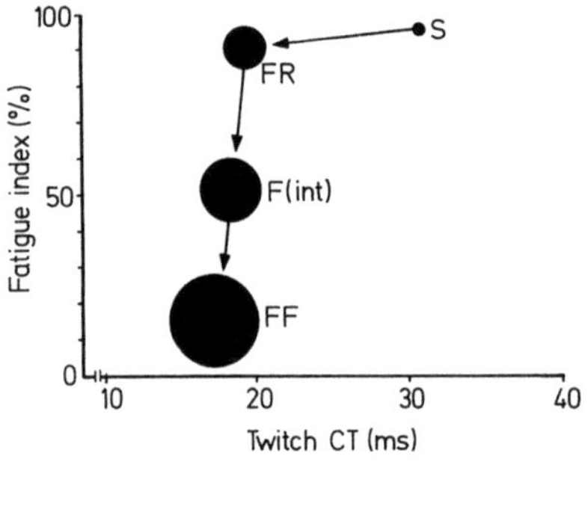

Fig. 1. Plot of fatigue resistance (fatigue index, %) vs twitch contraction time (ms) for 80 motor units of the cat's m. peroneus longus. Mean values are shown for motor units of the categories S, FR, F(int) and FF, respectively, as classified according to the criteria of [4]. *Plotted symbols* have diameters in direct proportion to the maximum tetanic tension of the respective group of units (mean maximum tension for FF units: 27.9 g). *Arrows* indicate presumed order of recruitment in most kinds of contractions (cf. [3]). (Data from results of Kernell et al. [28]; reproduced from Kernell [22])

indicated that the latter factor may be of great importance. In direct comparisons which we have made between the anatomy and electrophysiology of single motoneurones, we found evidence indicating that small and thin-axoned alpha-motoneurones tend to have a greater specific membrane resistance than that of cells with larger axons and cell bodies [25]. The existence of such differences between different types of alpha-motoneurones has recently been reported also from other laboratories [6, 13]. Such findings imply that, even if the active synapses were randomly distributed within a motoneurone pool (which is, of course, not necessarily the case), excitation would tend to produce larger depolarizations in small and slow cells than in larger and faster ones. As a consequence, the small and slow motoneurones would be expected to be more easily recruited than the large and fast ones.

Which problems are there then with respect to the other gradation strategy: the *rate-modulation* of muscle force? By stimulating a motor unit at different impulse rates, one may determine its relation between contractile tension and activation frequency. As was already shown for whole muscles long ago by Cooper and Eccles [8], this 'tension-frequency curve' is markedly sigmoid in shape (cf. Fig. 3 A). An effective gradation of force by a modulation of stimulus rate occurs only within a rather limited range of impulse frequencies, i.e. within the range covering the steep portion of the tension-frequency curve. This is true also for single motor units. The slower the twitch is of a muscle or motor unit, the further to the left lies its tension-frequency-curve, i.e. the slower are the rates needed for an optimum modulation of force production [5, 8, 26, 29].

The rate modulation of force is, of course, critically dependent on the way in which the impulse rates of motoneurones are modulated by synaptic activity. It is, in this connexion, of importance to realize that the synaptic excitation converging onto a motoneurone does not directly set its impulse rate. Asynchronous synaptic activity produces a relatively steady post-synaptic current, and excitatory currents of this kind will, if strong enough, elicit a repetitive discharge in the post-synaptic neurone. Thus, it works somewhat similarly to the way in which the repetitive discharge of a tonic mechanoreceptor is produced by a generator current. The discharge rate actually produced in the motoneurone very much depends on its own intrinsic membrane properties which determine how it will respond to steady excitation (for reviews and references, see [9, 21]). At a just suprathreshold intensity of current, the cell will discharge at its minimum rate of maintained firing. Over a fairly wide range of intensities above threshold, an increased current will produce an enhanced firing rate. For normal hindlimb muscles of cats, the minimum rate of maintained firing is matched to the twitch-speed of the corresponding muscle fibres. Thus, owing to its own intrinsic membrane properties, a barely recruited motoneurone will typically start to modulate its discharge at rates covering the steep portion of the tension-frequency curve of its muscle fibres [18–20]. Such a "speed-match" does, of course, lead to optimum conditions for the use of rate-modulation of motor unit force.

Short-Term Usage-Dependence of Gradation Mechanisms

I will now discuss some of the ways in which preceding activity changes the functional properties of the motoneurone-muscle system. Short-term changes take place in muscle fibres as well as in motoneurones. With respect to the muscle, it is well known that a constant pattern of test-stimulation may cause a gradual increase or decrease of force to occur over a time period of tens of seconds to minutes. These phenomena are commonly referred to as *potentiation* and *fatigue*, respectively. Besides such usage-dependent effects on contractile force, preceding activation may also cause a slowing-down of the twitch [7]. In comparison to these various muscle reactions, which are all likely to be of importance in normal motor behaviour, the activity-dependent reactions of motoneurones are less well known. I will therefore describe these cell properties in somewhat greater detail.

We studied the usage-dependent reactions of individual motoneurones by aid of intracellular stimulation and recording. During the intracellular application of steps of constant-current stimuli, motoneurones typically display a decrease in firing rate (referred to as *adaptation;* see [12, 17, 21]). This adaptation takes place in two phases. After a brief initial phase, which is usually practically completed already after a few hundred ms, there follows a prolonged period of gradual decline in firing rate which we have called late adaptation. During this phase of adaptation, the sensitivity to changes in current intensity stays almost constant. We have been studying the late adaptation in motoneurones of the cat's gastrocnemius [23, 24]. We found that even when the cells were activated by a weak constant current of only 5 nA above the threshold for rhythmic firing, they commonly showed a very marked decline of firing rate during the initial ½–1 min. If the discharge was made to start at a higher rate, as achieved by increasing the stimulus intensity, the decline in rate was also more marked. Such observations fit well with the view that the late adaptation is primarily caused by cumulative after-effects of many consecutive spikes (activation of electrogenic sodium-pump?; cf. [11]): the higher the discharge rate, the more spikes and after-effects per unit time. Hence, one would also expect that, for corresponding intensities of supra-threshold current, cells working at low rates would show less late adaptation than cells working at higher rates. This was indeed found to be the case. Typical examples are demonstrated in Fig. 2A, B. The two graphs show the relative changes in discharge rate (A) and force (B) of a slow-twitch (S) and a fast-twitch (FF) motoneurone-motor unit combination during 1 min of constant intracellular stimulation. In each case the motoneurone was activated by a current of 5 nA above the threshold for rhythmic firing. Slow-twitch motoneurones tended to maintain a fairly constant rate and force throughout the period of constant stimulation. In motoneurones of fast-twitch units, however, which were working at higher absolute levels of discharge frequency (cf. Legend of Fig. 2), the same supra-threshold intensity of stimulation caused a marked drop in rate as well as in force during the initial minute of stimulation (Fig. 2).

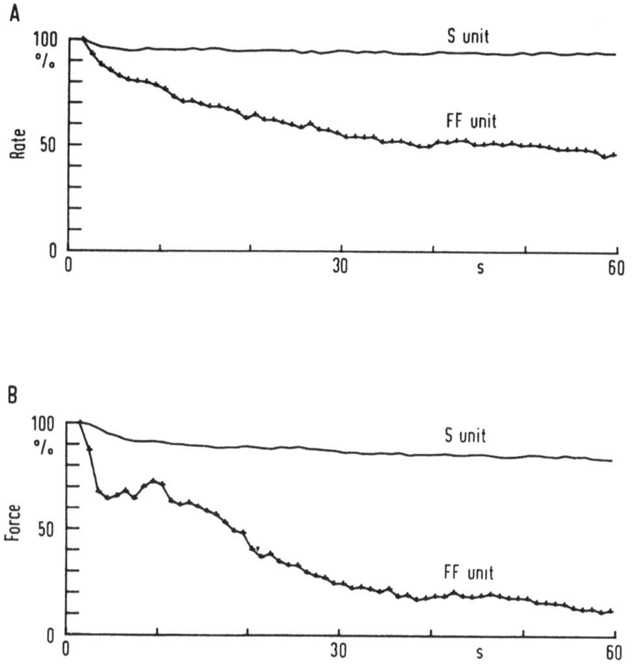

Fig. 2 A, B. Late adaptation in two individual motoneurones from m. gastrocnemius medialis. The cells were activated by currents injected through an intracellular microelectrode. Force production of their units was simultaneously monitored from the muscle (tendon connected to sensitive force transducer, muscle kept at optimum length for a twitch). The muscle unit of the S motoneurone had a twitch contraction time of 53 ms and a fatigue index of 97%. The muscle unit of the FF motoneurone had a twitch contraction time of 25 ms and a fatigue index of 7%. Values plotted in the illustration were derived from repetitive discharges elicited by constant injected currents lasting >1 min. **A** Plot of firing rate (%) versus time (s) for 1st min of discharges produced by current of 5 nA above threshold for rhythmic firing. For consecutive seconds of discharge, mean firing rates have been connected by ***straight lines***. Firing rates given as percentage of rate during 2nd second of discharge (= first value plotted = 16.1 Hz for the S unit and 28.9 Hz for the FF unit). **B** Contractile force produced by the discharges in **A.** Force given as percentage of value for 2nd second of discharge. For non-fused contractions, plotted values refer to mean force, as calculated over time periods of 1 s. Control measurements indicated that most of the force-decline of the FF unit was indeed caused by the decreased firing rate and not by muscle fatigue. (Kernell and Monster [24])

The findings of Fig. 2 illustrate that, in a maintained muscle contraction making use of fast units, the late adaptation of the motoneurones might cause a marked progressive fall in discharge rate and force to occur if the respective cells were driven by constant-intensity post-synaptic currents. Thus, a constant muscle force would often only be maintained by aid of a progressively increasing synaptic drive to the motoneurone pool. Such an increase in synaptic drive would presumably make itself felt as an increase in "effort" (cf. [31]). Thus, the late adaptation of motoneurones is presumably one of the mechanisms of importance for central aspects of motor fatigue.

Long-Term Usage-Dependence of Gradation Mechanisms

Besides the short-term after-effects of activity, there are also long-term effects which take place during the course of days to weeks. In the muscle, such alterations may concern the endurance as well as the speed and force of contraction. One of the most direct experimental methods for producing and studying such long-term effects of activity-changes is to evoke extra activity by means of direct electrical stimulation of muscles or muscle nerves (e.g. [30, 33]; for reviews and references, see [32, 34]). We have used this technique for comparing the long-term effects of different activity patterns on contractile properties [10, 27]. The chronic stimulation was applied to the left-side common peroneal nerve of cats by aid of a portable and remotely controlled ministimulator. Prior to the start of chronic stimulation, the nerve was deafferented; hence, the stimulation caused no behavioural reactions (no evidence for pain) and it evoked no reflex discharges. In the absence of chronic stimulation, the preceding operations (left-side dorsal rhizotomy and hemispinalization) did not cause any prolongation of the twitch or slowing-down of the tension-frequency curve. In an initial series of experiments, great amounts of stimulation were applied per day ("tonic" patterns, $\geqq 50\%$ of total time taken up by activity), and the total duration of treatment was 8 weeks [10]. As might be expected on the basis of preceding findings [32, 34], the fast muscle studied (m. peroneus longus) then became markedly slower, weaker and less fatigue-sensitive. All these activity-related changes are in accordance with the fact that the most easily recruited units of a muscle tend to be comparatively slow, weak, and fatigue-resistant (cf. Fig. 1; [3, 14]); such units might actually possess these various properties (mainly) as a consequence of the fact that they are so much used. Interestingly enough, however, we found no evidence indicating that the effects of long-term activation on twitch-speed would depend on the utilized pulse-rates (cf. [10, 27]; see also [1, 16]). Thus, in our type of preparation (mixed hindlimb muscles with normal innervation) the slowing effects of "extra" activity were apparently dependent on the daily amount of activation as well as, perhaps, on the diurnal distribution of the activity periods (cf. [1]).

Besides our studies of "tonic" stimulation patterns, we have also investigated the effects of smaller quantities of extra activation [27]. To our surprise, even amounts as small as 0.5% of extra activity-time per day (i.e. 7.2 min per 24 h taken up by applied stimulation) had a significant effect on contractile speed. Even though activation was in this case given as brief high-frequency bursts of 100 Hz (burst-duration 0.1 s), the muscle became significantly slower with respect to its twitch as well as to its tension-frequency relation. On average, the twitch contraction time was, in these cases, prolonged by $35 \pm 10\%$ (mean $\pm$ SD for ipsi- vs contra-lateral twitch; $n = 5$). The mean effect on the tension-frequency relation is demonstrated in Fig. 3 A. This graph shows contractile force plotted versus the pulse frequency of bursts of test stimuli. The left-side curve (triangles) refers to chronically stimulated peroneus muscles and the right-side curve (crosses) came from the nonstimulated control muscles of the contra-lateral side. The steep region of the tension-frequency curve was clearly situated over a range of pulse rates that were slower for the chronically stimulated muscles than for the control

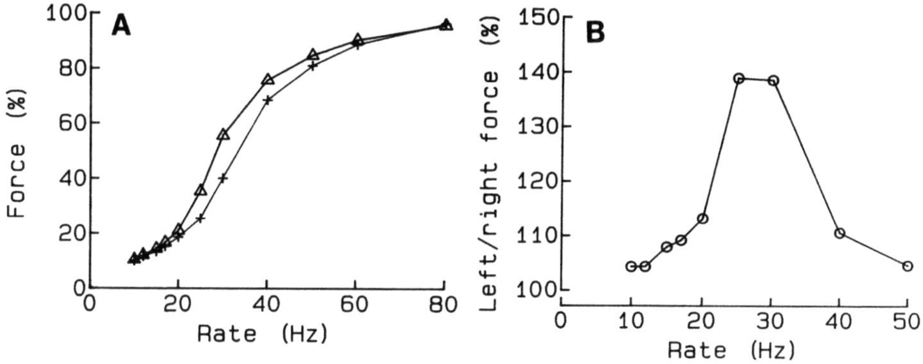

Fig. 3 A, B. Effects of small amounts of chronic stimulation on the gradation-properties of fast hindlimb muscles. **A** Tension-frequency plot for mean data from peroneus longus muscles (*PerL*) that had been subjected to chronic activation (stimulation of left-side common peroneal nerve; *triangles*) and for the contra-lateral control muscles of the same 5 animals (*crosses*). Chronic stimulation had in these cases been given as 0.1 s bursts of 100 Hz, repeated once every 4 s during 3 daily training periods of 96 min each. The total amount of daily activation was thus 7.2 min (= 0.5% of 24 h). The treatment lasted for 4 weeks. In a final acute experiment under pentobarbitone anaesthesia, the PerL muscles of both hindlimbs were dissected free and connected to a force transducer for the measurement of isometric contractile properties. These measurements, including those shown in the diagram, were performed with the muscle at the optimum length for a twitch. During recordings, the muscles were covered with liquid paraffine at 37–38 °C. The tension-frequency relation was determined by aid of test-bursts of 1 s duration, and the plotted forces (% of maximum tetanic force) refer to the *mean* tension obtained during the latter half of each test burst (cf. [29]). **B** Ratios between the forces of **A** as produced, at each rate of test stimulation, by the group of chronically activated muscles (*triangles*) and by their contra-lateral controls (*crosses*), respectively. (Kernell and Eerbeek, unpublished work)

muscles (Fig. 3 A). The interval of test-stimulation needed for producing 50% of the maximum tetanic force showed an experimental/control (i.e. left/right) ratio of 116±11% (cf. Fig. 3 A); this ratio was significantly greater than that seen in cats that had been deafferented and hemispinalized but not chronically stimulated (ratio 93±6%; $n=5$; $P<0.01$). Maximum contractile force was not significantly changed in the chronically stimulated muscles shown in Fig. 3.

The difference between the two tension-frequency curves of Fig. 3 A may seem relatively slight. What matters in the present context is, however, the extent to which a given motoneurone discharge would produce a greater force after a period of applied "training" than before. That effect may be quantified by calculating, for each stimulus rate, the ratio between the forces of the "trained" muscle and the contralateral control muscle. For the tension-frequency curves of Fig. 3 A, these ratios have been plotted in Fig. 3 B. As is demonstrated by this latter graph, even a very modest-looking shift of the steep region of a tension-frequency curve will produce quite an appreciable effect on the force evoked by physiologically relevant discharge rates (e.g. 25–30 Hz; Fig. 3 B).

Experiments such as those depicted in Fig. 3 indicate that also rather minor changes in long-term activity patterns may produce significant alterations in muscle properties of importance for force-gradation. If all units of a muscle were changed as indicated by Fig. 3, contractions of moderate force would have to be produced by a lower motoneuronal discharge rate after the "training-period" than before. Thus, if the motoneurones stayed unchanged, the effect of the "training" would be to decrease the amount of pool-excitation needed for producing moderate amounts of contractile force in the muscle. Such adjustments of excitatory drive would presumably be impossible without the use of sensori-motor correction mechanisms. The nature of these correction mechanisms is dealt with by several other contributions in this volume.

Summary

A brief but fairly general survey is given of functional properties of the motoneurone-muscle system, as studied for the cat's hindlimb. Firstly, mechanisms involved in the normal recruitment- and rate-gradation of muscle force are described and discussed. Secondly, the article deals with some of the ways in which these gradation-mechanisms may be influenced by preceding activity. With respect to short-term effects (time scale: seconds-minutes) special attention is given to the gradual decline in discharge rate that takes place in a motoneurone during a period of direct stimulation at a constant intensity (late adaptation). Long-term after-effects of activity are described with respect to the changes in muscular gradation-properties that one finds after a number of weeks of chronic stimulation. It is pointed out that even rather small amounts of extra activity per day (e.g. 0.5% of total time) may then be sufficient for producing significant alterations in the way in which muscle force is graded by a change in activation rate.

Acknowledgment. The investigations were supported in part by the Foundation for Medical Research (FUNGO), which is subsidized by the Netherlands Organization for the Advancement of Pure Research (ZWO).

References

1. Al-Amood WS, Lewis DM, Rowlerson A (1985) The effects of chronic stimulation at low rates on denervated rat soleus muscle. J Physiol (Lond) 361:36P
2. Bawa P, Binder MD, Ruenzel P, Henneman E (1984) Recruitment order of motoneurons in stretch reflexes is highly correlated with their axonal conduction velocity. J Neurophysiol 52:410–420
3. Burke RE (1981) Motor units: anatomy, physiology and functional organization. In: Brooks VB (ed) Handbook of physiology – the nervous system II, part 1. American Physiological Society, Bethesda, pp 345–422
4. Burke RE, Levine DN, Tsairis P, Zajac FE (1973) Physiological types and histochemical profiles in motor units of the cat gastrocnemius. J Physiol (Lond) 234:723–748
5. Burke RE, Rudomin P, Zajac FE (1976) The effect of activation history on tension production by individual muscle units. Brain Res 109:515–529

6. Burke RE, Dum RP, Fleshman JW, Glenn LL, Lev-Tov A, O'Donovan MJ, Pinter MJ (1982) An HRP study of the relation between cell size and motor unit type in cat ankle extensor motoneurons. J Comp Neurol 209:17–28
7. Close RI (1972) Dynamic properties of mammalian skeletal muscles. Physiol Rev 52:129–197
8. Cooper S, Eccles JC (1930) The isometric responses of mammalian muscles. J Physiol (Lond) 69:377–385
9. Crill WE, Schwindt PC (1984) Ionic mechanisms underlying the excitation-to-frequency transduction: studies by voltage clamp methods. Arch Ital Biol 122:31–41
10. Eerbeek O, Kernell D, Verhey BA (1984) Effects of fast and slow patterns of tonic long-term stimulation on contractile properties of fast muscle in the cat. J Physiol (Lond) 352:73–90
11. Grafe P, Rimpel J, Reddy MM, ten Bruggencate G (1982) Changes of intracellular sodium and potassium ion concentrations in frog spinal motoneurons induced by repetitive synaptic stimulation. Neuroscience 7:3213–3220
12. Granit R, Kernell D, Shortess GK (1963) Quantitative aspects of repetitive firing of mammalian motoneurones, caused by injected currents. J Physiol (Lond) 168:911–931
13. Gustafsson B, Pinter MJ (1984) Relations among passive electrical properties of lumbar alpha-motoneurones of the cat. J Physiol (Lond) 356:401–431
14. Henneman E, Mendell LM (1981) Functional organization of motoneuron pool and its inputs. In: Brooks VB (ed) Handbook of physiology – the nervous system II, part 1. American Physiological Society, Bethesda, pp 423–507
15. Henneman E, Somjen G, Carpenter DO (1965) Functional significance of cell size in spinal motoneurons. J Neurophysiol 28:560–580
16. Hudlická O, Tyler KR, Srihari T, Heilig A, Pette D (1982) The effect of different patterns of long-term stimulation on contractile properties and myosin light chains in rabbit fast muscles. Pflügers Arch 393:164–170
17. Kernell D (1965a) The adaptation and the relation between discharge frequency and current strength of cat lumbosacral motoneurones stimulated by long-lasting injected currents. Acta Physiol Scand 65:65–73
18. Kernell D (1965b) The limits of firing frequency in cat lumbosacral motoneurones possessing different time course of afterhyperpolarization. Acta Physiol Scand 65:87–100
19. Kernell D (1979) Rhythmic properties of motoneurones innervating muscle fibres of different speed in m. gastrocnemius medialis of the cat. Brain Res 160:159–162
20. Kernell D (1983) Functional properties of spinal motoneurons and gradation of muscle force. In: Desmedt JE (ed) Motor control mechanisms in health and disease. Raven, New York, pp 213–226
21. Kernell D (1984) The meaning of discharge rate: excitation-to-frequency transduction as studied in spinal motoneurones. Arch Ital Biol 122:5–15
22. Kernell D (1986) Organization and properties of spinal motoneurones and motor units. Prog Brain Res 64:21–30
23. Kernell D, Monster AW (1982a) Time course and properties of late adaptation in spinal motoneurones in the cat. Exp Brain Res 46:191–196
24. Kernell D, Monster AW (1982b) Motoneurone properties and motor fatigue. An intracellular study of gastrocnemius motoneurones of the cat. Exp Brain Res 46:197–204
25. Kernell D, Zwaagstra B (1981) Input conductance, axonal conduction velocity and cell size among hindlimb motoneurones of the cat. Brain Res 204:311–326
26. Kernell D, Ducati A, Sjöholm H (1975) Properties of motor units in the first deep lumbrical muscle of the cat's foot. Brain Res 98:37–55
27. Kernell D, Eerbeek O, Donselaar Y, Verhey BA (1983a) Effects of moderate amounts of fast and slow rates of chronic stimulation on the contractile properties of a fast hindlimb muscle in the cat. Proc Int Union Physiol Sci 15:189
28. Kernell D, Eerbeek O, Verhey BA (1983b) Motor unit categorization on basis of contractile properties: an experimental analysis of the composition of the cat's m. peroneus longus. Exp Brain Res 50:211–219
29. Kernell D, Eerbeek O, Verhey BA (1983c) Relation between isometric force and stimulus rate in cat's hindlimb motor units of different twitch contraction time. Exp Brain Res 50:220–227

30. Lømo T, Westgaard RH, Engebretsen L (1980) Different stimulation patterns affect contractile properties of denervated rat soleus muscles. In: Pette D (ed) Plasticity of muscle. de Gruyter, Berlin, pp 297–309
31. McCloskey DI (1978) Kinesthetic sensibility. Physiol Rev 58:763–820
32. Salmons S, Henriksson J (1981) The adaptive response of skeletal muscle to increased use. Muscle Nerve 4:94–105
33. Salmons S, Vrbová G (1969) The influence of activity on some contractile characteristics of mammalian fast and slow muscles. J Physiol (Lond) 201:535–549
34. Vrbová G, Gordon T, Jones R (1978) Nerve-muscle interaction. Chapman & Hall, London

Activity of Motoneurons in Man under Stationary Conditions

R. DENGLER [1] and W. WOLF [2]

The activity of motoneurons can be assessed electromyographically by recording the action potentials of the muscle fibers they supply (motor unit potentials). Under stationary conditions, e.g., stable isometric contractions, motor units (MUs) fire tonically if they have once been recruited [2, 7, 10, 12, 14]. The major advantage of stationary conditions is that they are appropriate for a more detailed analysis of the discharge sequence of MUs, with particular respect to their regularity. If several MUs can be recorded simultaneously, their interaction can be studied by correlating the times of their discharges. The following contribution, which deals with the discharges of single MUs, will refer to both their regularity and their temporal relations.

The first dorsal interosseus muscle of the hand was investigated during sustained (20–120 s) isometric contractions (10% of maximum force; visual feedback) in normal subjects and patients with various central motor disorders. The discharges of up to 6 single MUs could be simultaneously recorded using two bipolar needle electrodes. MU action potentials were automatically recognized by a computer system using a specific segmentation algorithm and a classifier consisting of correlation techniques (see also W. Wolf and R. Dengler, this volume).

The regularity of the discharge sequence of MUs can be assessed by various methods [5, 8]. A quick survey is provided by instantaneous frequency curves as illustrated in Fig. 1. Under the above conditions, MUs fire with a rather constant frequency for time periods of up to 3 min. While some minor frequency oscillations are occasionally seen, a consistent increase or decrease is rather unusual, at least as long as the force is kept constant. Mean firing frequencies at forces of 10% ranged in normal subjects between 8 and 13 Hz, whereas in patients with parkinsonism and chorea occasionally higher frequencies of up to 16 Hz were observed.

More accurate results are obtained by constructing interval histograms (IH). Normal IH exhibit a narrow basis and are bell-shaped, reflecting a normal distribution. IH of patients with central motor disorders are frequently broader and of irregular shape (Fig. 2). For the calculation of mean intervals and standard deviations (SD), it has proven useful to reject very long intervals (longer than mean +2 SD) and to correct for long-term frequency oscillations applying a so-called floating mean [1]. Following these procedures, SDs in normal subjects are lower

[1] Neurologische Klinik und Poliklinik der Technischen Universität München, Möhlstr. 28, D-8000 München 80, FRG.
[2] Abteilung Elektrotechnik der Universität der Bundeswehr München, Werner-Heisenberg-Weg 39, D-8014 Neubiberg, FRG.

Clinical Aspects of Sensory Motor Integration
Ed. by A. Struppler and A. Weindl
© Springer-Verlag Berlin Heidelberg 1987

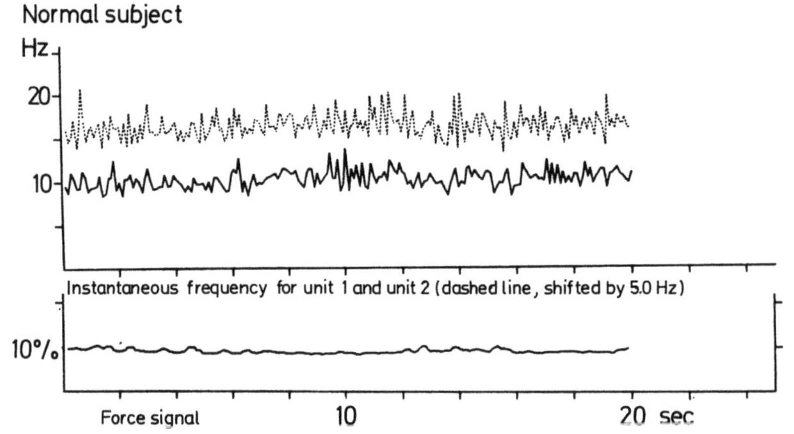

Fig. 1. Instantaneous frequencies of 2 simultaneously recorded motor units. The *bottom trace* shows the corresponding force kept constant at 10% of the maximum

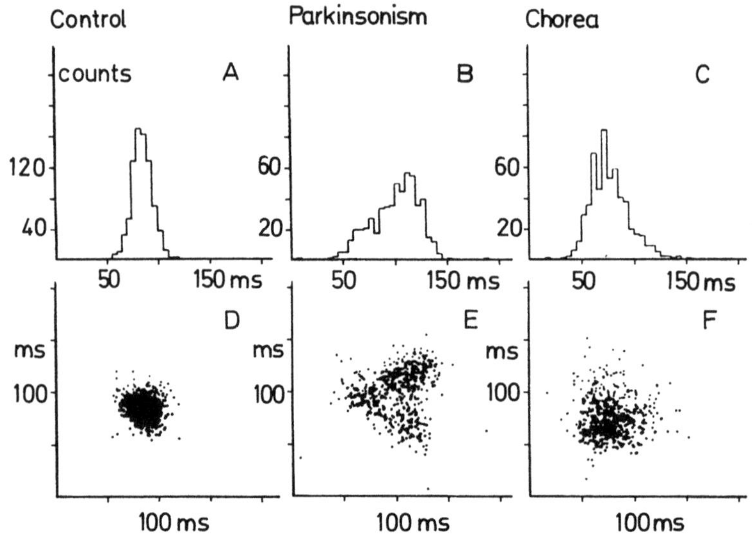

Fig. 2 A–F. Interval histograms (**A–C**) and corresponding joint interval histograms (**D–F**) of single motor units. Characteristic changes are seen in parkinsonism (**B, E**) and chorea (**C, F**)

than 15% of the mean interval, whereas higher values are found in about one-half of patient MUs. This changes, however, seem to be rather unspecific and cannot be used for a reliable discrimination between different central motor disorders.

Joint interval histograms (JIH) reflect the relation between neighboring intervals. For each MU discharge, the preceeding interval is plotted on the abscissa and the following on the ordinate (Fig. 2). JIH of normal MUs show a densely packed crowd of points because neighboring intervals are about equally long without signs of interdependence. In parkinsonism, JIHs are frequently triangle-

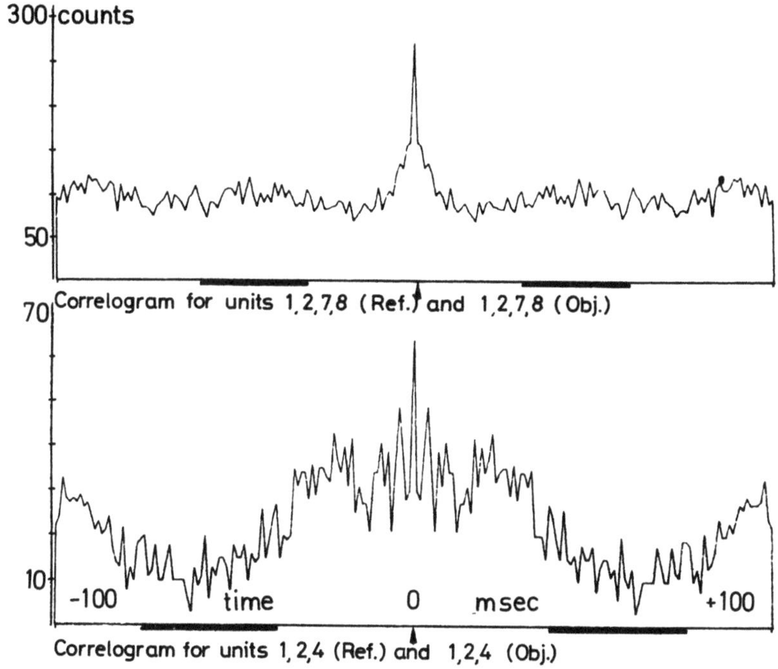

Fig. 3. Multiunit cross-correlograms. The *upper* is constructed from the times of discharges of 4 MUs of a normal subject. The narrow peak above "0" indicates "short-term synchronization". The *lower* is constructed from the discharges of 3 MUs of a patient with parkinsonism and reveals "braod-peak synchronization"

shaped, which points to long-short sequences. Choreic conditions are often associated with rather irregular patterns. Statistical treatment of JIHs, e.g., calculation of regression lines, variances, and correlation coefficients, discriminates, similarly to the IH, between normal subjects and patients, but not between different central motor disorders.

Although MUs fire rather independently of each other, a correlation of the times of their discharges frequently reveals a weak tendency to synchronization [6, 11, 13]. This so-called "short-term" synchronization is thought to be caused by the synchronous occurrence of EPSPs in several motoneurons sharing axonal branches from a common input cell. In corresponding correlograms (Fig. 3), a narrow peak above "0" (width 5–10 ms) indicates a more than random synchronization. We have seen the "short-term" synchronization in about 50% of pairs of MUs analyzed and nearly always in multiunit correlograms (Fig. 3). Very recently we could show that these synchronies are not evenly distributed but tend to form clusters of 5–10 subsequent events [3].

Another type of synchronization corresponding to the so-called "broad-peak" synchronization [4, 9] is regularly found in multiunit correlograms from patients with parkinsonism (Fig. 3). It lasts about 10–50 ms (width of the peak above "0") and is probably due to synchronization of the input to the motoneurons. We have also seen it in patients with mild parkinsonism, as well as in the absence of a clinically manifest tremor. There is generally an underlying rhythmicity of 5–7 Hz (Fig. 3) associated with force oscillations visible in high-gain records. Thus, it appears that MU synchronization of the "broad-peak" type is a rather consistent finding in parkinsonism although it might be clinically latent. So far, we have not observed it in normal subjects up to ages of 70 years, although it might occur in activated physiological tremor and in senile tremor.

Studies as described above require the application of computer techniques. Computers, however, will soon be part of the equipment of most EMG laboratories. Thus, it appears worthwhile to investigate whether such studies might be of value in the diagnostic assessment of central motor disorders.

Summary

The discharge pattern of single motor units (MUs) in stationary isometric contractions was studied in normal subjects and in patients with parkinsonism or chorea. Interval histograms, joint interval histograms and cross-correlograms of several simultaneously recorded MUs were used. MUs of normal subjects fired rather regularly whereas the discharge sequences of many MUs of both patient groups were grossly irregular. Cross-correlograms of normals and of choreics generally showed a synchronization of the "short-term" type. All patients with parkinsonism, however, revealed a synchronization of the "broad-peak" type. The diagnostic value of these investigations is discussed.

Acknowledgment. We thank Mrs. C. Klesius for her expert technical assistance.

References

1. Andreassen S, Rosenfalck A (1980) Regulation of the firing pattern of single motor units. J Neurol Neurosurg Psychiatry 43:897–906
2. Clamann HP (1970) Activity of single motor units during isometric tension. Neurology 20:254–260
3. Dengler R, Wolf W, Struppler A (1984) Synchronous discharges in pairs of steadily firing motor units tend to form clusters. Neurosci Lett 47:167–172
4. Dengler R, Birk P, Wolf W, Struppler A (1984) Temporal relations of simultaneously active motor units: a study in normal subjects and patients with central motor disorders. Neurology 34 [Suppl 1]:361
5. Dengler R, Wolf W, Birk P, Struppler A (1985) Möglichkeiten der computer-assistierten Signalerkennung in der Elektromyographie zentralmotorischer Störungen. In: Gänshirt H, Berlitz P, Haack G (Hrsg) Verhandlungen der deutschen Gesellschaft für Neurologie, Bd 3. Springer, Berlin Heidelberg New York, S 875–900
6. Dietz V, Bischofsberger E, Wita C, Freund HJ (1976) Correlation between discharges of two simultaneously recorded motor units and physiological tremor. Electroencephalogr Clin Neurophysiol 40:97–105
7. Freund HJ (1983) Motor units and muscle activity in voluntary motor control. Physiol Rev 63:387–436

8. Freund HJ, Wita C (1971) Computeranalyse des Intervalmusters einzelner motorischer Einheiten bei Gesunden und Patienten mit supraspinalen Störungen. Arch Psychiat Nervenkr 214:56–71
9. Kirkwood PA, Sears TA, Stagg D, Westgaard RH (1982) The spatial distribution of synchronization of intercostal motoneurons in the cat. J Physiol 327:137–155
10. Kranz H, Baumgartner G (1975) Human alphamotoneurons discharge, a statistical analysis. Brain Res 67:324–329
11. Milner-Brown HS, Stein RB, Lee RG (1975) Synchronization of human motor units: possible roles of exercise and supraspinal reflexes. Electroencephalogr Clin Neurophysiol 38:245–254
12. Person RS, Kudina LP (1972) Discharge frequency and discharge pattern of human motor units during voluntary contraction of muscle. Electroencephalogr Clin Neurophysiol 32:471–483
13. Sears TA, Stagg D (1976) Short-term synchronization of intercostal motoneurone activity. J Physiol 263:357–381
14. Tanji J, Kato M (1973) Firing rate of individual motor units in voluntary contraction of abductor digiti minimi muscle in man. Exp Neurol 40:771–783

Automatic Sorting and Analysis of Multiunit EMG Recordings

W. Wolf [1] and R. Dengler [2]

Introduction

In the clinical environment, invasive EMG recordings are currently used to acquire information about either the shape of the motor unit action potential (MUAP) waveform [3–9] or the motor unit discharge pattern [4–7]. Despite the use of selective bipolar needle electrodes, records mostly comprise MUAPs of different motor units, and, therefore, the application of a sorting procedure which separates the MUAPs of the single motor units for analysis is indicated. In the case of analyzing one single unit at a time, simple methods (like amplitude window) which can be performed on both a hardware and a software level are satisfactory, since only the extraction of a very prominent MUAP from the recorded signal is required. If, however, a simultaneous observation of several units is required for studying temporal relationships between them [5], a decomposition of the EMG signal into its constituent MUAP trains is necessary. Such a decomposition, however, represents a very complex pattern recognition task, even in the simple case of EMG recordings taken during stable weak isometric contractions; therefore, it can only be accomplished by computer software, and there are reports about efficient programs from several authors [1, 8, 10, 11]. As part of our investigations concerning motor-unit behavior in central disorders (see R. Dengler and W. Wolf, this volume) we have also developed such a decomposition program, which will be outlined by this report.

Signal Acquisition

EMG signals were recorded from the first dorsal interosseus muscle, as shown in Fig. 1, and, together with the force signal from a strain gauge, stored on digital tape for off-line evaluation by a medium-scale computer PE 8/32. The sampling rate was 16 kHz, and the length of the records varied from 20 to 120 s, depending on the performance of the test subject.

[1] Abteilung Elektrotechnik der Universität der Bundeswehr München, Werner-Heisenberg-Weg 39, D-8014 Neubiberg, FRG.

[2] Neurologische Klinik und Poliklinik der Technischen Universität München, Möhlstr. 28, D-8000 München, FRG.

Clinical Aspects of Sensory Motor Integration
Ed. by A. Struppler and A. Weindl
© Springer-Verlag Berlin Heidelberg 1987

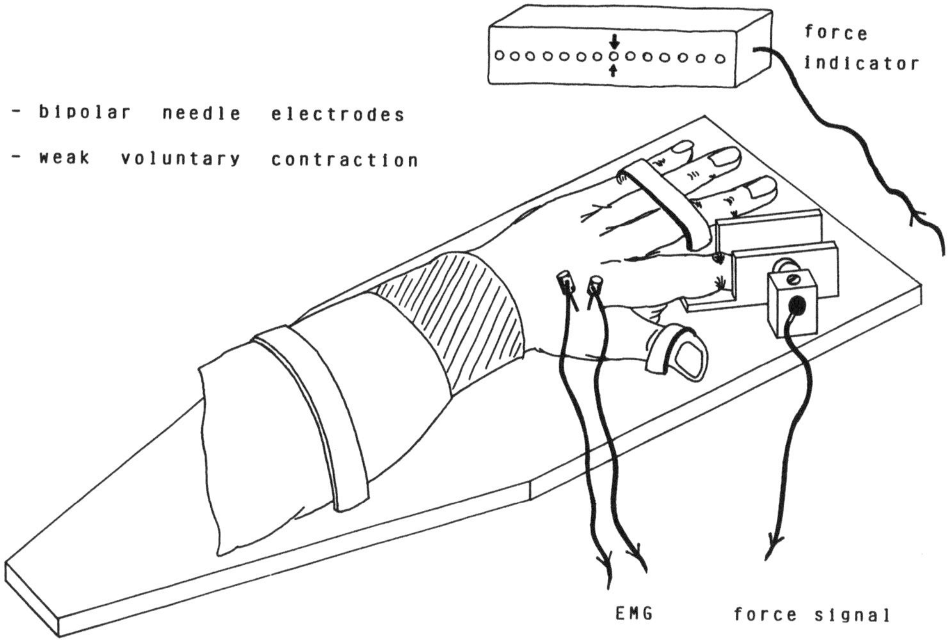

Fig. 1. Experimental setup: the subject has to develop a constant isometric force against a strain gauge device by his index finger, which is guided by a laterally moveable rail. The actual force is visualized for feedback by a row of light emitting diodes ("force indicator"). EMG signals are recorded by two bipolar needle electrodes from different sites within the first dorsal interosseus muscle

Decomposition Technique

Decomposition of the EMG signal is performed in three steps: (a) segmentation, (b) template selection, and (c) classification. Since performance and computer time required to decompose a record automatically are opposing parameters, a "performance" adjustment is possible at each stage in order to shorten the evaluation process by leaving unsolved problems to the decision of a highly trained operator.

Segmentation

The segmentation procedure has to detect all useful MUAPs contained in the signal trace. Signal segments which only show "noise" (consisting of instrumentation noise and background activity) are removed from the trace, and only those segments which meet the MUAP criteria are stored, together with a time label, for further processing. This means a great reduction of storage requirements. But the salient point in segmentation is that a very sensitive criterion produces a

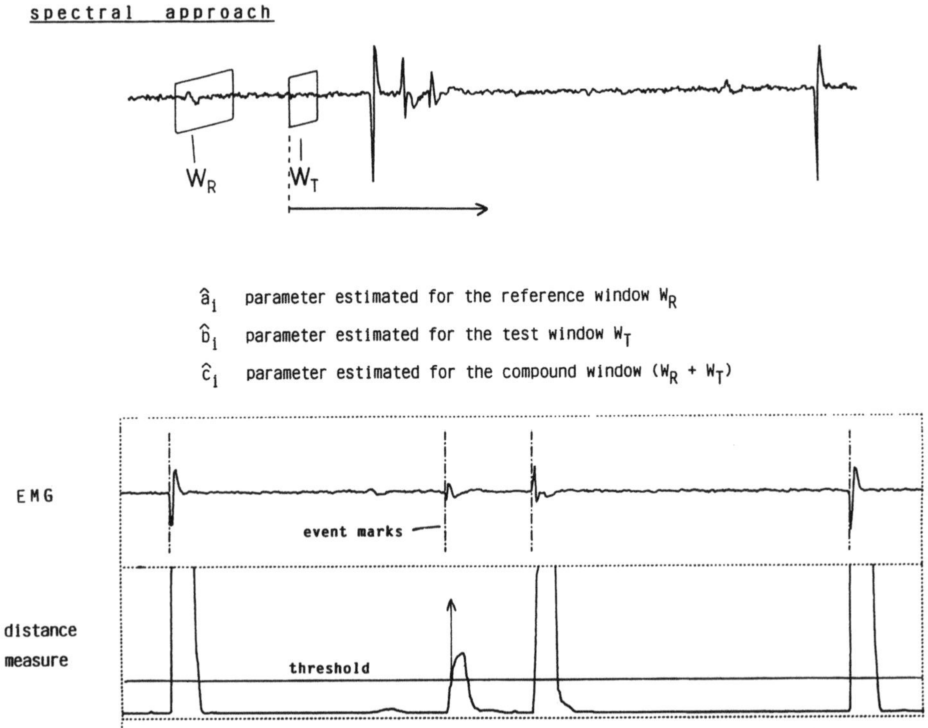

Fig. 2. The distance measure given in the *bottom trace* is derived from a comparison of the EMG signals within a fixed reference window W_R and a moving test window W_T, and specifies the actual dissimilarity between both. Segment boundaries of the EMG are set when the distance measure exceeds the threshold value

number of false alarms, and these require a large amount of signal processing for subsequent cancelling. On the other hand, use of a strong criterion will cause the program to overlook significant events which can hardly be recovered later on in the evaluation process. An amplitude threshold criterion, as used in most of the literature, is not effective for small signals contaminated by noise. For this reason, we developed a spectral approach to this problem using the autoregressive signal description. A detailed description of the theory is given in [2], but the basic idea will be outlined roughly (Fig. 2). First, we took a data segment – the so-called reference window W_R, which is fixed in time – and described it by a set of parameters a_i. It is important to select a segment which contains no useful MUAPs but is representative for the background noise. Then, we took the adjacent data segment – the so-called test window W_T, which will be shifted in time – and determined the parameters b_i for it as well. A third set of parameters c_i fitted the composite window $W_R + W_T$. If the parameter set c_i describes each of the windows as well as the individually adapted sets a_i and b_i, respectively, the distance measure is small and no segmentation boundary is indicated; if not, the signal characteristics (energy and spectrum) of W_T and W_R differ and a boundary is set.

Template Selection

As a basis for the classification procedure, an MUAP template must be determined for each motor unit being considered. For this purpose, a cluster program which uses the cross correlation factor as a criterion sorts all the segments into different groups. Groups with high correlation factors within the group and low correlation factors to all members of other groups are regarded as "template groups". The means over the central elements of these template groups represent the MUAP templates.

Classification

Pattern recognition can be performed by two different approaches: (a) parametric methods using predetermined features (slope, amplitude, duration, etc.) and (b) nonparametric methods based on correlation techniques. Both methods are implemented on our computer system, and results show that correlation techniques exceed the parametric method in performance. Especially, the decomposition of waveforms produced by superposition of two or more simultaneously occurring MUAPs can only be achieved by the correlation method. Our presently used correlation procedure shows three operational levels: 1. The stored EMG segment is correlated with each of the templates, allowing a slight shift to achieve correct alignment and, consequently, maximum correlation. If the maximum correlation of the segment to one of the templates is very high but low to all other templates, this MUAP is classified accordingly and used for updating the template. This continuous adaptation of the templates to slow changes in MUAP shape (e.g., due to slight needle slipping) improves the performance. 2. If the maximum correlation is not as high as before but still good, only classification is done without updating. 3. If the correlation of the segment to none of the templates is outstanding, a superposition is assumed. For decomposing this segment, correlation is performed with all possible combination of templates as described in [6].

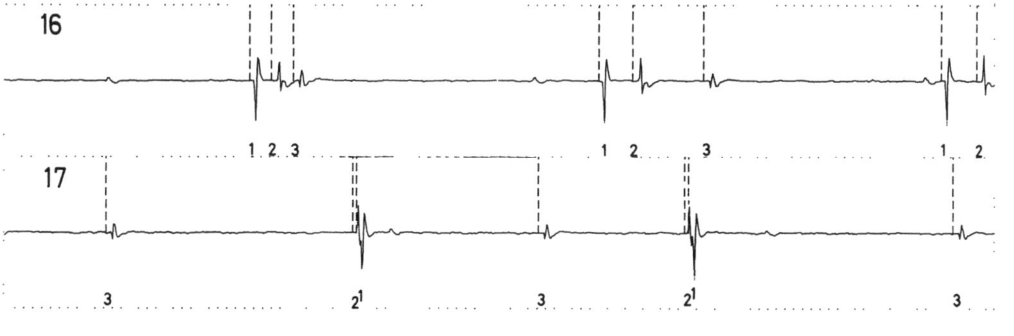

Fig. 3. Plot of the final classification results which shows the original EMG signal with the motor unit action potentials labelled by the unit number. Each trace represents an EMG segment of 250 ms

The performance of the decomposition procedure highly depends on the number of different MUAPs contained in the EMG record and on the simularity of the MUAP waveforms of the different motor units. For records which comprise MUAPs of two motor units and show high recording quality we have obtained classification error rates <1%. Figure 3 shows a control output of the program with the MUAPs marked by the vertical dashed lines and their classifications given by the small numbers below. A drawback of the correlation techniques is their high computational costs, but at present computer technology is developing so fast that this represents no problem in the long-term perspective.

Statistical Evaluation

For statistical evaluation of the discharge behavior, we implemented several commonly known procedures such as instantaneous frequency, interdischarge interval histograms, joint interval histogram, and cross-correlation (see also [5]). Furthermore, we have developed special procedures for analyzing the temporal relationship between the discharges of different motor units, especially the synchronies. Since the corresponding force signal is also available, the relation of motor-unit firing behavior to the time course of the force as well as to power spectrum properties can be studied.

Conclusion

Automatic MUAP sorting and analysis provide all the information currently used by clinicians, as well as additional information on the motor-unit firing pattern, and, therefore, represent a useful tool for clinical studies as well as for physiological investigations. But it should be pointed out that the effectiveness of decomposition by computer highly depends on the skill of the clinician: indeed, he implicitly performs the first partial decomposition of the EMG when he observes the signal on the scope while searching for a proper electrode position.

Summary

Decomposition of the EMG signal into its constituent motor unit action potential (MUAP) trains is a prerequisite for studying the activity pattern of motor units. The paper describes the basic principles of a computer program for automatic EMG decomposition and the statistical treatment of the MUAP data. The performance of the decomposition algorithm is demonstrated, and the usefulness of its clinical application is discussed.

References

1. Andreassen S (1983) Computerized analysis of motor unit firing. In: Desmedt JE (ed) Computer-aided electromyography. Karger, Basel (Progress in clinical neurophysiology, vol 10, pp 150–163)
2. Appel U, Von Brandt A (1983) Adaptive sequential segmentation of piecewise stationary time series. Information Sciences 29:27–56
3. Buchthal F (1962) The electromyogram. World Neurol 3:16–34
4. Clamann HP (1969) Statistical analysis of motor unit firing patterns in a human skeletal muscle. Biophys J 10:1233–1251
5. Dengler R, Wolf W, Birk P, Struppler A (1984) Synchronous discharges in pairs of steadily firing motor units tend to form clusters. Neurosci Lett 47:167–172
6. Figueiredo RJP, Gerber A (1983) Separation of superimposed signals by a cross-correlation method. IEEE Trans Acoustics, Speech and Signal Processing ASSP 31:1084–1089
7. Freund HJ (1983) Motor unit and muscle activity in voluntary motor control. Physiol Rev 63:387–436
8. Guiheneuc P (1985) Computer pattern recognition of motor unit potentials. In: Struppler A, Weindl A (eds) Electromyography and evoked potentials. Springer, Berlin Heidelberg New York, pp 114–121
9. Hausmanowa-Petrusewicz I, Kopec J (1983) Quantitative EMG and its automation. In: Desmedt JE (Ed) Computer-aided electromyography. Karger, Basel (Progress in clinical neurophysiology, vol 10, pp 164–185)
10. Mambrito B, De Luca CJ (1984) A technique for the detection, decomposition and analysis of the EMG signal. Electroencephalogr Clin Neurophysiol 58:175–188
11. Studer RM, Figueiredo RJP, Moschytz GS (1984) An algorithm for sequential signal estimation and system identification for EMG signals. IEEE Trans Biomedical Eng BME-31, no 3, pp 285–295

Functional Implications of Structure and Synaptology of Motor Neurons in Motor Neuron Disease

S. Conradi [1]

Motor neuron disease, or amyotrophic lateral sclerosis (ALS) is by no means uncommon. The number of new cases per year is more than half that reported for multiple sclerosis although the prevalence is low because of the rapid and fatal course of the disease. It is – and has been for the last 100 years – a major challenge to the neurologist to search for the cause and for an effective treatment of this disease. Despite the prominent symptoms such as progressive generalized paresis in striated muscle, the neuronal damage responsible for paresis is far from generalized, as it is concentrated in lower motor neurons in the spinal cord and brain stem and in neurons in descending tracts of the spinal cord, especially in the pyramidal tract.

In fact, the neuronal damage in ALS is more focal than that. Motor neurons of extraocular muscles have long been known to escape the disease process, as have external sphincter motor neurons. In the motor nerve roots, only part of the axons are affected, since Sobue et al. [22] have described a practically selective loss of the large axons of alpha motoneurons, with gamma motor axons and preganglionic sympathetic axons being much less affected, if at all. The damage to descending fiber tracts at the spinal cord level seems rather diffuse, but at higher levels there is a practically selective affection of pyramidal tract fibers [11]. The morphological changes in lower motor neurons in ALS are mostly unspecific and cannot be used for a more detailed understanding of the pathophysiology of the damage to the neurons, although aggregations of neurofilaments in the motor neuron cell bodies and proximal axon segments has been observed in some early cases [9]. Similar neuronal changes have been found in diseases affecting slow axonal transport mechanisms.

To gain a better understanding of the disease's pathophysiology it is important to explain this selective neuronal damage in most alpha motoneurons. The first opportunity to understand the mechanisms underlying a widespread lower motor neuron affection in a disease process was given by the description of retrograde axoplasmic transport in peripheral motor axons of foreign macromolecules which had been injected in muscle [13]. This transport route short-circuits the blood-brain and blood-nerve barrier at the motor end-plate and gives access for certain macromolecules in plasma to be transported into the CNS. This phenomenon has been called "the Trojan horse" route. It had already been discovered that poliovirus is transported via this route into lower motor neurons (cf. [3]); rabies virus is also transported in this manner into the spinal cord after uptake by the motor end-plates by means of binding to presynaptic nicotinic acetylcholine receptors (cf. [24]).

[1] Department of Neurology, Karolinska Sjukhuset, Box 60500, S-10401 Stockholm, Sweden.

Clinical Aspects of Sensory Motor Integration
Ed. by A. Struppler and A. Weindl
© Springer-Verlag Berlin Heidelberg 1987

To get an idea of the extent of this transport mechanism, Broadwell and Brightman [4] studied the uptake of horseradish peroxidase in neuronal cell bodies after intravenous injection and found labelling of the nerve cell bodies of large and small lower motor neurons, preganglionic autonomic neurons, spinal ganglion neurons and also in some neurons located near regions in the CNS lacking blood-brain barrier, such as the area postrema. Obviously, the blood-brain barrier can be bypassed in most of the peripheral nervous system neurons. Certain macromolecules can also be transported transsynaptically in the CNS, as was first shown in experiments with intraneuronal injection of dye for anatomical purposes where the dye was found to have spread to adjacent neurons through synapses [25]. Thus, material taken up through peripheral nerves can travel long distances in the CNS.

As seen, it is important to know whether there is evidence for the presence in the blood plasma of ALS patients of cytotoxic substances which could affect unprotected parts of the peripheral nervous system. There are several demonstrations of this. Wolfgram and Myers [23] reported neuronal changes in cultured spinal cord tissue after exposure to ALS but not to control serum. Roisen et al. [19] recently confirmed this finding in a large experimental series using a detailed quantitative description of the neuronal changes. Incubation of cultured spinal cord tissue to ALS plasma gives also an increased binding of immunoglobulins to the culture, as compared to controls [8]. Cytotoxic effects of ALS plasma might well affect other kinds of cells than neurons. Our own group has demonstrated that ALS plasma regularly induces an increased hemolysis of normal erythrocytes at incubation, which can be measured by spectrophotometry [5, 6]. This property of ALS plasma, which was now been found in 75 consecutive cases, is not found in the plasma of neurologically diseased controls having Charcot-Marie-Tooth disease, Guillain-Barré syndrome, or traumatic paraplegia. The cytotoxic activity toward erythrocytes is retained in isolated Immunoglobin G (IgG) and Immunoglobin A (IgA) fractions of ALS plasma [20]. Remarkably, the effect is still present after dilution of ALS plasma more than 1000 times, which indicates a very large excess of cytotoxic substance in comparison with the target organ. We now study the question whether the immunoglobulins responsible for cytotoxic effects on erythrocytes can also bind to the motor end-plate. In case it can be demonstrated that the interaction of immunoglobulins of ALS plasma with erythrocytes can be used as a model system for the effect on the motor end-plate, further experiments with erythrocytes can be rewarding. For example, hemolysis can easily be described quantitatively by means of spectrophotometry; immunosuppressive treatment of ALS patients has thereby been shown regularly to reduce the induced hemolysis considerably.

Thus, in ALS patients toxic material such as immunoglobulins could damage peripheral motor and sensory nerve endings, either by being taken up and transported retrogradely or by interfering with uptake and transport mechanisms occurring normally. Damage to descending pathways, including the pyramidal tract neurons, would then be caused by retrograde transsynaptic spread. Pyramidal tract neurons make monosynaptic contacts with lower motor neurons in primates (but not in other mammals) (see [14]).

Against this background it must be analyzed whether there are any factors which could explain why alpha motoneurons are more vulnerable than others to noxious influence mediated by retrograde axoplasmic flow mechanisms in the PNS. Such factors exist and they have to do with the anatomy and synaptic connections of the motor neurons. Due to the monosynaptic contact between dorsal root I A fibers and alpha motoneurons in the stretch reflex arch, the alpha motoneurons occupy an unfortunate position in that they are subjected to a double attack when a noxious influence is exerted via retrograde flow mechanisms in the PNS and has a transsynaptic spread. Gamma motoneurons, on the contrary, lack monosynaptic contact with dorsal root fibers, as do preganglionic autonomic neurons. Dorsal ganglia neurons do not receive any synapses at all, at least on the cell bodies. Actually, the position of many alpha motoneurons is even more unfortunate, since they receive monosynaptic recurrent alpha motor axon collaterals, as described some years ago [7]. This implies a third transport route from the periphery.

The presence of dorsal root synapses on alpha motoneurons must also have implications for the metabolism and surface chemistry of these neurons, in comparison with gamma motoneurons. In contrast to the latter, alpha motoneurons have to synthetize receptors for the transmitter substance of the dorsal root fibers in the monosynaptic pathway. Such receptors could, like acetylcholine (ACh) receptors, be widespread on the motor neuron surface. They could therefore exert an influence on uptake mechanisms at the motor end-plate, in analogy with what was mentioned above about virus uptake. In fact, the presence of dorsal root synapses on the motor neuron surface has an additional consequence, since there is one more morphological type of synapse on alpha motoneurons, which appears to have an occurrence and distribution on the motor neuron surface which parallels that of the large dorsal root synapses [12]. These large synapses of the so-called C-type have a highly characteristic appearance with a subsynaptic cistern, and they are not found on gamma motoneurons ([15]; see also A. H. Pullen, this volume). The origin of these synapses is not exactly known, although they are considered to originate from spinal interneurons [17, 18]. The presence of these synapses could likewise exert an influence on the surface chemistry and metabolism of alpha motoneurons, especially since larger stacks of endoplasmic reticulum are regularly found just below the subsynaptic cistern of these synapses.

According to this line of reasoning, the question of the presence of dorsal root synapses on those alpha motoneurons escaping in ALS is of great interest. In oculomotor neurons, there is no clear physiological evidence of a monosynaptic reflex from muscle spindle afferents, although trigeminal afferent terminals have been claimed morphologically to contact these neurons [16]. Type C synapses are not found on trochlear motor neurons [2]. Regarding the neurons in Onuf's nucleus, which innervate the external sphincter muscles, the information is still somewhat unprecise. There is no monosynaptic stretch reflex on most of these neurons and C-type synapses are apparently lacking ([21a)].

Now, with the claim that transport mechanisms in sensory fibers are involved in ALS, is there any evidence of damage to sensory fibers? Ultrastructural changes in sural nerve fibers of ALS patients have been described [10], and one group of neurons in the spinal cord which receives a heavy synaptic input from muscle spindle afferents (namely the neurons in Clarke's column) have been reported to be severely damaged in a high proportion of ALS patients [1].

As seen above, there are some aspects of the morphology and physiology of spinal cord neurons which have hitherto received little attention in relation to disease, although they might play an important role for the pathophysiology and distribution of neuronal damage in ALS. An analysis of the same type might be fruitful in other degenerative neuronal diseases.

Summary

Just as the synaptic connexions of motor neurons regulate the neuronal activity in the normal state, they could be equally important in the regulation of pathophysiological processes. In motor neuron disease (ALS), there are several arguments in support of the hypothesis that the degeneration of lower motor neurons is induced in peripheral axonal endings by interference with retrograde axoplasmic transport mechanisms. A theory is presented which explains the known differences in the vulnerability of various types of motor neurons (alpha, gamma, oculomotor, autonomic, etc.) in ALS on the basis of the differences in their synaptic connexions with other neurons of the peripheral nervous system, whose retrograde axoplasmic transport mechanisms might be affected in a similar way as in motor neurons. An analysis of the same type might be fruitful for other degenerative neuronal diseases.

Acknowledgments. Our studies are supported by the Swedish Medical Research Council (Proj. No 5178), Karolinska Institutet, Stockholms Läns Landsting, MS-fonden, Magnus Bergvalls Stiftelse, and the Vivian L. Smith Foundation for restorative neurology.

References

1. Averback P, Crocker P (1982) Regular involvement of Clarke's nucleus in sporadic amyotrophic lateral sclerosis. Arch Neurol 39:155–156
2. Bak IJ, Choi Wb (1974) Electron microscopic investigation of the trochlear nucleus in the cat. Cell Tissue Res 150:409–423
3. Bodian D, Howe HA (1941) The rate of progression of poliomyelitis virus in nerves. Bull Johns Hopkins Hosp 69:79–85
4. Broadwell RD, Brightman MW (1976) Entry of peroxidase into neurons of the central and peripheral neurons system from extracerebral and cerebral blood. J Comp Neurol 166:257–284
5. Conradi S, Ronnevi L-O (1982) Cytotoxic factor in plasma from ALS patients provokes haemolysis of normal erythrocytes. Acta Neurol Scand [Suppl 90] 65:246–247
6. Conradi S, Ronnevi L-O (1985) Cytotoxic activity in the plasma of ALS patients against normal erythrocytes. Quantitative determinations. J Neurol Sci 68:135–145
7. Cullheim S, Kellerth J-O, Conradi S (1977) Evidence for direct synaptic interconnections between cat spinal alpha motoneurons via the recurrent axon collaterals. A morphological study using intracellular injection of horseradish peroxidase. Brain Res 132:1–10

8. Digby J, Harrison R, Jehanli A, Lunt GG, Capildeo R, Clifford-Rose F (1984) Immunological changes in motor neurone disease. In: Clifford-Rose F (ed) Research progress in motor neuron disease. Pitman, London, pp 368–378
9. Hirano A (1982) Aspects of the ultrastructure of amyotrophic lateral sclerosis. In: Rowland LP (ed) Human motor neuron diseases. Raven, New York, pp 75–86
10. Dyck PJ, Stevens JC, Mulder DW, Espinosa RE (1975) Frequency of nerve fiber degeneration of peripheral motor and sensory neurons in amyotrophic lateral sclerosis. Neurology 25:781–785
11. Ikuta F, Makifuchi T, Ichikawa T (1979) Comparative studies of tract degeneration in ALS and other disorders. In: Tsubako T, Toyokura T (eds) Amyotrophic lateral sclerosis. University Park Press, Baltimore, pp 177–200
12. Kellerth J-O, Conradi S, Berthold C-H (1983) Electron microscopic studies of serially sectioned cat spinal alpha motoneurons: motoneurons innervating fast-twitch (type FF) units of the gastrocnemius muscle. J Comp Neurol 214:451–458
13. Kristensson K (1970) Transport of fluorescent protein tracer in peripheral nerves. Acta Neuropathol 16:293–300
14. Kuypers HGJM (1981) Anatomy of the descending pathways. In: Brooks VB (ed) The nervous system. Handbook of physiology, sect 1, vol II. American Physiological Society, Washington, pp 597–666
15. Lagerbäck PÅ (1985) On ultrastructural studies of cat lumbosacral gamma motor neurons after retrograde labelling with horseradish peroxidase. J Comp Neurol 240:256–264
16. Manni E, Bortolami R (1982) Proprioception in eye muscles. In: Lennerstrand G, Zee DS, Keller EL (eds) Functional basis of ocular mobility disorders. Pergamon, Oxford New York, pp 53–64
17. McLaughlin B (1972) Propriospinal and supraspinal projections to the motor nuclei in the cat spinal cord. J Comp Neurol 144:475–500
18. Pullen AH, Sears TA (1978) Modification of "C"-synapses following partial central deafferentation of thoracic motorneurons. Brain Res 145:141–146
19. Roisen FJ, Bartfeld H, Donnenfeld H, Baxter J (1982) Neuron-specific in vitro cytotoxicity of sera from patients with amyotrophic lateral sclerosis. Muscle Nerve 5:48–53
20. Ronnevi L-O, Conradi S, Karlsson E (1984) Cytotoxic effects of immunoglobulins in amyotrophic lateral sclerosis (ALS). Acta Neurol Scand [Suppl 98] 69:182–183
21. Schröder HD (1980) Organization of the motoneurons innervating the pelvis muscles of the male rat. J Comp Neurol 192:567–587
21a. Schröder HD (1985) Anatomical and pathoanatomical studies on the spinal efferent systems innervating pèlvic structures. J Anton Nerv Syst 14:23–48
22. Sobue K, Matsouka Y, Mukai E, Takayanagi T, Sobue I, Hashizume Y (1981) Spinal and cranial motor nerve roots in amyotrophic lateral sclerosis and X-linked recessive bulbospinal muscular atrophy. Morphometric and teased-fiber study. Acta Neuropathol 55:227–235
23. Wolfgram F, Myers L (1972) Toxicity of serum from patients with ALS for anterior horn cells in vitro. Transam Neurol Assoc 97:19–23
24. Wunner WH (1982) Is the acetylcholine receptor a rabies virus receptor? Trends in Neurosciences 5:413–415
25. Zieglgänsberger W, Reiter C (1974) Interneuronal movement of Procion Yellow in cat spinal neurons. Exp Brain Res 20:527–530

Muscle Thixotropy and Its Effect on Spindle and Reflex Responses to Stretch

K.-E. HAGBARTH [1], J. V. HÄGGLUND [1], M. NORDIN [1], and E. U. WALLIN [1]

A thixotropic substance is characterized by the fact that its stiffness or viscosity is dependent on the past history of movement. The word is commonly used for gels, which become fluid when shaken or stirred and which then gradually regain their original high viscosity after a period of rest. Stirring forces in such solutions tend to disrupt bonds between molecules, which then reform when the stirring forces cease.

As early as 1951, Buchthal and Kaiser [4] demonstrated that isolated resting frog muscle fibres have thixotropic properties. A single ramp stretch of such a muscle fibre is sufficient to limber it up, to reduce its short-range stiffness so that it becomes more compliant to succeeding stretch stimuli. After a stretch it has to rest for a number of minutes to regain its original stiffness. The short-range stiffness of the resting muscle is believed to depend on bonds between actin and myosin filaments in extrafusal fibres, bonds which are broken by stretch movements beyond a certain extent and which then need some time to form again once the motion ceases [7, 8, 14].

During the last decades, it has become increasingly clear that the intrafusal muscle fibres of the cat also possess thixotropic properties which can give rise to long-term variations in the inherent stiffness of the fibres, in turn influencing the stretch sensitivity of the primary spindle endings [1–3, 5, 15–17]. Morgan et al. [13] recently arrived at the conclusion that in resting intrafusal fibres there are filament bonds which in general terms behave like bonds in extrafusal fibres. They advance the thesis that both stretch and contraction can break such bonds, and if the fibre is held in a stretched position for a while new bonds will be formed, now adapted to the new length. When it returns to the rest length, the fibre will be slack and it may take up to half an hour before this slack is taken up spontaneously by the formation of new bonds adapted to the short length.

Lakie et al. [11] recently reported that human musculo-tendinous structures possess thixotropic properties reminiscent of those previously described for isolated muscle fibres of the frog. Prompted by their findings, we have searched in human subjects for signs of thixotropy in the relaxed finger flexor muscles. In particular, we wanted to see whether it is possible to demonstrate signs of thixotropy in man, not only in extrafusal but also in intrafusal fibres.

The experimental set-up is illustrated in Fig. 1. The arm and hand were resting on a support with the extended fingers under a cover, free to move around the metacarpo-phalangeal (MCP) joints. The stiffness of the relaxed finger flexor muscles was explored by measuring the amplitude of the angular finger excur-

[1] Department of Clinical Neurophysiology, Uppsala University, Akademiska Sjukhuset, S-751 85 Uppsala, Sweden.

Clinical Aspects of Sensory Motor Integration
Ed. by A. Struppler and A. Weindl
© Springer-Verlag Berlin Heidelberg 1987

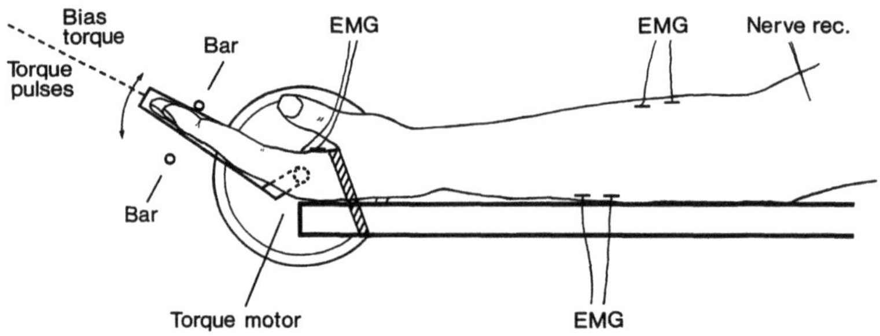

Fig. 1. Schematic drawing of the experimental set-up. (From Hagbarth et al. [6])

sions induced by small extension torque pulses of a given strength, delivered by a torque motor with its axis corresponding to the axis of rotation at the MCP joints.

A series of initial experiments were made on subjects with the hand and arm muscles totally paralysed by nerve blocks, and EMG recordings were made to check that the muscles really were totally inactive. With the muscles paralysed and without any bias torque from the motor, the fingers usually came to rest in a semi-flexed position as shown in Fig. 1. However, after a single passive stretch of the fingers they did not return spontaneously, but stayed in a more extended position; and after a single imposed flexion they stayed in a more flexed position. The resting position of the fingers also changed when a series of weak extension torque pulses were delivered by the motor. The recurrent pulses pushed the fingers in a step-wise fashion towards a more extended position where they tended to stay when the train of torque pulses ceased.

In order to study the dynamic stiffness of the system these drifts in the resting position had to be avoided. A stable resting position was obtained by placing a stop-bar on the volar side of the fingers and by letting the torque motor deliver a bias flexion torque of sufficient strength to bring the fingers back to this bar after each extension test pulse or conditioning manoeuvre. When this technique was used on subjects with paralysed arm and hand muscles it was found that single large finger extension movements were followed by prolonged reduction of short-range stiffness, as tested by the recurrent pulses (Fig. 2, *left*). The loosening of the system was maximal immediately following the conditioning stretch movements. As the test pulses were then repeated at regular intervals, it could be seen how the system gradually regained some of its stiffness without fully returning to the control level within observation periods of 10–15 min. After-effects opposite to those following large extension movements were seen following large passive finger flexions. After such movements the system became stiffer, as shown by the marked reduction of the finger deviations induced by the torque pulses (Fig. 2, *right*).

Similar results with "limbering-up" effects after stretching and "stiffening-up" effects after finger flexor shortening were also obtained in subjects who without nerve blocks succeeded in remaining relaxed in their arm and hand muscles during the tests. These subjects were also asked to perform voluntary isometric

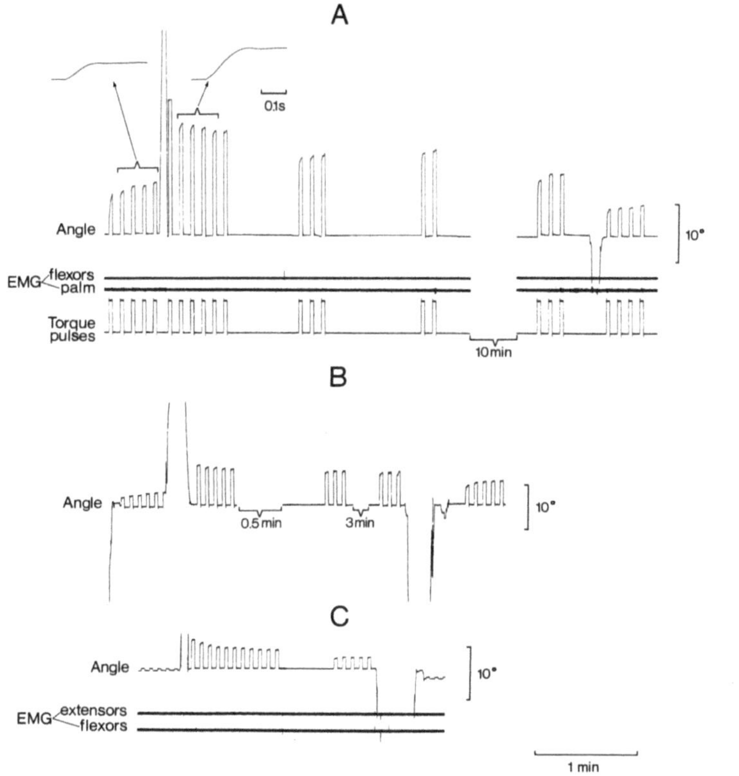

Fig. 2 A–C. Stiffness changes induced by passive large finger extensions (*left*) and finger flexions (*right*). **A** during total median nerve block with local anaesthetics, **B** in a patient with severe brachial plexus lesion, and **C** during total finger flexor and extensor paralysis caused by a cuff inflated around the upper arm. In each experiment the torque pulses were of constant strength, and the change in stiffness is evidenced by the angular deflections from the resting position. The inserts with expanded time scale in **A** were obtained by averaging the angular movements indicated by the square brackets. Besides increased amplitude there was also an enhanced velocity of the extension movements after the conditioning extension manoeuvre. (Hagbarth et al. [6])

finger flexor contractions of a moderate strength, lasting for a few seconds. Such contractions were followed by a sustained decrease in stiffness similar to that occurring after passive stretch. However, when the finger flexor contractions were not isometric but the subject instead was allowed to make an active finger flexor movement, there was a post-contraction increase in stiffness similar to that following passive finger flexions.

Several subjects could not prevent stretch reflex contractions from appearing in response to the extension test pulses. These reflex responses were clearly seen in the EMG recordings from the finger flexors, and in the goniometer traces one could also see the reflex jerks as more or less distinct downgoing deflections. Figure 3 shows a representative example of how the stretch reflex responses were altered by the various conditioning manoeuvres. Those manoeuvres, which reduced the passive stiffness of the system, i.e. stretch and isometric finger flexor

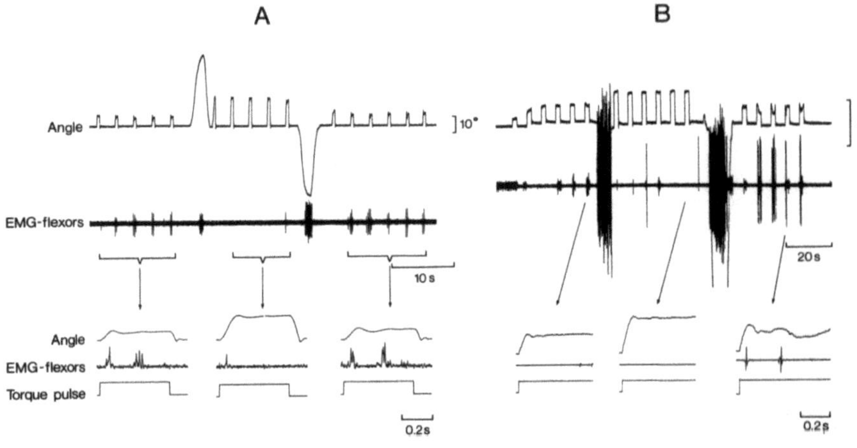

Fig. 3 A, B. Reflex changes accompanying variations in stiffness due to different conditioning manoeuvres. **A** shows how a passive extension movement caused disappearance of reflex EMG responses, which then reappeared after a passive flexion. The inserts with expanded time scale were obtained by averaging the marked 5 trials. **B** illustrates reduced reflex responses after an isometric voluntary contraction despite increased angular deflections (*left*). By contrast, both the inherent stiffness and the reflex responses were increased following a resisted shortening voluntary contraction (*right*). The inserts in this case depict the response to individual torque pulses. The insert EMG is integrated in **A**. See text for further comments. (Hagbarth et al. [6])

contractions, also led to a reduction of the stretch reflex responses. Those manoeuvres which enhanced the passive stiffness, i.e. passive and active finger flexions, caused a potentiation of the stretch reflexes. In other words, the passive inherent and the active reflex stiffness were affected in similar ways and seemed to vary in parallel after the different conditioning manoeuvres.

These observations led to the question whether the stretch reflex changes were a manifestation of thixotropic events in intrafusal muscle fibres, altering the stretch sensitivity of the primary spindle endings. To check this possibility the multiunit afferent stretch responses to the test pulses were recorded with tungsten microelectrodes inserted into finger flexor muscle nerve fascicles. Tests were made to assure that the compound afferent volleys originated from intramuscular stretch receptors with response characteristics (to electrically induced muscle twitches) typical for primary muscle spindle afferents.

As illustrated in Fig. 4, the afferent stretch responses to the extension test pulses were reduced following conditioning large amplitude stretch movements, indicating that a reduction in muscle spindle stretch sensitivity had occurred; this could explain why the stretch reflexes were less pronounced. Similar signs of reduced muscle spindle stretch sensitivity were also found after isometric voluntary finger flexor contractions, whereas the afferent stretch responses were enhanced following passive and active finger flexions.

Signs of reduced muscle spindle stretch sensitivity following a conditioning stretch movement were also found in experiments where the muscle nerve was blocked by lidocaine proximal to the recording site. This indicates that the effects were not due to fusimotor-induced changes in spindle stretch sensitivity but to al-

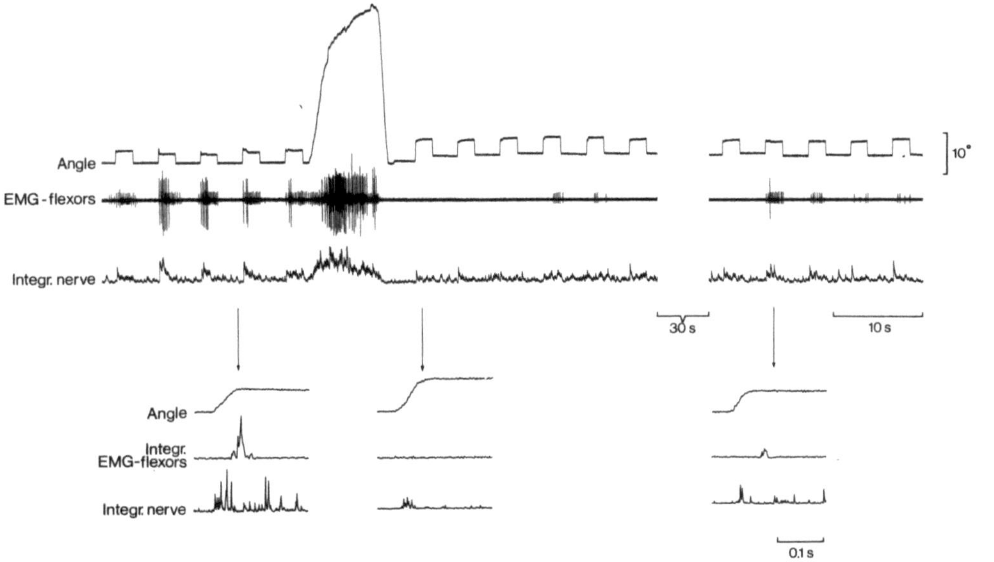

Fig. 4. Reduction of stiffness, afferent stretch responses and reflex EMG responses following a conditioning passive extension movement. During the observation period a slow recovery occurred but the initial levels were not reached. EMG was recorded from the forearm finger flexors and multiunit neural activity from a muscle nerve fascicle supplying the same muscles. The inserts with expanded time scale show the responses to selected individual torque pulses. After the large passive extension an enhancement could also be observed of the velocity of the extension movements. As in previous figures, the torque pulses were unaltered during the trial. (Hagbarth et al. [6])

terations in the inherent stiffness of the intrafusal fibres, changes of the type to be expected if these fibres, like the extrafusal ones, possessed thixotropic properties.

The results have some practical, clinical implications. Among physiotherapists it is well-known that one can use different types of stretching manoeuvres to limber up stiff muscles and to reduce spasticity. Many physiotherapists are also convinced of the beneficial effects of the so-called proprioceptive neuromuscular facilitation (PNF) technique [9, 10], in which voluntary isometric contractions are combined with stretching manoeuvres (however, cf. [12]). To judge by the present results, it seems that such manoeuvres can reduce not only the passive, inherent stiffness of the muscles but also the active reflex stiffness. In healthy subjects the limbering-up effects of stretch and contraction are probably mainly due to the inherent thixotropy of the extrafusal fibres, possibly combined with thixotropy also in tendinous structures. In spastic patients, on the other hand, with their enhanced reflex stiffness it may be the thixotropic properties of the intrafusal fibres which are mainly responsible for the beneficial effects.

Summary

The inherent, short-range stiffness of relaxed human finger flexor muscles was examined by measuring the speed and amplitude of angular finger extensions induced by recurrent weak extension torque pulses. The stiffness (tested with the fingers in a resting, semi-flexed position) was enhanced following transient, large-amplitude finger flexions (active or passive) and reduced following transient, large-amplitude finger extensions or isometric voluntary finger flexor contractions. The after-effects gradually declined during observation periods of several minutes.

As judged by simultaneous recordings of neural afferent and reflex EMG responses to the torque pulses, muscle spindle and stretch reflex sensitivity varied in parallel with the changes in inherent muscle stiffness.

The in-parallel changes in inherent and reflex stiffness can be explained in terms of thixotropic behaviour of extra- and intrafusal muscle fibres. Loosening of thixotropic bonds in extra- and intrafusal muscle fibres may contribute to the beneficial effects of limbering-up manoeuvres commonly employed by physiotherapists.

References

1. Baumann TK, Emonet-Dénand F, Hulliger M (1983) Temporal characteristics of the sensitivity-enhancing after-effects of fusimotor activity on spindle IA afferents. Brain Res 258:139–143
2. Brown MS, Goodwin GM, Matthews PBC (1969) After-effects of fusimotor stimulation on the response of muscle spindle primary afferent endings. J Physiol (Lond) 205:677–694
3. Brown MC, Goodwin GM, Matthews PBC (1970) The persistence of stable bonds between actin and myosin filaments of intrafusal muscle fibres following their activation. J Physiol (Lond) 210:9–10
4. Buchthal F, Kaiser E (1951) The rheology of the cross-striated muscle fibre with special reference to isotonic conditions. Danske Biologiske Meddeleser 21:7, 1–318
5. Emonet-Dénand F, Hunt CC, Laporte Y (1983) Persistent effects of fusimotor activity and muscle stretch on responses of primary endings to dynamic gamma stimulation in cat soleus spindles. J Physiol (Lond) 345:101P
6. Hagbarth K-E, Hägglund JV, Nordin M, Wallin EU (1986) Thixotropic behaviour of human finger flexor muscles with accompanying changes in spindle and reflex responses to stretch. J Physiol (Lond) 368:323–342
7. Hill DK (1968) Tension due to the interaction between the sliding filaments in resting striated muscle. The effect of stimulation. J Physiol (Lond) 199:637–684
8. Joyce GC, Rack MH, Westbury DR (1969) The mechanical properties of cat soleus muscle during controlled lengthening and shortening movements. J Physiol (Lond) 204:461–474
9. Kabat H (1950) Studies on neuromuscular dysfunction XIII: new concepts and techniques of neuromuscular reeducation for paralysis. Permanente Foundation Med Bull 8:121–143
10. Knott M, Voss DE (1968) Proprioceptive neuromuscular facilitation: techniques. Harper & Row, New York, pp 91–100
11. Lakie M, Walsh EG, Wright GW (1984) Resonance at the wrist demonstrated by the use of a torque motor, an instrumental analysis of muscle tone in man. J Physiol (Lond) 353:265–285
12. Moore MA, Hutton RS (1980) Electromyographic investigation of muscle stretching techniques. Med Sci Sports Exerc 12:322–329

13. Morgan DL, Prochazka A, Proske U (1984) The after-effects of stretch and fusimotor stimulation on the responses of primary endings of cat muscle spindles. J Physiol (Lond) 356:465–477
14. Nichols TR, Houk JC (1976) Improvement in linearity and regulation of stiffness that results from action of stretch reflex. J Neurophysiol 39:119–142
15. Poppele RE, Quick DC (1981) Stretch-induced contraction of intrafusal muscle in cat muscle spindle. J Neurosci 1:1069–1074
16. Proske U (1975) Stretch-evoked potentiation of responses of muscle spindles in the cat. Brain Res 88:378–383
17. Smith JL, Hutton RS, Eldred E (1974) Postcontraction changes in sensitivity of muscle afferents to static and dynamic stretch. Brain Res 78:193–202

Cytochemical Reevaluation of Location and Translocation of Acetylcholinesterase in the Motor End-Plate

G.W. KREUTZBERG [1] and L. TÓTH [2]

Introduction

The localization of acetylcholinesterase (AChE, EC 3.1.1.7.) in the motor end-plate region has been investigated for almost three decades with light and electron microscopy and with several biochemical techniques. The extensive literature has been critically reviewed by Silver [44] and by Friedenberg and Seligman [17] with special emphasis on the cytochemical methods used. Although data in the literature vary with the different methods used, there is general agreement on the localization of the enzyme in the synaptic cleft of the neuromuscular junction. However, activity related to other structures, e.g., the subneural endoplasmic reticulum, the L-tubules of the sarcoplasmic reticulum, the sarcolemmal T-system, the teloglial Schwann cells, the axolemma and the axonal terminal, have been mentioned occasionally but questioned or even denied by other investigators [2, 3, 7, 12, 27, 49, 50, 51].

With the emergence of the concept of AChE as a secretory and mobile protein released in the peripheral and central nervous system [6, 8, 9, 18, 29, 30, 34, 40, 41], the various locations of the enzyme can be better understood. In this study, reevaluation of AChE localization in the diaphragm by means of high-resolution cytochemistry has now revealed the presence of this enzyme in locations not described earlier and which suggest an extracellular transport of AChE.

Material and Methods

Investigations were carried out on adult, male Wistar rats weighing 250 g. The animals were anesthetized by Nembutal and then perfused with a mixture of 2% glutaraldehyde, 4% paraformaldehyde and 4% saccharose buffered with sodium cacodylate, pH 7.4. The solution contained 2×10^{-3} M calcium chloride. After perfusion diaphragms were removed and fixed in the same aldehyde solution for another six hours, then rinsed overnight in cacodylate buffer solution (pH 7.4). The motor end-plate area was cut out and 100 μm sections were prepared on a vibratome. For demonstration of AChE the sections were incubated for 3–5 min and processed further according to Lewis and Shute [32, 33]. In modifying this method we have replaced sodium acetate in the incubation medium with 0.05 M sodium succinate, and adjusted the pH to 5.3. The substrate is acetylthiocholine

[1] Department of Neuromorphology, Max-Planck-Institute for Psychiatry, Munich, FRG.
[2] Department of Anatomy, University Medical School, Szeged, Hungary.

Clinical Aspects of Sensory Motor Integration
Ed. by A. Struppler and A. Weindl
© Springer-Verlag Berlin Heidelberg 1987

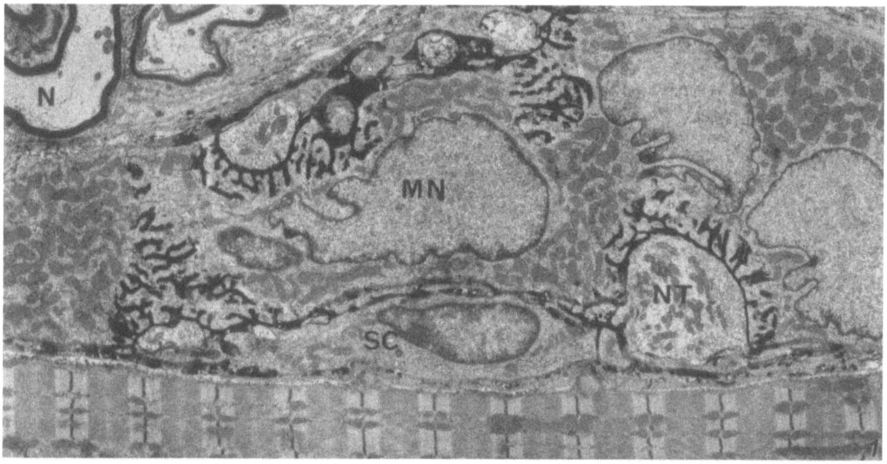

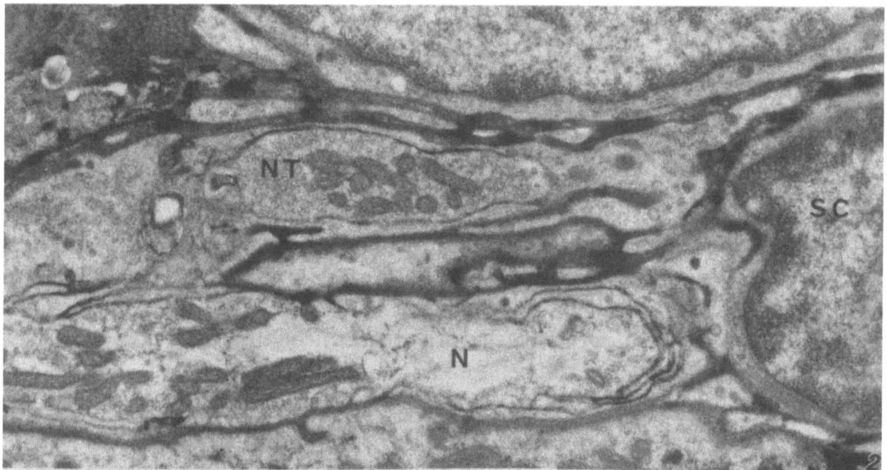

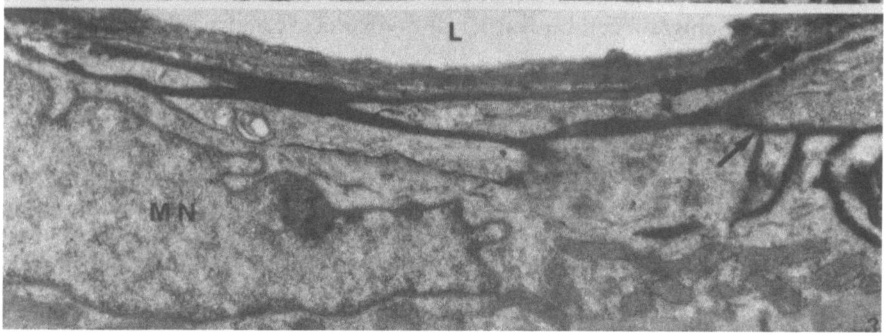

Fig. 1. Low-power overview of AChE distribution in the motor end-plate region of a red muscle fiber, rat diaphragm. *N*, nerve; *MN*, muscle cell nucleus; *SC*, Schwann cell; *NT*, nerve terminal. ×5700

Fig. 2. Electron-dense reaction product indicates enzyme activity in the synaptic clefts extending along the basal lamina of adjacent Schwann cells (*SC*). *NT*, nerve terminal; *N*, nerve. ×18000

Fig. 3. There is a continuity of AChE activity extending from the neuromuscular junction (*arrow*) to the basal lamina of Schwann cells and of a capillary (*top*). Red fiber; *L*, lumen of capillary; *MN*, muscle cell nucleus. ×18000

iodide. After incubation the sections were washed and poststained with osmium tetroxide before being dehydrated and embedded in araldite (Durcupan®). During dehydration the blocks were briefly (10–20 min) stained with uranyl acetate. Thin sections were poststained with lead *citrate according* to Reynolds [39]. Sections were viewed on an EM 9 or EM 10 Zeiss electron microscope. For specificity of the cytochemical reaction the incubation medium always contained 2×10^{-4} m ethopropazine in order to inhibit unspecific cholinesterases (EC 3.1.1.8). Control incubations lacking the substrate acetylcholine iodide or containing the AChE inhibitor BW 284 C 51 were carried out and revealed no reaction products in cells and extracellular spaces.

Observations

A section through the motor end-plate region of the diaphragm shows the general distribution of the electron-dense reaction product which indicates the activity of AChE (Fig. 1). It is seen in the synaptic cleft of neuromuscular junctions, in the junctional folds and in a patchy fashion in the basement membrane of surrounding Schwann cells or so-called teloglial cells (Fig. 3). The latter is a non-junctional activity and deserves special interest. Figures 1–3 demonstrate continuity of the basal laminae of the neuromuscular junctions and of the teloglial Schwann cells. Continuity exists also between these basement membranes and the intercellular spaces of the Schwann cell sleeves, which form several cytoplasmic layers on the axonal preterminals, analogous to the organization of the node of Ranvier. As demonstrated in Fig. 4, Schwann cell and axonal terminal may form desmosome-like connections. In the extracellular space of this junction AChE activity is especially strong. No enzyme activity has ever been found inside the satellite cells. This negative finding supports the idea that extracellular AChE in the motor end-plate region is not of Schwann cell but rather of myogenic or neuronal origin.

Fig. 4. In the center a desmosome-like junction (*arrow*) between axon termonal (*NT*) and Schwann cell (*SC*): high AChE activity in the extracellular space of the junction and of the Schwann cell sleeves. × 61 500

Fig. 5. AChE activity in the synaptic cleft (*arrow*) is strong but not uniform. The inner lamina densa shows less activity than the two outer laminae rarae. In the subneural region a weak activity is seen in the perinuclear cisterna of the myocyte nucleus (*MN*), in the Golgi complex (*G*) and in the elongated thin cisternae of the rough sarcoplasmic reticulum (*SR*). The smooth sarcoplasmic vesicles are larger, comparatively coarse and contain higher activity. × 51 000

Fig. 6. In the primary synaptic cleft one single and five aligned, unstained circular discs of ca. 40 nm in diameter can be seen (*arrow*) (see also Fig. 5). Interposed between this region and the axon terminal a double Schwann cell finger with AChE activity in the intercellular clefts is demonstrated. × 51 000

Fig. 7. Sarcoplasmic vesicles (*SV*) are strongly AChE-positive and are seen associated with the secondary folds but never with the primary synaptic gutter. In the *upper left* a synaptic vesicle with AChE activity is seen with a "connection" to the presynaptic membrane area. Red fiber. × 54 000

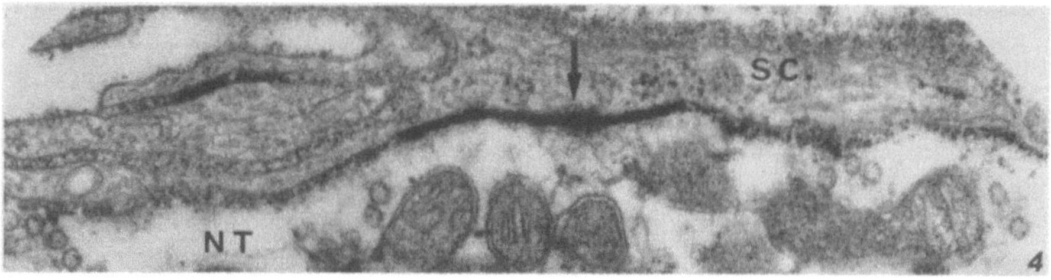
SC
NT
4

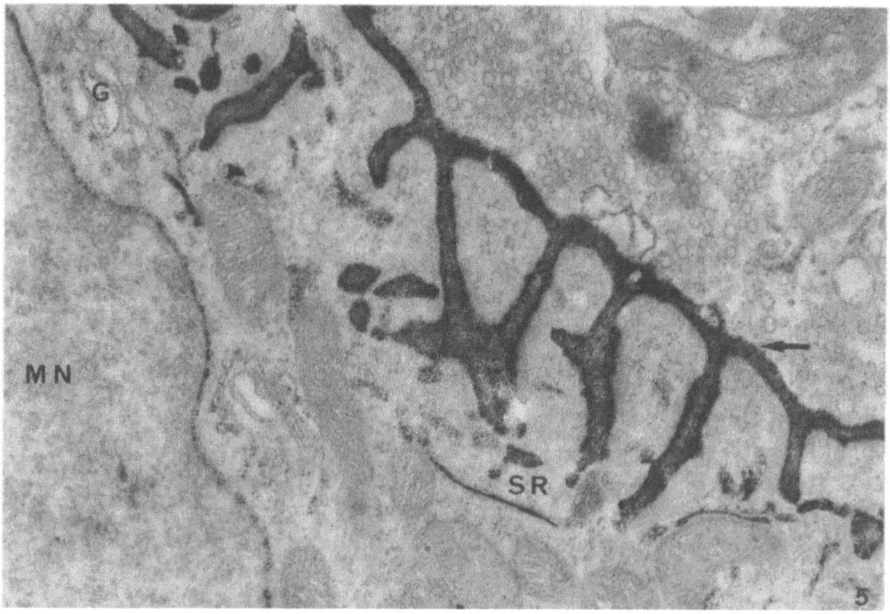
G
MN
SR
5

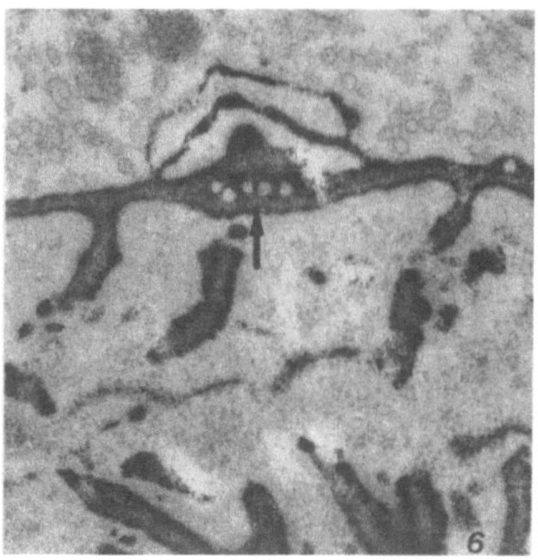
6

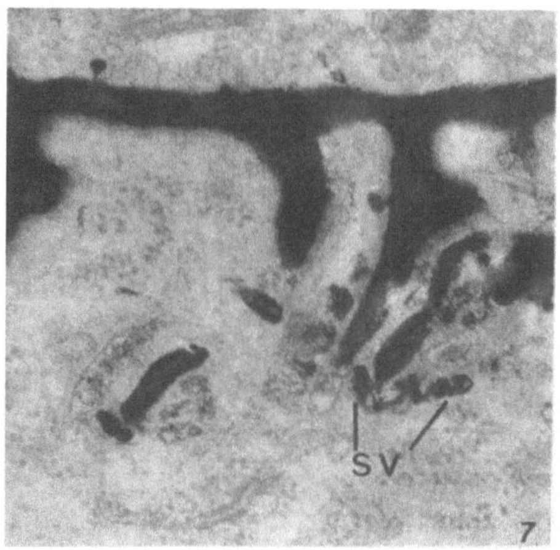
SV
7

AChE activity is also frequently observed in the basal laminae of capillaries adjacent to the motor end-plate. Continuity between the junctional and the vascular basal lamina is clearly seen in Fig. 3. No enzymatic activity has been encountered inside vascular elements; suggesting that they cannot produce the enzyme and thus are not the source of the activity visualized by the reaction product in their basal lamina.

With the short incubation period of 3–5 min, activity is rarely seen in nerve fibers (Fig. 1). Occasionally some AChE-positive segments of the axolemma can be seen in the myelinated motor fibers and at the axolemma of the preterminal axons (Fig. 2). However, no activity is seen within axonal terminals with the exception of a few synaptic vesicles at the presynaptic membrane which seem to be contaminated by enzyme from the synaptic cleft (Fig. 7).

AChE in the Subneural Apparatus

The junctional folds penetrate deeply into the subneural sarcoplasm. This area of the muscle fiber consists of lobulated nuclei and a ground cytoplasm enriched with mitochondria, free polysomes, and endoplasmic reticulum usually termed sarcoplasmic reticulum (SR). AChE activity is seen in the subneural region in various locations:

1. Occasionally in the perinuclear cisternae of the lobulated nuclei of the subneural region (Fig. 5).
2. Very weak activity is occasionally seen in the Golgi apparatus of the muscle cells (Fig. 5).
3. Some cisternae of the granular sarcoplasmic reticulum show deposits of enzyme reaction product (Fig. 5).
4. Weak to very strong activity is found in comparatively wide rounded or elongated cisternae of the smooth sarcoplasmic reticulum (Figs. 5–10). In close proximity to the junctional folds these organelles seem to become spherical in shape and are therefore called postsynaptic or sarcoplasmic vesicles [38]. Various sorts of relations between the sarcoplasmic vesicles and the synaptic folds can be recognized (Figs. 5–10); shorter and longer stalk connections, no visible conection, and cisternae-to-fold fusion. Frequently a rather short stalk connection between a cisterna and a junctional fold is seen (Figs. 8–10). Sarcoplasmic vesicles are not seen in connection with the primary synaptic gutter. They prefer association with the secondary folds of the neuromuscular junction in all three types of muscle fibers, red (Figs. 7 and 8), white (Fig. 9), and intermediate (Fig. 10), the latter demonstrating the most extensive appearance of AChE-positive postsynaptic vesicles. In all types of subneural apparatus we have seen rounded or irregular-shaped sarcoplasmic vesicles with AChE reaction product deposited mainly on the inside of the vesicular membrane, whereas the core is almost unstained. These "short vesicles" are occasionally linked to the secondary folds by stalk-like connections, suggesting that they have possibly extruded the AChE-positive contents into the synaptic cleft (Fig. 8). In the primary and secondary cleft, AChE activity is seen associated with the pre- and postsynaptic membrane. Although we have mostly applied

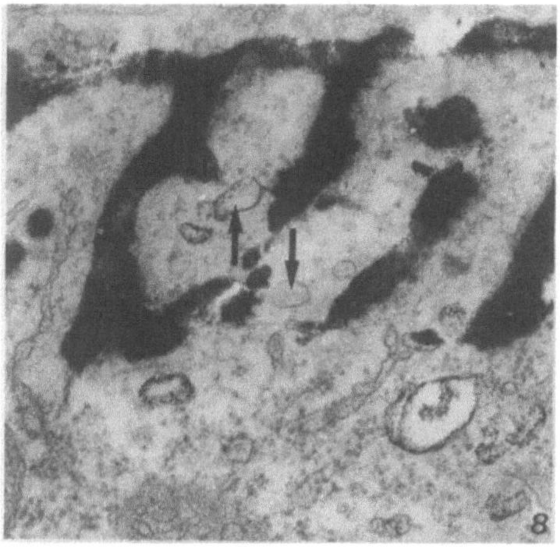

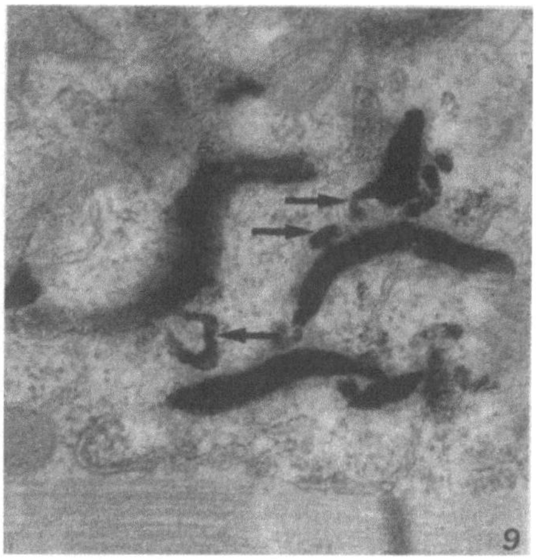

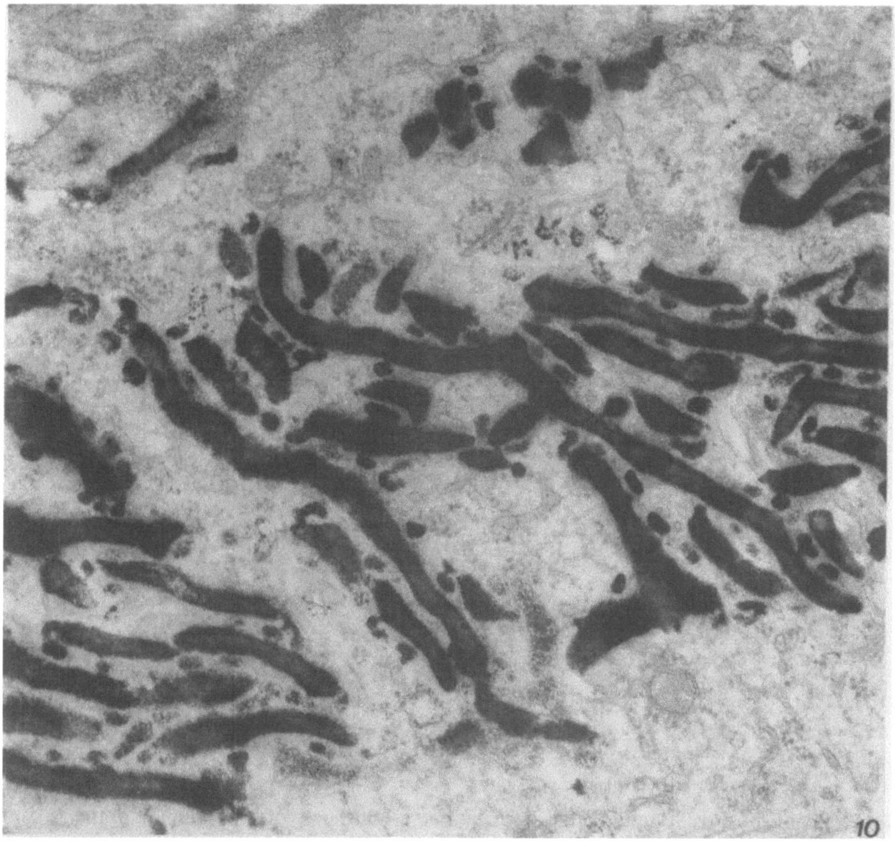

Fig. 8. In the subneural apparatus of this red fiber several "empty" sarcoplasmic vesicles (*arrows*) can be seen. One of them (*center*) still has a stalk-like connection to the secondary fold. ×54000

Fig. 9. In this white fiber subneural apparatus several AChE-positive sarcoplasmic vesicles can be seen connected to the secondary folds (*arrows*). ×54000

Fig. 10. Neuromuscular junctions of the intermediate fiber type show the highest frequency of AChE-positive postsynaptic or sarcoplasmic vesicles with a variety of dockings to the deep secondary folds. ×51000

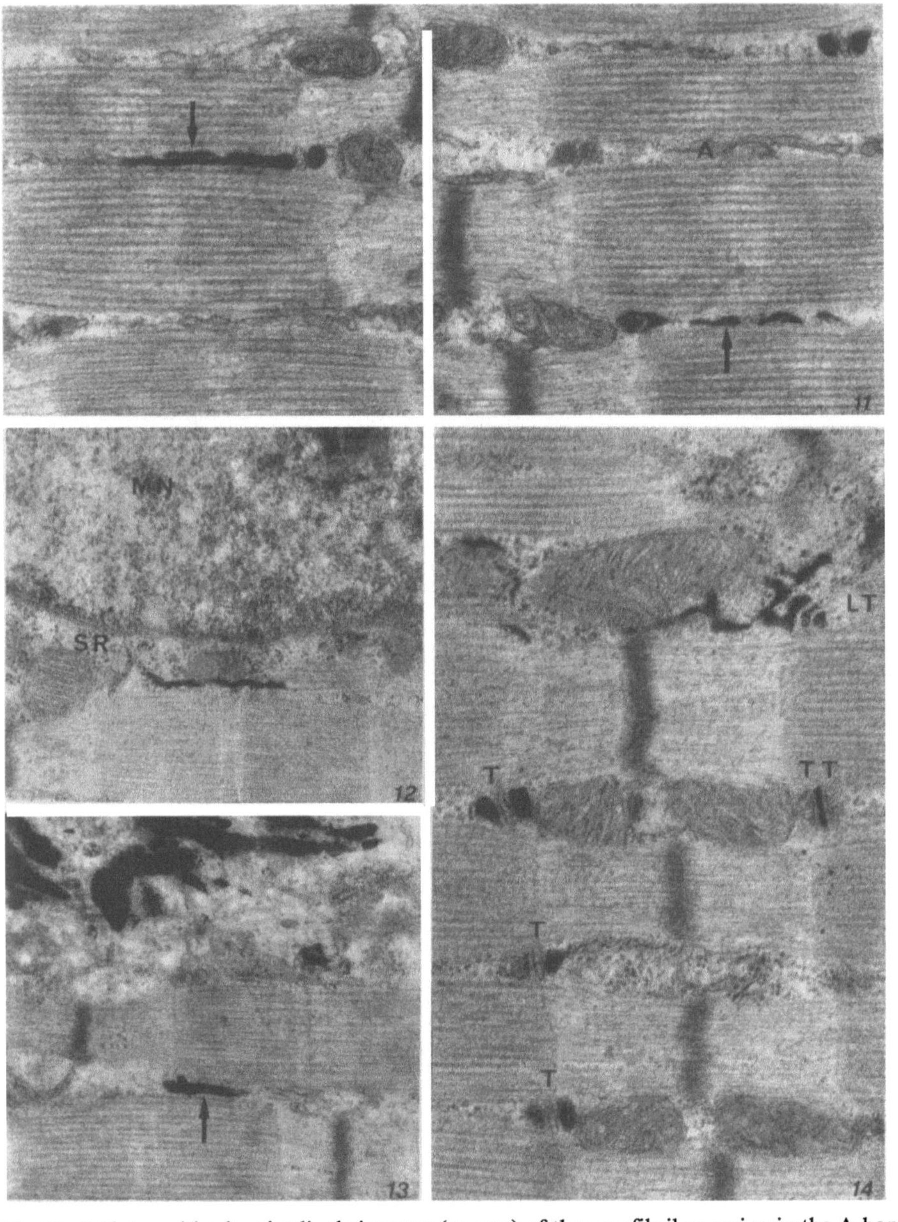

Fig. 11. AChE-positive longitudinal cisternae (*arrows*) of the myofibrils running in the A-bands (*A*). ×51 000

Fig. 12. AChE activity in a partially granular sarcoplasmic reticulum (*SR*) cisterna on the side of the perikaryon directed toward the myofibrils. Red fiber; *MN*, muscle cell nucleus. ×36 000

Fig. 13. Inter-fibrillary longitudinal cisternae (*arrow*) in the vicinity of the subneural apparatus (*top*) show AChE activity in this white fiber type. ×45 000

Fig. 14. Branching longitudinal cisternae (*LT*) occasionally contain AChE activity (*top right*) which can continue into the lateral sacs of the triades (*T; left*) localized in the junctions between the A and I band. *TT*, T-tubule. ×54 000

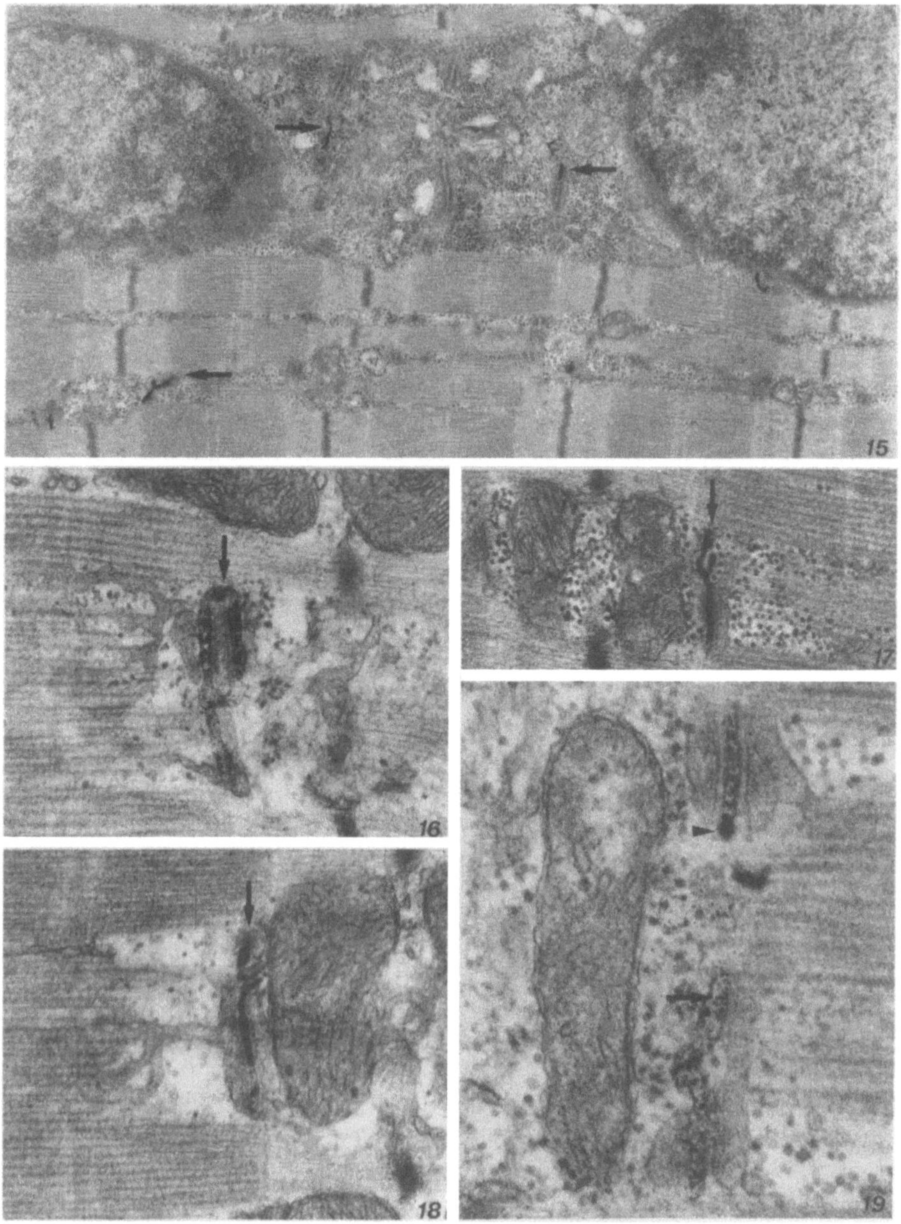

Fig. 15. Low-power demonstration of several AChE positive T-tubules (*arrows*) in the perikaryon and the myofibrils. × 5700

Figs. 16–18. AChE activity in the double or bifurcating T-tubules of triads and pentads at the A/I band junction (*arrow*). 16 & 18 = × 54000, 17 = × 51000

Fig. 19. In the upper part a T-tubule in a triad (*arrowhead*), below a larger transversally running intertriad tubule with scattered AChE activity (*arrow*). × 102000

very short incubation times (3–5 min) the tissue has to be cut especially thin to reveal this distribution (Figs. 5 and 6). Schwann cell fingers, which occasionally run under the axonal terminals and touch the basal lamina of the synaptic cleft, also show AChE activity in their extracellular spaces (Figs. 5 and 6). Unstained circular discs of the size of synaptic vesicles (ca. 40 nm diameter) can be detected single, in doublets, and in rows in the synaptic clefts (Figs. 5 and 6).

AChE in the Interfibrillar Sarcoplasmic Reticulum

AChE activity is seen in various channels of the sarcoplasmic reticulum ensheathing the myofibrils, e.g., in a number of longitudinal cisternae as they run parallel to the A-band (Fig. 11). Sarcoplasmic reticulum cisternae positive in AChE are also detected on the myofibril side of the perikaryon of a red muscle (Fig. 12). In white fibers, where the motor end-plate region almost touches the superficial myofibrils, strong AChE activity is seen in interfibrillary longitudinal cisternae of this region (Fig. 13). Activity often continues into the terminal cisterna forming the lateral sacs of the triads at the A-I junctional zone (Figs. 11 and 14). The fenestrated collar system seems not to contain AChE although in some cases we have seen branching L-tubules with activity (Fig. 14).

AChE in the T-System

AChE reaction product can be seen in a number of transversely and obliquely running tubules of the T-system (Figs. 15–19) as well as in the medial tubule of some triads (Figs. 15, 17, and 19). In the diaphragm two triads per sarcomere are encountered. They are located at the junctions between the A and I band (Fig. 14). A peculiarity of the diaphragm is the formation of pentads located in the perikaryon (Fig. 15) and at the A-I junctions (Fig. 16). In several instances the T-tubules of pentads and of bifurcating T-tubules demonstrate AChE activity (Figs. 15–18). Enzyme-positive structures of the T-system are more frequent than those of the interfibrillar L-system. AChE activity is not seen in all T-tubules. Enzyme-positive tubules of L- and T-systems are located mainly in several myofibrils in the vicinity of the subneural apparatus. In triads the T-tubule or the lateral sacs can contain the enzyme (Fig. 14). We have not seen a triad where both systems were simultaneously enzyme-positive.

Discussion

The thiocholine-iodide method of Lewis and Shute [32] combined with good perfusion fixation technique allows high resolution cytochemistry of AChE activity. The specificity of the reaction is checked by the use of appropriate inhibitors and by substrate lacking controls. Thus, in accordance with Lewis and Shute [32, 33], who checked various incubation conditions and never found reaction product in locations suspicious for diffusion, the appearance of electron-dense reaction product in our preparations is assumed to demonstrate the site of AChE activity (cf. Miledi [37]).

The enzyme is demonstrated in a number of different compartments intracellularly and extracellularly including subneural endoplasmic and sarcoplasmic reticulum L-system, synaptic membranes, sarcolemmal system (T-tubules), the basal laminae of Schwann cells and capillaries and in intracellular spaces of teloglial cells containing axons or other Schwann cells.

Locally there are two possible sources for the enzyme: the axon and the muscle cell. The neuronal enzyme seems to reach only rather low concentrations at the terminals. It cannot be demonstrated with the short incubation periods used in our experiments. It is generally accepted that the main source of AChE at the neuromuscular junction is the muscle. Its production seems to be under the control of the nerve (cf. Guth [20], Lentz [31]). However, even after denervation the muscle continues to produce AChE, though on a lower level. Upon observation of a neuromuscular junction under normal conditions as well as after deafferentation [1, 7, 48] in inhibition experiments with phosphorus compounds [19, 53], during ontogenesis [26, 45, 46, 52], or in vitro [42], it appears that the bulk of the enzyme is produced locally in the endoplasmic reticulum of the subneural region.

The origin of the junctional AChE has been under debate for many years [44]. Since motor nerves regularly contain AChE in their axolemma and axoplasmic reticulum [12, 28] it cannot be ruled out that the neuronal enzyme ends in the membrane of the junction. The situation becomes even more disturbing in light of the fact that junctional AChE decreases to 20% after deafferentation [10, 11], which could be interpreted to mean that 80% of the junctional AChE is provided by the nerve. However, the persistent though diminished enzyme activity in a denervated muscle proves local AChE synthesis in the muscle. This postulate has further been substantiated by molecular analysis of the AChE types present in the muscle. Hall [22] has shown that three different forms of AChE, 4S, 10S and 16S, exist in the muscle but only the 4S and 10S variety is recognized in the nerve fibers [14, 15]. We think that the cytochemical localization of AChE activity in the subneural apparatus and in sarcoplasmic organelles strongly supports the assumption of a myogenic origin of most of the junctional AChE. Enzyme activity can occur in the perinuclear cisternae, in the channels of the interfusal sarcoplasmic reticulum including the L-tubules, the fenestrated collar and the lateral sacs of the triads. These channels are extensions of the sarcoplasmic reticulum in which protein synthesis occurs (cf. [16]). An enzyme which is produced in the SR should also be encountered in its extensions. In fact, these parts of the SR show scattered AChE activity in a patch-like fashion.

The strongest activity of AChE is seen in vesicular and cisternal structures accompanying the secondary folds of the subneural appartus. We consider these organelles as sites of synthesis of AChE. These enzyme-positive structures are located mainly in close vicinity to the end-plate which might reflect the induction influence of the nerve terminals on the myogenic enzyme production. Most of the cisternae are agranular but some show ribosomes bound to the membranes. The cisternae approach the secondary folds and seem to fuse with the subneural sarcolemma at an attachment point formed by a short stalk seen in many of the vesicle-like cisternae. Obviously the content of these vesicles is released into the synaptic cleft and the enzyme has to find its anchorage sites in the basal lamina of the synaptic cleft. The distribution of the reaction product in the junctional basal lamina shows that an high enzyme activity is exposed toward the postsynaptic and the presynaptic membrane [36].

The molecular configuration of AChE has been studied and it has been established that the enzyme exists in a globular and in an asymmetric form [35, 54]. The hydrophobic, collagen-like stalk is insinuated into the basal lamina and a polymer of 4 subunits is exposed to the synaptic cleft [13, 43]. At the neuromuscular junction a release of AChE has been obverved in vitro [23, 25, 47] and after treatment with proteases [4, 5, 21, 45]. The cytochemical data presented here show that the enzyme can indeed be recognized in a number of extracellular sites to which it must have been secreted. We assume that the enzyme or active subunits, as have been characterized biochemically, first appear in the synaptic cleft and then travel along the basal laminae thus reaching sites on the Schwann cells and the vasculature. Enzyme moving from the synaptic cleft along the subneural sarcolemma can easily reach the external opening of the transverse tubular system. This explains enzyme activity present in the T-tubules [50]. Penetration of enzymes into the T-system has also been demonstrated by extracellular injection of horseradish peroxidase [27] of the macromolecular tracer ferritin [27], and of lanthanum [55].

In conclusion, high resolution cytochemistry of AChE at the neuromuscular junction and its vicinity stresses local synthesis and offers new insights into the mobility of the enzyme. It is likely that the myogenic enzyme is synthesized mainly in sarcoplasmic reticulum of the subjunctional sarcoplasm and probably in the tubular system of the smooth sarcoplasmic reticulum. Translocation occurs at the secondary folds where the enzyme is released to the synaptic cleft. Within this cleft the enzyme is supposed to stick with its collagen-like tail in the lamina densa of the basal lamina, orienting the enzymatically active units toward the pre-, and postsynaptic membrane. The loosely attached enzyme or its subunits can be detached from these anchorage places and then becomes visible in the basal lamina of Schwann cells and vascular cells in the intercellular spaces of Schwann cells and in the sarcolemmal T-system, which is known to be open to the extracellular space.

Summary

The localization of acetylcholinesterase (AChE) activity has been studied by cytochemical electron microscopy at neuromuscular junctions of the rat diaphragm. Short incubation times revealed activity almost only in the muscle cells and the special extracellular spaces. In the subneural region activity was seen in the perinuclear cisternae, in the Golgi complex and the cisternae of the sarcoplasmic reticulum. Postsynaptic or sarcoplasmic vesicles contain high activity. They merge with the secondary folds of neuromuscular junctions of white, red and intermediate types of muscle fibers. Thus, these vesicles would seem to represent the secretory organelle of the muscle cell. In the specialized extracellular spaces of the neuromuscular region AChE is seen in the primary and secondary synaptic clefts, at Schwann cell surfaces, at basal lamina of the local capillaries and the sarcolemma. Furthermore, there is activity in the longitudinal and transverse tubular system of the muscle fibers. The localization of enzyme activity suggests a production site in the muscle cells, a release at the secondary folds of the neuromuscular junction and spreading to various extracellular sites.

References

1. Barat A, Gómez Barriocanal J, Ramírez G (1984) Two classes of collagen-tailed, asymmetric molecular forms of acetylcholinesterase in skeletal muscle: differential effects of denervation. Neurochem Int 6:403–412
2. Barrnett RJ (1962) The fine structural localization of acetylcholinesterase at the myoneural junction. J Cell Biol 12:247–262
3. Bergmann RA, Ueno H, Morizono Y, Hanker JS, Seligman AM (1967) Ultrastructural demonstration of acetylcholinesterase activity of motor end-plates via osmiophilic diazothioethers. Histochemie 11:1–2
4. Betz W, Sakmann B (1971) "Disjunction" of frog neuromuscular synapses by treatment with proteolytic enymes. Nature 232:94–95
5. Betz W, Sakmann B (1973) Effects of proteolytic enzymes on function and structure of frog neuromuscular junctions. J Physiol (Lond) 230:673–688
6. Brimijoin S, Skau K, Wiermaa MJ (1978) On the origin and fate of external acetylcholinesterase in peripheral nerve. J Physiol (Lond) 285:143–158
7. Brzin M, Kiautá T (1979) Cholinesterase of the motor end-plate of the rat diaphragm. Progr Brain Res 49:313–322
8. Chubb IW, Smith AD (1975) Release of acetylcholinesterase into the perfusate from the ox adrenal gland. Proc R Soc Ser B 191:263–269
9. Chubb IW, Goodman S, Smith AD (1976) Is acetylcholinesterase secreted from central neurones into cerebrospinal fluid? Neuroscience 1:57–62
10. Collins PL, Younkin SG (1982) Effect of denervation on the molecular forms of acetylcholinesterase in rat diaphragm. J Biol Chem 257:13 638–13 644
11. Crone HD, Freeman SE (1972) The acetylcholinesterase activity of denervated rat diaphragm. J Neurochem 19:1207–1208
12. Davies R, Koelle GB (1967) Electron microscopic localization of acetylcholinesterase and nonspecific cholinesterase at the neuromuscular junction by the gold-thiocholine and gold-thiolacetic acid methods. J Cell Biol 34:157–171
13. Dudai Y, Herzberg M, Silman I (1973) Molecular structures of acetylcholinesterase from electric organ tissues of electric eel. Proc Nat Acad Sci USA 70:2473–2476
14. Fernandéz HL, Duell MJ, Festoff BW (1979) Neurotrophic control of 16S acetylcholinesterase at the vertebrate neuromuscular Junction. J Neurobiol 19:441–454
15. Fernandéz HL, Patterson MR, Duell MJ (1980) Neurotrophic control of 16S acetylcholinesterase from mammalian skeletal muscle organ culture. J Neurobiol 11:557–570
16. Franzini-Armstrong C (1973) Membranous systems in muscle fibers. In: Bourne GH (ed) The structure and function of muscle, vol 2. Academic, New York London, pp 532–619
17. Friedenberg RM, Seligman AM (1972) Acetylcholinesterase at the myoneural junction: cystochemical ultrastructure and some biochemical considerations. J Histochem Cystochem 20:771–792
18. Greenfield S, Cheramy A, Leviel V, Glowinski J (1980) In vivo release of acetylcholinesterase in cat substantia nigra and caudate nucleus. Nature 284:355–357

19. Grubic Z, Sketelj J, Klinar B, Brzin M (1981) Recovery of acetylcholinesterase in the diaphragm, brain, and plasma of the rat after irreversible inhibition by soman: a study of cytochemical localization and molecular forms of the enzyme in motor end-plate. J Neurochem 37:909–916
20. Guth L (1968) Trophic influences of nerve on muscle. Physiol Rev 48:645–687
21. Hall ZW, Kelly RB (1971) Enzymatic detachment of end-plate acetylcholinesterase from muscle. Nature New Biology 232:62–63
22. Hall ZW (1973) Multiple forms of acetylcholinesterase and their distribution in end-plate and non-end-plate regions of rat diaphragm muscle. J Neurobiol 4:343–361
23. Hollunger EG, Niklasson BH (1973) The release and molecular state of mammalian brain acetylcholinesterase. J Neurochem 20:821–836
24. Huxley HE (1964) Evidence for continuity between the central elements of the triads and extracellular space in frog sartorius muscle. Nature 202:1067–1071
25. Inestrosa NC, Ramírez BU, Fernandéz HL (1977) Effect of denervation and of axoplasmic transport blockage on the in vitro release of muscle end-plate acetylcholinesterase. J Neurochem 28:941–945
26. Karcsú S, Tóth L, Csillik B (1975) Electron cytochemistry of acetylcholinesterase in the neuromuscular junction during ontogenesis. Z mikrosk-anat Forsch 89:447–459
27. Karnovsky MJ (1964) The localization of cholinesterase activity in rat cardiac muscle by electron microscopy. J Cell Biol 23:217–232
28. Kasa P (1968) Acetylcholinesterase transport in the central and peripheral nervous tissue: the role of tubules in the enzyme transport. Nature 218:1265–1267
29. Kreutzberg GW, Tóth L (1974) Dendritic secretion: a way for the neuron to communicate with the vasculature. Naturwissenschaften 61:37
30. Kreutzberg GE, Tóth L, Kaiya H (1975) Acetylcholinesterase as a marker for dendritic transport and dendritic secretion. In: Kreutzberg GW (ed) Physiology and pathology of dendrites. Advances in neurology, vol 12. Raven, New York, pp 269–281
31. Lentz TL (1971) Nerve trophic function: in vitro assay of effects of nerve tissue on muscle cholinesterase activity. Science 171:187–189
32. Lewis PR, Shute CCD (1966) The distribution of cholinesterase in cholinergic neurons demonstrated with the electron microscope. J Cell Sci 1:381–390
33. Lewis PR, Shute CCD (1969) An electron-microscopic study of cholinesterase in the rat adrenal medulla. J Microscopy 89:181–193
34. Lucas CA, Kreutzberg GW (1985) Regulation of acetylcholinesterase secretion from neuronal cell cultures. 1. Actions of nerve growth factor, cystoskeletal inhibitors and tunicamycin. Neuroscience 14:349–360
35. Massoulié J, Bon S (1982) The molecular forms of cholinesterase and acetylcholinesterase in vertebrates. Ann Rev Neurosci 5:57–106
36. McMahan UJ, Sanes JR, Marshall LM (1978) Cholinesterase is associated with the basal lamina at the neuromuscular junction. Nature 271:172–174
37. Miledi R (1964) Electron-microscopical localization of products from histochemical reactions used to detect cholinesterase in muscle. Nature 204:293–295
38. Padykula HA, Gauthier GF (1970) The ultrastructure of neuromuscular junctions of mammalian red, white and intermediate skeletal muscle fibers. J Cell Biol 46:27–41
39. Reynolds ES (1963) The use of lead citrate at high pH as an electron-opaque stain in electron microscopy. J Cell Biol 17:208–212
40. Rotundo RL, Fambrough DM (1980 a) Synthesis, transport and fate of acetylcholinesterase in cultured chick embryo muscle cells. Cell 22:583–594
41. Rotundo RL, Fambrough DM (1980 b) Secretion of acetylcholinesterase: relation to acetylcholine receptor metabolism. Cell 22:595–602
42. Rubin LL, Schuetze SM, Weill CL, Fischbach GD (1980) Regulation of acetylcholinesterase appearance at neuromuscular junctions in vitro. Nature 283:264–267
43. Silman I, Lyles JM, Barnard EA (1978) Intrinsic forms of acetylcholinestase in skeletal muscle. FEBS Lett 94:166–170
44. Silver A (1974) The Biology of cholinesterase. In: Neuberger A, Tatum EL (eds) Frontiers of biology. Elsevier, North-Holland, Amsterdam
45. Sketelj J, Brzin M (1979) Attachment of acetylcholinesterase to structures of the motor endplate. Histochemistry 61:239–248

46. Sketelj J, Brzin M (1980) 16S Acetylcholinesterase in end-plate-free regions of developing rat diaphragm. Neurochem Res 5:653–658
47. Smilowitz H (1980) Routes of intracellular transport of acetylcholine receptor and esterase are distinct. Cell 19:237–244
48. Tennyson VM, Kremzer LT, Brzin M (1977) Electron microscopic-cytochemical and biochemical studies of acetylcholinesterase activity in denervated muscle of rabbits. J Neuropath Exp Neurol. 36:245–275
49. Teräväinen H (1967) Electron-microscopic localization of cholinesterases in the rat myoneural junction. Histochemie 10:266–271
50. Teräväinen H (1969a) Distribution of acetylcholinesterase in extraocular muscle fibres of the rat. Histochemie 18:174–176
51. Teräväinen H (1969b) Localization of acetylcholinesterase in the rat myoneural junction. Histochemie 17:162–169
52. Tóth L, Karcsú S (1979) Über die ultrastrukturelle Lokalisation der Acetylcholinesterase-(AChE-) Aktivität im Diaphragma des Rattenembryo. Acta Histochem (Jena) 64:148–156
53. Tóth L, Karcsú S, Poberai M, Savay GY (1981) A light- and electron-microscopic histochemical study on the mechanism of DFP-induced acute and subacute myopathy. Neuropathol Appl Neurobiol 7:399–410
54. Younkin SG, Rosenstein C, Collins PL, Rosenberry TL (1982) Cellular localization of the molecular forms of acetylcholinesterase in rat diaphragm. J Biol Chem 257:13630–13637
55. Zacks SI (1970) Direct connections between the T-system and the subneural apparatus in mouse neuromuscular junctions demonstrated by lanthanum. J Histochem Cytochem 18:302–304

Control of Transmitter Release at Cholinergic and Glutamatergic Nerve Terminals

J. DUDEL [1]

It is generally assumed that on depolarization of a nerve terminal calcium enters, and that the rise in the intraterminal calcium concentration ([Ca_1]) triggers the release of transmitter quanta [5–7]. Facilitation of release in consecutive depolarizations, then, is caused by residual calcium from the preceding calcium inflow [6, 8]. In this scheme the time course of release of quanta and that of facilitation after a depolarization pulse both are controlled by the time course of $[Ca]_i$. Release, however, stops within 10 ms after depolarization, while facilitation lasts for seconds. This discrepancy and some other results not presented here have led us to propose a more complicated scheme (Fig. 1) for the control of release [2, 3, 9, 10]. Depolarization (*d*) in addition to the calcium inflow triggers an activator A* with two inactive presteps in the membrane, and, further, a repressor R* allows release only after re- or hyperpolarization (*h*). The following description of control of release in nerve terminals on muscles of frog or crayfish will refer to this scheme.

Figure 2 shows the dependence of amplitude (*m*, number of quanta per stimulus) and time course of release on the amplitude of depolarization. If the depolarization pulse is increased from −0.5 μA to −1 μA, which corresponds approximately to 50 and 100 mV depolarization, respectively, release rises 170-fold. In spite of this large increase, the time course of release is unaltered (see also [1]), contrary to expectations for a process controlled by the rise and removal of $[Ca]_i$

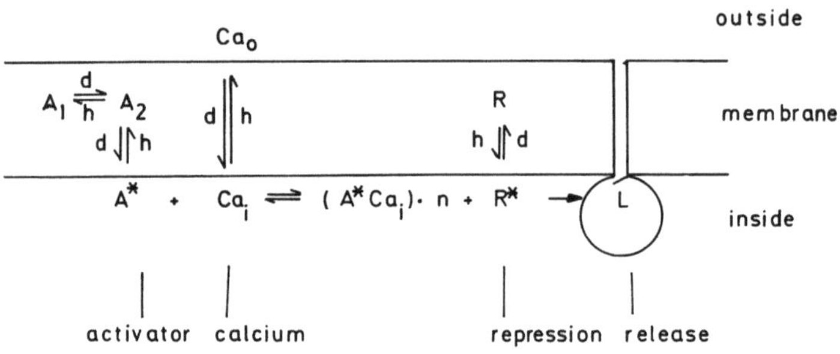

Fig. 1. Scheme of proposed reactions leading to release. Inactive stages A_1 and A_2 shift to the activator A^* due to depolarization (*d*) and back due to re- or hyperpolarization (*h*). Similarly, extracellular *Ca* (Ca_o) crosses the membrane (Ca_i). At the intracellular side of the membrane, the complex (A^*Ca_i) reacts with the cooperativity *n* with a deactivated repressor R^* and effects release (*L*) of a quantum. The repressor *R* is activated by depolarization (*d*), but is deactivated to R^* rapidly after repolarization (*h*)

[1] Physiologisches Institut der Technischen Universität München, Biedersteiner Str. 29, D-8000 München 40, FRG.

Clinical Aspects of Sensory Motor Integration
Ed. by A. Struppler and A. Weindl
© Springer-Verlag Berlin Heidelberg 1987

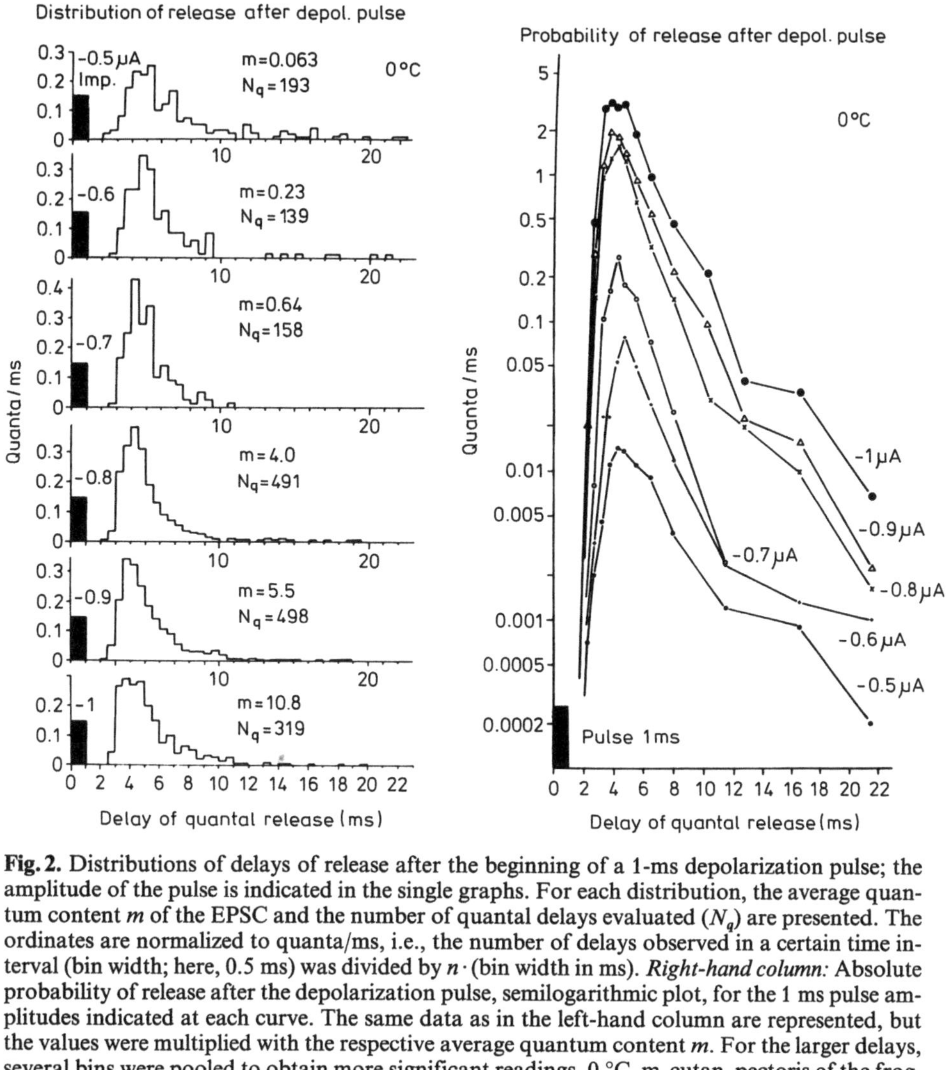

Fig. 2. Distributions of delays of release after the beginning of a 1-ms depolarization pulse; the amplitude of the pulse is indicated in the single graphs. For each distribution, the average quantum content m of the EPSC and the number of quantal delays evaluated (N_q) are presented. The ordinates are normalized to quanta/ms, i.e., the number of delays observed in a certain time interval (bin width; here, 0.5 ms) was divided by $n \cdot$ (bin width in ms). *Right-hand column:* Absolute probability of release after the depolarization pulse, semilogarithmic plot, for the 1 ms pulse amplitudes indicated at each curve. The same data as in the left-hand column are represented, but the values were multiplied with the respective average quantum content m. For the larger delays, several bins were pooled to obtain more significant readings. 0 °C, m. cutan. pectoris of the frog. (Dudel [2])

[11]. We conclude that the time course of release is determined by that of the activator A*, while the amount of release is controlled by a complex (A^*Ca_i) (Fig. 1), and thus in addition by calcium inflow. Also, the duration of the depolarization of the terminal does not essentially influence the time course of release, but the amount of release changes very steeply. Figure 3 presents a double logarithmic plot of the dependence of release on the duration of depolarization. For levels of depolarization between −0.4 and −1.2 µA release varies in a wide range, and the slopes of the dependence on duration reach 7–8. This would mean that release should rise with a power n of 7–8 of (A^*Ca_i) (Fig. 1).

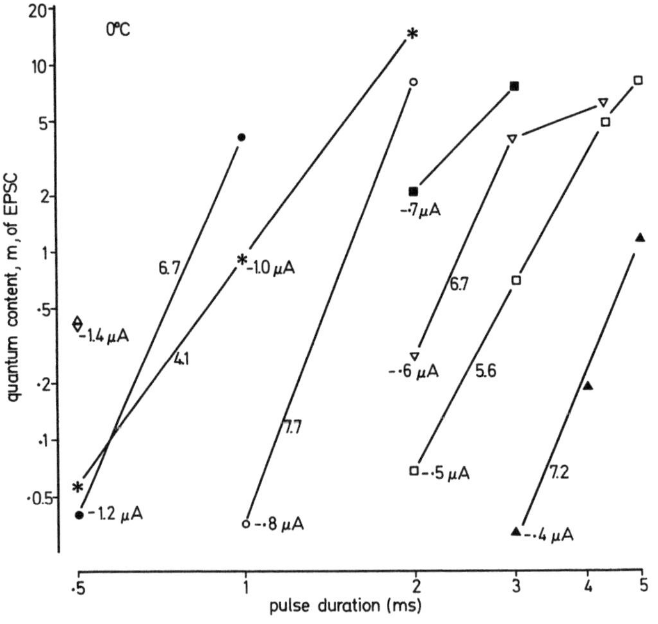

Fig. 3. Double logarithmic plot of average quantum content (*m*) of EPSCs in dependence on pulse duration, at one terminal. Values obtained at one pulse amplitude are connected by *lines* and the respective current pulse amplitudes are indicated. In addition, the (double-logarithmic) slopes of 4.1–7.7 are inserted at the respective connections. 0 °C. m. cutan. pectoris of the frog. (Dudel [2])

Release is not only determined by amplitude and duration of depolarization, but also by the amplitude and timing of de- or hyperpolarizing pre- or post-pulses which by themselves are too small to trigger release. Figure 4 shows the effects of pre-pulses: depolarizing ones increase release up to 10-fold, while, unexpectedly, hyperpolarizing ones depress release to a similar extent. The effects decrease with the lengthening of the interval between the pre-pulse and the releasing pulse, with a time course which is very similar to that of release (Fig. 2). This indicates that these pre-pulses act on the process $A_1 \rightleftharpoons A_2 \rightleftharpoons A^*$ (Fig. 1). Accordingly, such subthreshold pre-pulses, but also post-pulses placed directly after the releasing pulse, strongly modulate release, but do not affect facilitation of a following excitatory postsynaptic current (EPSC) [3, 10].

A last finding which cannot be explained by control of release by $[Ca]_i$, not even with the addition of an activator, is the „latency shift" with changes of pulse duration: the first releases appear after the end of the depolarization even if this lasts up to 5 ms [2, 5]. The shift in latency can also be reached by small (10 mV) depolarizing post-pulses. As seen in Fig. 5, the latency increases from 2 to 4 ms when post-pulses of increasing duration are given. The effect would increase with larger post-pulses (not illustrated). Comparison with the control shows that for a few ms the post-depolarization prevents releases that would have already been triggered by the test pulse. However, the post-pulses strongly increase total release, obviously due to activation of A*. The temporary prevention of release by

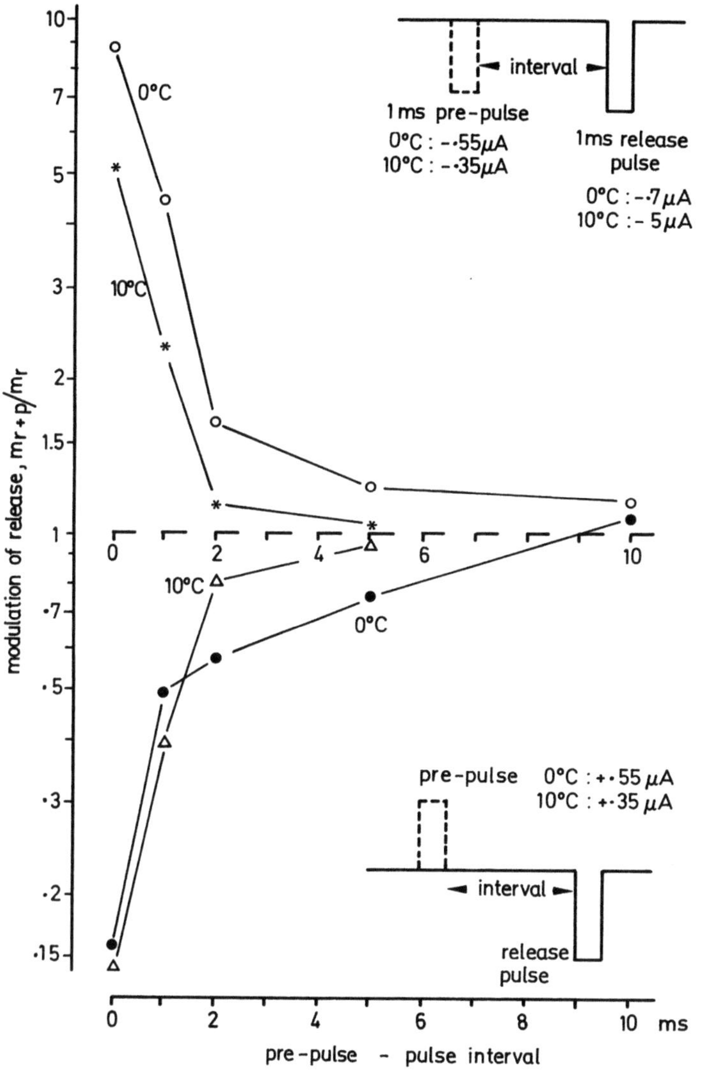

Fig. 4. An EPSC with the quantum content m_r is elicited by a releasing pulse of fixed amplitude and duration. Prepulses of 1 ms duration and fixed amplitude (see *insert*) modulate the EPSC which results in the quantum content m_{r+p}. Modulation m_{r+p}/m_r is plotted in the *ordinate* (logarithmic scale) in dependence on the interval between pre-pulse and releasing pulse (*abscissa*). The pulse amplitudes were different for 0 °C and 10 °C to compensate partly for the effects of temperature on release. In the *upper half* of the figure depolarizing pre-pulses enhanced, and in the *lower half* hyperpolarizing pre-pulses depressed release. Frog, m. cutan. pectoris. (Dudel [3])

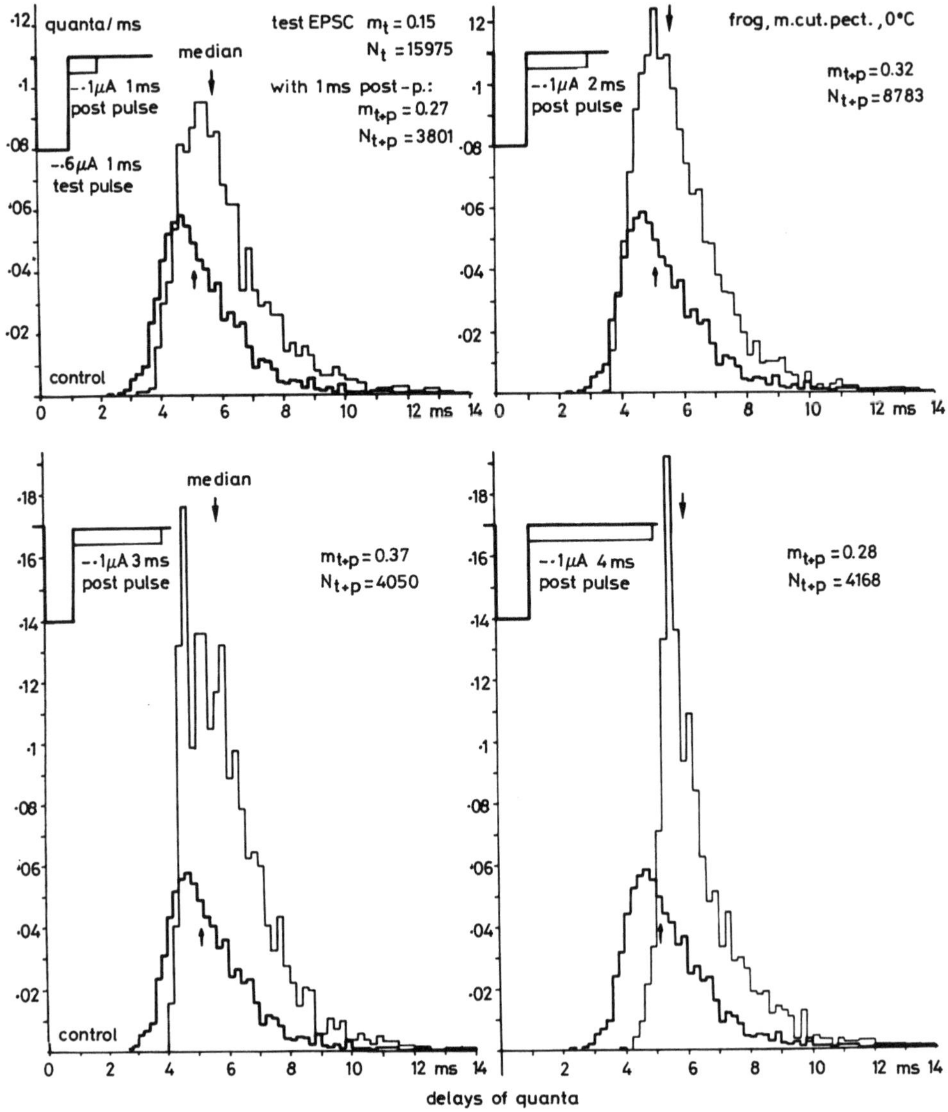

Fig. 5. Time course of release after a fixed test pulse of −0.6 μA and 1 ms duration: heavily drawn control distribution in each plot. If a depolarizing −0.1 μA post-pulse of 1–4 ms duration is added (see *insets*), the lightly drawn distributions of release result: release is increased but also repressed initially. The differences in the initial distributions are highly significant. Ordinate probability of release like in the right hand part of Fig. 2. m_t is the average number of quanta released by the test pulse, which is increased to m_{t+p} on addition of the respective post-pulse. N_t and N_{t+p} are the numbers of trials evaluated for the respective distribution

small depolarizations must be attributed to a further step in the control of release, which was named „repression“ and is removed rapidly by re- or hyperpolarization (Fig. 1).

The characteristics of neuromuscular transmission in the frog briefly presented here have been found to be very much the same in crayfish, with small quantitative variations. Excitatory neuromuscular transmission in crustaceans is organized differently from that in vertebrates: synapses are distributed over the whole muscle fibre, not all terminals are reached by action potentials, the probability of release is low at single terminal areas, and the transmitter is glutamate. Thus, these synapses are similar to the ones in the CNS of vertebrates. The great similarities in the mechanisms of control of release despite very different animal species and functional organizations make it probable that the proposed mechanisms are fairly general properties of synaptic terminals.

The powerful control of release by amplitude and duration of depolarization is not realized in neuromuscular transmission in vertebrates – with the possible exception of pathological conditions. However, amplitude and duration of depolarization of the terminal may be varied in other systems, for instance by presynaptic inhibition as seen in crayfish [4]. In the CNS terminals receiving axo-axonic synaptic inputs, subthreshold modulatory de- or hyperpolarizations also seem possible and may enrich the repertoire of synaptic functions. The functional role of the temporary repression of release after depolarization seems difficult to understand.

Summary

A new model of the control of transmitter release from synaptic terminals is proposed, in which in addition to the rise in intracellular Ca concentration ($[Ca]_i$), another activator A* is produced by depolarization, and the complex (Ca_iA^*) cooperatively triggers release. In addition, depolarization represses release causing the synaptic delay. This model is able to explain (1) the very large variations of release on changes in amplitude and duration of depolarization, while the time course of release is little affected, (2) modulation of release due to a depolarization pulse by de- or hyperpolarizing small pre- and post-pulses, and (3) latency shifts of release produced by small depolarizing post-pulses.

References

1. Datyner NB, Gage PW (1980) Phasic secretion of acetylcholine at a mammalian neuromuscular junction. J Physiol (Lond) 303:299–314
2. Dudel J (1984a) Control of quantal transmitter release at frog's motor nerve terminals: I Dependence on amplitude and duration of depolarization. Pflügers Arch 402:225–234
3. Dudel J (1984b) Control of quantal transmitter release at frog's motor nerve terminals: II Modulation by de- or hyperpolarizing pulses. Pflügers Arch 402:235–243

4. Dudel J, Kuffler SW (1961) Presynaptic inhibition at the crayfish neuromuscular junction. J Physiol (Lond) 155:543–562
5. Katz B, Miledi R (1967) The release of acetylcholine from nerve endings by graded electric pulses. Proc R Soc (Lond) B 167:23–38
6. Katz B, Miledi R (1968) The role of calcium in neuromuscular facilitation. J Physiol (Lond) 195:481–492
7. Llinás R, Steinberg IZ, Walton K (1981) Relationship between presynaptic calcium current and postsynaptic potential in squid giant synapse. Biophys J 33:3232–3252
8. Parnas H, Dudel J, Parnas I (1982) Neurotransmitter release and its facilitation in crayfish. I Saturation kinetics of release, and of entry and removal of calcium. Pflügers Arch 393:1–14
9. Parnas I, Dudel J, Parnas H (1984) Depolarization dependence of the kinetics of phasic transmitter release at the crayfish neuromuscular junction. Neurosci Lett 50:157–162
10. Parnas I, Parnas H, Dudel J (1986) Neurotransmitter release and its facilitation in crayfish. VIII Modulation of release by hyperpolarizing pulses. Pflügers Arch 406:131–137
11. Stockbridge N, Moore JW (1984) Dynamics of intracellular calcium and its possible relationship to phasic transmitter release and facilitation at the frog neuromuscular junction. J Neurosci 4:803–811

Neurotrophism – Another Approach

W. W. HOFMANN [1]

Introduction

The mechanism by which a motor nerve axon "nourishes" or maintains a "trophic influence" on the muscle fiber it reaches remains unknown, despite much investigation and interest. A large body of literature, including several excellent reviews [12, 13, 15, 16] suggests that the presynaptic element exerts its control in one of three ways: (a) by release of the normal transmitter, acetylcholine (ACh), (b) by some aspect of the stimulus-evoked postsynaptic response itself, i.e., in muscle, the contraction, and (c) by the release of other substances delivered to the presynaptic axon terminals by centrifugal axoplasmic flow from the motor neuron soma.

When the trophic influence is lost, either through actual disconnection (denervation) or by blockade of axoplasmic transport, the well-known results in the case of striated muscle include: (a) partial depolarization, (b) spontaneous fibrillatory activity, (c) increased membrane ohmic input resistance (R_m), (d) a remarkable increase in the number of available ACh receptors (AChRs), spreading from the normal end-plate region to cover the whole fiber, and (e) despite this increased chemosensitivity, a failure of intrafiber metabolism, with an increase in net protein breakdown, leading to atrophy. The latter is particularly interesting, since the muscle appears to be "starving in a sea of plenty."

The most extensively studied proposal is that it is simply the loss of ACh itself which leads to the effects of denervation, since both stimulus-linked and spontaneous quantal release of the transmitter essentially cease when tested in standard ringer within a few hours after surgical denervation [11, 32]. Ingenious experiments [4, 7, 14, 28, 31], however, have failed to prove this point, and the role of the ionotrophic transmitter-receptor interaction in maintaining muscle metabolism, contractile properties, and structure remains to be determined.

A denervation change considered most likely to result directly from loss of ACh is the increase in AChR, a receptor "up-regulation" superficially analagous to that seen in other systems where the ligand concentration affects the availability of binding sites [27]. However, since similar increases in AChR can be shown to occur while ACh output seems to be normal [2, 19], even this notion is contested. The redistribution of proteins known to govern the ionic channels for electrogenesis might explain other denervation effects such as depolarization and fibrillation if some transmitter were available. Without its ligand, however, AChR seems inert, and the various electrophysiological changes after denervation [1, 26] cannot be attributed simply to the presence of more transmitter binding sites.

[1] Department of Neurology (127), Veterans Administration Medical Center, Palo Alto, CA 94304, USA.

Clinical Aspects of Sensory Motor Integration
Ed. by A. Struppler and A. Weindl
© Springer-Verlag Berlin Heidelberg 1987

Perhaps the strongest argument against the ACh theory is found in the detailed experiments in which denervated muscle has been stimulated in vivo and in vitro with patterns of impulses mimicking normal innervation [21, 23, 24, 29]. In these studies, reduction in post-denervation chemosensitivity, restoration of electrical properties, and short-term preservation of normal contraction parameters have all been convincingly demonstrated by direct stimulation in the absence of ACh. The inference has been that muscle activity is the major trophic element. Interpretation of these elegant experiments has been made difficult by the observation that changes similar to those of denervation can be induced in a fully active muscle with normal neuromuscular transmission. Muscles whose nerves have been treated with colchicine may, as if denervated, become slightly depolarized, universally chemosensitive, spontaneously active, and may lose up to 20% in weight, though they remain normally active [2, 19].

The hypothesis derived from the colchicine experiments was that axoplasmic centrifugal convection delivered a "trophic" substance to the muscle, and lack of this material, even with normal ACh release and contractions, caused the observed changes. This last theory has itself been challenged on the ground that colchicine may exert some of its effects by direct action on the muscle membrane or that nerve conduction may have been transiently impaired, lending to a period of muscle inactivity, which markedly accentuated the effects of nerve damage [6].

In summary, the ACh theory has met with a sizeable number of objections, the mechanism underlying the remarkable effects of direct muscle stimulation is still unknown, and the axoplasmic flow theory is in need of support from additional experiments in which the blocking agent does not directly affect muscle membranes.

Further search for trophic factor(s) might be aided by defining the features that should be expected in an "ideal" regulatory substance. Such properties would include:

1. The ability to maintain the normal resting membrane potential. This property would have little to do with the AChR-operated, voltage-dependent, passive ion fluxes responsible for excitability. For steady-state conditions, the trophic effect would be mediated through the energy-dependent, active Na^+/K^+ exchange pump responsible for normal ionic concentration gradients.
2. The ability to preserve membrane structure and all intrafiber contents by regulating the uptake of protein-building materials, by increasing protein synthesis, and by inhibiting protein breakdown. This property would maintain muscle bulk and strength, and the nature of the proteins synthesized under control of this substance could determine the type of conversion from fetal to mature state as well as the final histochemical character of the muscle.
3. The ability to regulate the uptake of the most readily available fuel, namely, glucose. This control should be quickly adjustable to changes in contraction requirements or the needs of repair and growth systems.
4. The ability to regulate the turnover, and thus the availability, of transmitter receptor units of the end-plate, as well as the rest of the fiber membrane. This property, again a form of metabolic regulation, might have nothing to do with intrasarcoplasmic biochemical events and might operate strictly within the membrane.

It is worth emphasizing that all the above functions could be classified as "metabotrophic" and that they would presumably have to begin at or within the membrane. It is the purpose of this report to indicate that a powerful metabotrophic element with all the properties required is present in abundance on the surface of all skeletal muscle, and probably most other cells. This substance is the insulin receptor, a molecular complex uniquely situated to supply the muscles' needs, and it is now found to be under neural control.

The ligand, insulin, required to activate the receptor is always available in the extracellular fluid, as are the various nutrients whose transfer to the muscle fiber interior is continually required. The system is set up to operate effectively without any necessary electrogenic signals from the motor nerve. To exert its repertoire of insulin-mediated "trophic" effects, the nerve needs only to deliver some compound able to govern either the number or the effectiveness of the insulin receptors (IR). To accommodate sudden changes in workload, there would need to be either a method by which to correlate delivery of the trophic material with the frequency of nerve action potentials or a method to sense and respond to increasing numbers of muscle contractions. From the present and earlier [20] studies, it appears that the IR is, in fact, under some sort of nervous control, and a property of the receptor is illustrated here which can quickly match muscle demands and energy supply.

Material and Methods

In all but a few experiments using young rat diaphragm, the investigation utilized soleus or diaphragm from adult mice [1]. The electrophysiological measurements were made with standard glass microelectrodes, and estimates of acetylcholine receptors (AChR) or insulin receptors (IR) were made with I^{125}-labelled α-bungarotoxin [2] or I^{125}-labelled rat insulin [3], respectively. For prolonged incubation of excised diaphragms, a standard culture medium arrangement [4] was used, in which infection was controlled with penicillin and streptomycin [5]. In tests of axoplasmic flow, highest purity colchicine was used [6].

In the electrophysiological tests, resting membrane potentials (RMP) were recorded in fresh and "cultured" mouse diaphragm fibers. The preparations were maintained at 21–22 °C in tissue culture medium for the periods shown in the various figures, the medium being changed every 24 h. Those specimens tested with insulin were maintained in culture with zinc-free porcine insulin sodium at

[1] Rats: Wistar strain, male, 50–70 g. Mice: male Swiss-Webster.
[2] α-Bungarotoxin [^{125}I] lyophilized, Sp. Act. 10–20 Ci/g. New England Nuclear, Boston, MA, USA.
[3] Insulin [^{125}I] rat, Sp. Act. 80–100 Ci/g, kindly supplied from the laboratories of Dr. Gerald Reaven, Department of Medicine, Stanford University, GREC, Veterans Administration Medical Center, Palo Alto, CA, USA.
[4] Minimum essential medium (MEM) with Hank's salts. Gibco Laboratories, Santa Clara, CA 95050, USA.
[5] Penicillin base, 2000 U/ml, streptomycin base 2000 mcg/ml.
[6] Colchicum purissimum, crystalline, Merck, EM Labs, Elmsford, NY 10523, USA.

a final concentration of 0.5 to 1.0 μM [7]. At the test time of interest, the muscles were transferred to the nutrient solution of Bretag [5] for microelectrode studies. RMPs from cultured preparations were compared with those in hemidiaphragm freshly removed, the latter also being studied with or without insulin immediately after excision. End-plate regions, easily visible in such preparations, were also examined for the presence of miniature end-plate potentials (mepps).

In other experiments, hemidiaphragms previously incubated in identical culture media were transferred to nutrient, gassed ringer containing either I^{125}-labelled bungarotoxin [2] or I^{125}-labelled rat insulin [3] to estimate the amount of AChR or IR, respectively. In determining the amount of specific ACh binding, two tests were used. The first was to compare in cultured and fresh muscle the radioactivity in the end-plate-containing 1 mm central strip of diaphragm with that counted in similar (non-end-plate) strips near the rib origin of fibers [7, 33]. The second method was to incubate segments cut from the various preparations with and without the addition of a large excess of "cold" ligand [7,8]. In the case of AChR, the "cold" BUTX was about 1 μM. With IR measurements, the unlabelled insulin [7] concentration was 8 μM. With diaphragm, as many as 8 useable segments could be obtained. When the test muscle was soleus, the excised preparation, either test or control, was divided longitudinally under a dissecting microscope by teasing into two similar portions. For each given test condition, one strip of soleus could then be incubated (for binding studies, 4 h) in a low concentration of radioactive ligand only, the other with tracer plus "cold".

In tissue-culture incubations, the diaphragms were always denervated at the very entry of the phrenic nerve to the muscle. In denervation experiments on soleus, the sciatic was sectioned in the thigh. Corrections were applied for isotope decay, for exposure dose, and for differences in specific activity where indicated, and plots of both AChR and IR binding activities were made after subtracting "non-specific" binding.

In vivo denervation experiments always utilized soleus, with each animal's unoperated leg acting as control. When it was desired to measure the effects of colchicine, two methods were employed. The first was with a small Silastic [9] cuff containing 1 mg/g colchicine applied to one sciatic nerve. The second was with 0.5–1.0 µl of 50 mM colchicine injected through a 27-gauge needle and microsyringe into one sciatic perineurium. The muscles from these mice were tested at 80–96 h.

Responses of the IR to neuromuscular stimulation were examined with three techniques. In the first, the cut sciatic nerve on one side of an anesthetized mouse was supramaximally stimulated with 5 trains of shocks at 20 Hz for 10 s each, just prior to removal of soleus, which had previously been exposed. The second approach was to deliver repeated trains of tetanic (40 Hz; 10 s; $\times$ 5) stimulation to hemidiaphragm through the phrenic nerve in vitro, after which the preparation was cut into several strips, each of which was then incubated in a specific concentration of cold plus tracer insulin to estimate binding affinity. To learn what

[7] Insulin, porcine, sodium, Lot 050YB7 kindly supplied by Dr. R. Chance, Research Division, Eli Lilly Co., Indianapolis, IN 46206, USA.
[8] α-Bungarotoxin (Bungarus multicinctus) Sigma Chemical Co., St. Louis, MO 14508, USA.
[9] "Silastic" Medical Grade Elastomer 382, Dow-Corning Co., Midland, MI 48460, USA.

aspect of stimulation produced the observed changes in insulin binding, other soleus and diaphragm strips were pre-incubated for 15 min in 50 mM [K^+] solution and then exposed for 4 h to labelled and various concentrations of unlabelled ligand. In these muscles, all of which appeared relaxed at the end of the final incubation, the stimulus was steady membrane depolarization for a short period prior to the binding assay. It was not possible to determine whether the potassium-treated muscles developed a brief contracture at the beginning of the depolarization.

Tests of the effect of botulinum toxin [10] on IR were made in two ways. In 6 young rats, the diaphragm was surgically exposed under ether anesthesia and 5 μl, containing 2×10^{-10} g/μl of the toxin, was directly injected under microscopic control into the underside of the muscle near the point of entry of the left phrenic nerve. For the rest, the soleus nerve-entry point in 16 mice was infiltrated with 5 μl of solution containing 2.5×10^{-11} g/μl toxin.

Results

1. Electrophysiological Effects of Insulin. Incubation of denervated diaphragm was followed by the expected mild depolarization of muscle fibers, unless the preparations were incubated with insulin, in which case the resting membrane potential (RMP) was relatively better maintained. In freshly removed diaphragm, insulin also increased the RMP, as has been observed in other studies [35]. These results are illustrated in Table 1. In the tissue-culture medium used,· the diaphragm specimens contained surprisingly numerous end-plate regions at which miniature end-plate potentials (mepps) of normal uniquantal form, amplitude, and frequency could be recorded as late as 48–72 h after isolation. The presence or absence of insulin did not appear to affect this spontaneous transmitter release. As shown in the typical experiment of Fig. 1, ACh binding had risen to about 30% over control as early as 15 h, while mepps were reduced by only about 10% in this medium. The relative depolarization noted in 2-day incubated control preparations was nearly absent if the muscle had been incubated in insulin, at a concentration of about 1 μM.

2. Effects of Denervation of the Insulin Receptor. Figure 2 (curve A) illustrates that, after isolation in culture medium or section of the appropriate motor nerve

Table 1. Effects of insulin on membrane potential in fresh and in denervated muscle preparations

	Control (mV)	(*n*)	24 h	(*n*)	48 h	(n)	72 h	(*n*)
Standard ringer	72.8±4.1	160	68.8±5.1	60	61.3±5.5	70	60.9±6.8	40
Plus insulin	78.7±3.4	115	70.1±5.3	50	67.9±4.8	75	68.5±7.6	45

[10] Botulinum toxin type A, Sigma Chemical Co., St. Louis, MO 14508, USA.

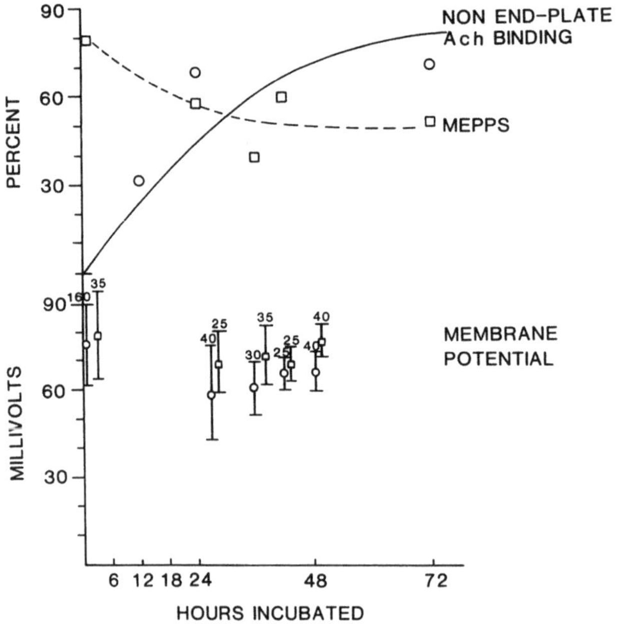

Fig. 1. Development of extrajunctional ACh receptors with quantal transmitter release still largely intact. (In tissue culture medium for times as shown prior to testing.) Curves fitted by eye to ACh binding values in non-end-plate regions, as estimated with I^{125} α-bungarotoxin. Values in *upper two curves* shown as percentage of those obtained in similarly treated hemidiaphragms from control side in each animal. *Lower plots* are of membrane potentials and ranges. Number of fibers tested above each point. Values from diaphragm incubated for times shown in standard culture medium without insulin (○) and from muscles incubated with ca. 1 μM insulin (□). Note early fall in RMP while mepps still present and definite repolarization in all cases with insulin

in vivo, both diaphragm and soleus muscles demonstrated increased insulin-displaceable insulin binding as compared with controls. In the diaphragm experiments, the control was a freshly removed muscle, while with soleus, results were compared with the muscle of the unoperated leg. In the soleus (in vivo) experiments, the increase of insulin binding was not adequate to prevent weight loss of approximately 10% in the denervated muscle at 48 h.

Since botulinum toxin (BOTX) mimics the effects of denervation [14], it was of interest to examine the responses of the IR on BOTX-treated muscles. Two preparations were used: mouse soleus (16 animals) whose motor nerve terminals had been infiltrated, and the diaphragm of young rats (50 g) injected 48–96 h previously through a surgical approach to the phrenic terminals on one side from the underside of the muscle (6 animals). Curves B and C in Fig. 2 illustrate the result: again, a rise in insulin-displaceable insulin binding. It can be seen that the increase in the maximum amount of binding is not as great as that following nerve section, but it occurs when the toxin is injected near the motor nerve to soleus or diaphragm. These findings plus the observation that BOTX reduces the flow of some intra-axonal materials [3] led to further experiments, a typical example of

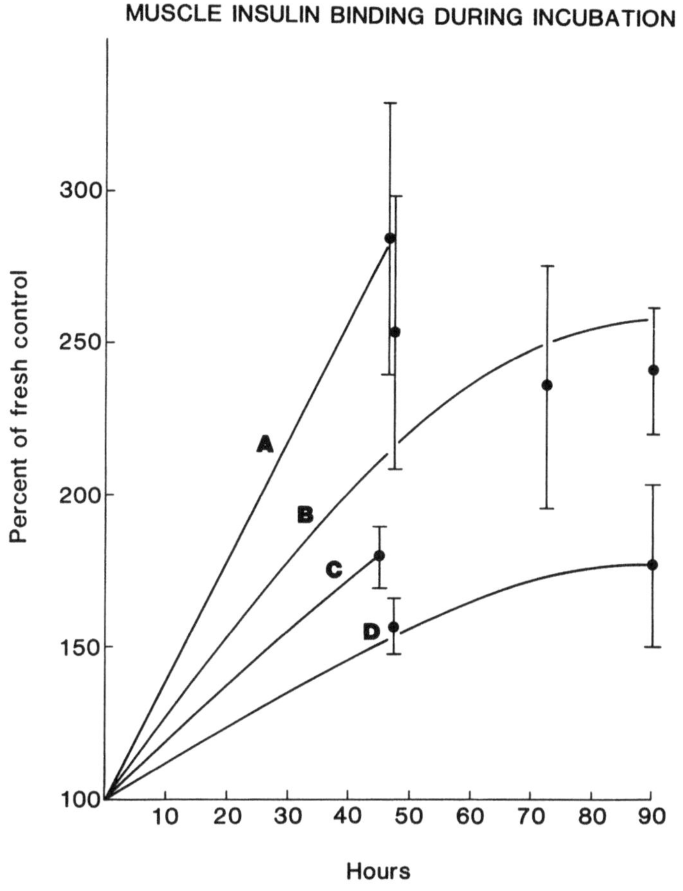

Fig. 2. Effects of various agents on maximum numbers of insulin binding sites on skeletal muscle. *Curves* fitted by eye to mean value points. *Bars* are ranges. *Curve A* surgically denervated soleii at times shown (48–50 h). Incubation in all cases as in text. All values expressed as percentages of those from opposite (control) soleii for each animal. *Curve B* soleus after botulinum toxin applied near motor nerve entry point. *Curve C* from diaphragm in young rats after botulinum toxin injected near phrenic endings. *Curve D* soleii at 2 and 4 days after injection of colchicine into sciatic nerve (see text). Some neural effect regulating the number of insulin binding sites is lost with denervation and is decreased after both partial botulinum poisoning and inhibition of axoplasmic flow

which is shown in curve D in Fig. 2. The question of interest was whether interference with axoplasmic flow in a given nerve would also affect the maximum insulin-binding capacity in the muscle it supplied. Colchicine was applied to the exposed sciatic nerve on one side by means of Silastic cuffs in 6 and by direct injection in 16 mice. Each animal served as its own control. Eight preparations were used to determine colchicine's effect on insulin-stimulated 2-deoxyglucose [11] uptake in vitro. For the cuff experiments, the colchicine concentration in the elas-

[11] Deoxy-D-glucose, 2[^{14}C (U)], Sp. Act. 282 mCi/mmol, New England Nuclear, Boston, MA, USA.

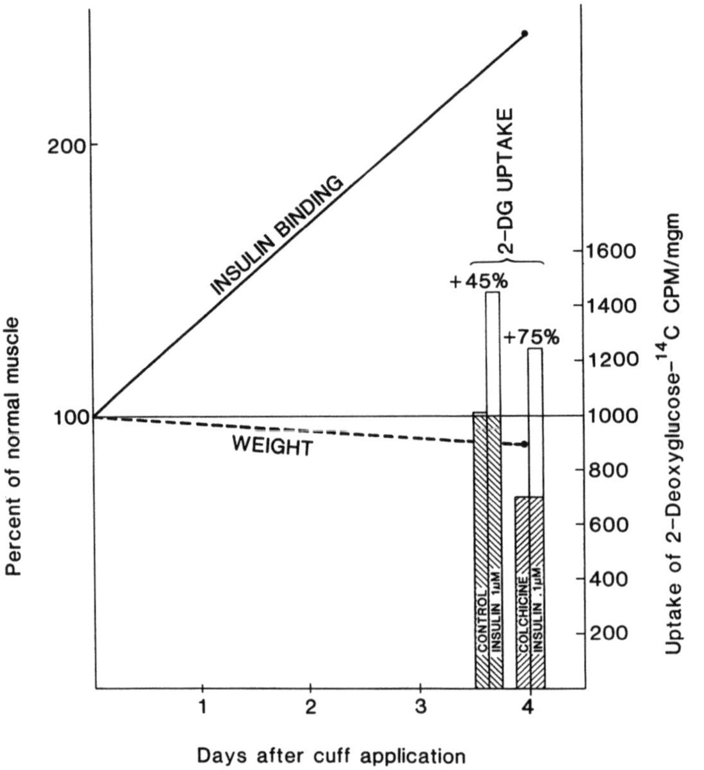

Fig. 3. Neural colchicine effects on ipsilateral soleus. 1 mg/g colchicine in Silastic cuff on sciatic. Tests of ipsilateral and control soleii in this experiment at 4 days. For ^{14}C-deoxyglucose uptake, control and soleus muscles split longitudinally to be tested with and without insulin in excess (1 μ*M*). Deoxyglucose concentration 3 μ*M*, incubation time 1 h at 21 °C. Insulin binding on similarly paired and treated soleii from other mice with identical colchicine dose. Insulin ^{125}I-binding recorded as insulin-displaceable cpm/mg muscle after 4-h incubation period. With colchicine-treated nerve, soleus has more insulin binding sites but less 2-deoxyglucose uptake, unless excess ligand is provided. New receptors have low affinity but can be saturated to produce increased metabotrophic effects

tomer was 1 mg/g, for the injection experiments, 0.5 μl of 50 m*M* drug was infiltrated under the sciatic perineurium in mid-thigh. The soleii on the treated side lost some weight, increased their insulin binding, and demonstrated increased insulin responsiveness, though the resting (non-insulin) level of deoxyglucose assimilation was reduced in a typical experiment by about one-half (Fig. 3). Inspection of curve D in Fig. 2 reveals that the colchicine effect on insulin binding was less on average than that of denervation or of botulinum toxin.

3. Effects of Stimulation on the Insulin Receptor. These experiments centered on the question whether the muscle insulin receptor could respond quickly to increased energy demands brought on by stimulation. Three methods were used to stimulate the muscles, which were either mouse soleus or diaphragm (see "Mate-

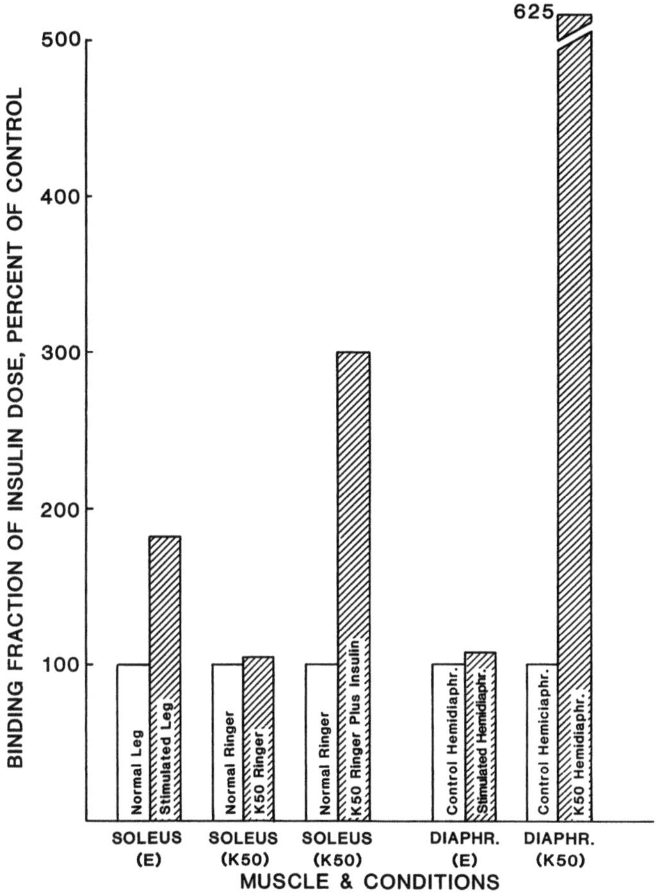

Fig. 4. Results of muscle electrical stimulation (see text) and potassium depolarization on maximum binding of labelled insulin by its receptor. *Clear columns* represent values for conditions listed in control muscle from same animal. Values are means of 4 or more experiments in each column. *Hatched columns* are from test muscles. [*E*, electrical stimulation (soleus) in vivo, or for diaphragm through phrenic nerve in vitro. *K50*, muscle incubated 15 min prior to binding assay (4 h) in presence of 50 mM K^+

rial and Methods"). The results are presented in Fig. 4. It can be seen that, after 4 h of incubation, soleus, electrically stimulated before removal, achieves a greater maximum binding of the administered insulin ^{125}I dose than control while 50 m*M* [K^+] stimulation (membrane potentials; mean 57.5 ± 6.2 mV; $n = 73$) is less effective. Nevertheless, insulin exposure of the potassium-depolarized muscles ($n = 6$) prior to testing gives rise to a great enhancement of maximum binding, suggesting that, in the presence of ligand, the voltage change can increase the number of binding sites. Diaphragm shows less (average) IR response to insulin after electrical, but greater (after) potassium, stimulation. Another view of the potassium effect on soleus is presented in Fig. 5. Here, the plots are of specific insulin binding with changes in the concentration of "cold" insulin in the incuba-

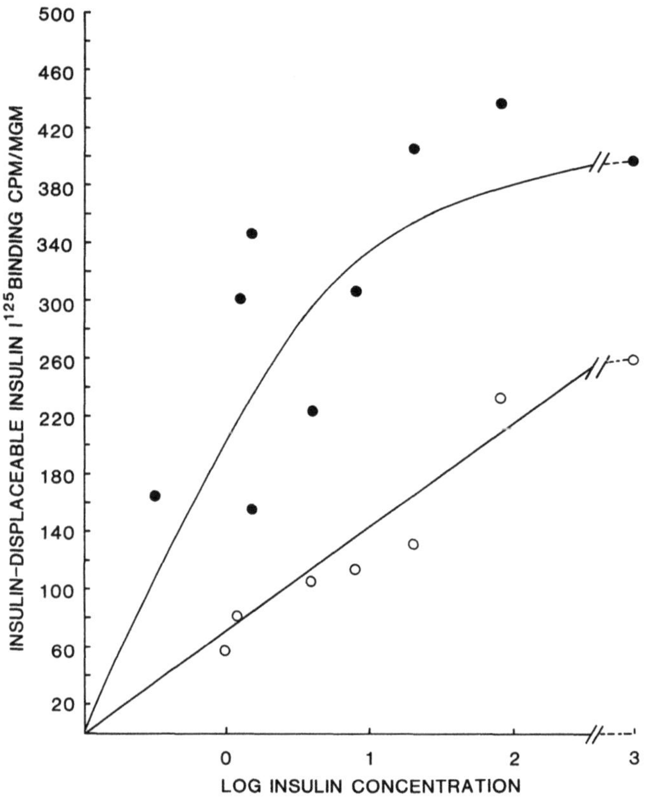

Fig. 5. Effects of muscle depolarization on insulin-binding affinity curves. Test muscles incubated 4 h in insulin-containing ringer at concentrations shown. Nonspecific binding subtracted from total in all cases to give plots of insulin-displaceable binding. *Curves* fitted by eye to values obtained in mouse hemidiaphragms. Results with 50 mM [K^+] in medium (●) and controls (○). The depolarization caused by potassium excess is associated with a shift-of-affinity curve, *left*, indicating that a lower concentration of insulin could have a greater effect

tion medium. It can be appreciated that, after potassium-depolarization stimulation (50 mM K^+ present in final incubating fluid), the binding affinity curve of soleus is shifted to the left. The maximum binding is also enhanced. These results suggest that one prompt and enduring effect of the stimulation of a muscle is increased complexing with insulin. This response need not involve ACh or contraction, as it is seen also with excess potassium, and at least one aspect includes an increase in affinity of the hormone receptor for its ligand.

4. *Effects of IR on AChR*. The contrast between the augmenting effects of depolarization on the insulin receptor and the well-known action of such stimulation to reduce extrajunctional AChRs [23] indicated that neuromuscular electrophysiological activity did not affect the two molecular populations in the same way. The question arose whether only one population of receptors normally re-

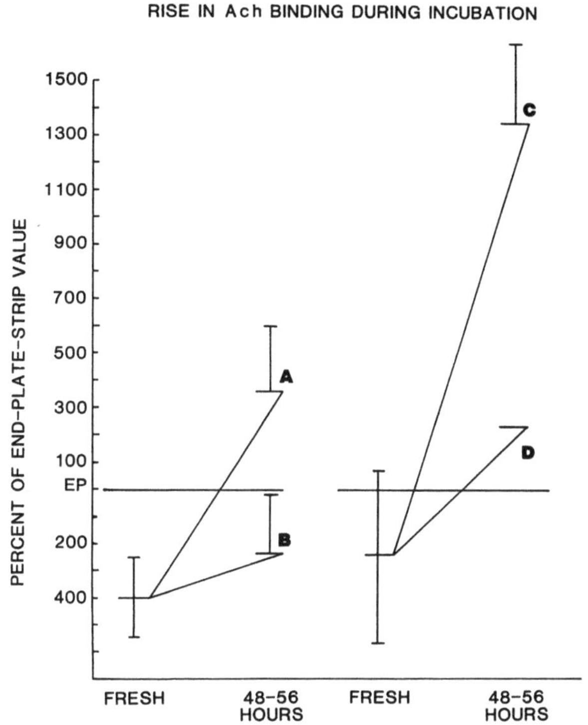

Fig. 6. Effects of insulin on development of extrajunctional AChR after denervation. *Curves* fitted by eye of mean values. *Bars* are ranges. *EP* represents end-plate values as 100%. Values below *EP* line indicate that non-end-plate region has less bungarotoxin binding, by the percentage shown, than end-plate area. Data in freshly removed diaphragm strips compared with those after incubation. *Curves* **A** and **B** by the end-plate strip method (see text). **C** and **D** are from identically treated strips, but with "specific" binding estimated by subtracting values in large excess of "cold" toxin from total. **B** and **D**, with insulin, ca. 1 μM. In all cases, incubation of denervated muscle with insulin clearly reduced the number of bungarotoxin-reactive binding sites

sponds to trophic effects and then regulates other membrane components (AChR, etc.) in the manner of a "master" regulator-receptor. To test this idea, extrajunctional AChR was allowed to develop in cultured muscles, either with or without insulin. These muscles also had the expected rise in IR, and the hypothesis was that, on exposure to an insulin excess, the latter receptors would become fully saturated and thus would function in the same way as if they had developed high binding affinity after stimulation. Any change in AChR under these conditions would suggest a direct regulatory effect of the increased insulin binding and metabolism. The results are seen in Fig. 6. Whether the "specific" binding of labelled bungarotoxin to AChR is measured by comparative counts of an end-plate-containing strip of diaphragm and a non-end-plate area or an end-plate strip exposed to excess "cold" toxin, the results are in the same direction. The non-end-plate area of fresh muscles always gave lower counts than the end-plate region, but, after incubation in tissue culture medium for from 2–4 days, AChR radioactivity in the extrajunctional portions of the muscle consistently rose to values greater than in the central strip. In confirmation of other workers' results [33], it was found in a few experiments at 96–100 h that the absolute number of counts per mg of the end-plate region also rose (not shown). The major effect of incubation in insulin (ca. 1 μ*M*) was to reduce the amount of bungarotoxin binding considerably. This change is about equal to the suppression of AChR seen in other studies using electrical stimulation of muscles [21, 23, 24].

Discussion

The effects of insulin observed in these experiments suggest that the hormone itself could be at least partly effective in substituting for the "trophic" influence of a motor nerve. There is, however, absolutely no reason to think that denervation, or any of the procedures used here, would have reduced the availability of circulating insulin, glucose, or any precursor materials needed for maintenance and growth. Moreover, since all insulin's actions appear to be receptor-mediated, the "trophic" effects found here for insulin should be viewed as those of another family of molecules, namely, the insulin receptors. A material capable of controlling these receptors would fulfill most of the requirements for a trophic factor listed above. The motor nerve thus would not need to bring "nourishment" to the muscle; it could simply regulate the delivery of what is needed from the extracellular fluid by means of a metabolic transducer located in abundance on the postsynaptic surface. The metabotrophic energy-gating system, on this view, operates in parallel with the purely ionotrophic (ACh) system and is designed to accommodate abrupt changes in requirements as a result of incoming impulses.

It would be attractive to consider that whatever substance from the nerve can regulate the insulin receptor might simply have joint effects on the AChR, both sets of glycoproteins being up- and down-regulated by a single material. Such an inference, however, cannot be made in the face of two essential observations. First, direct or indirect muscle stimulation has very different effects on AChR and IR. Second, insulin itself, by combining with its own receptor, can reduce the amount of extrajunctional AChR when no trophic substance is available. In terms of stimulation results, consistent reduction of AChR, as well as restoration of RMP, is well documented [34]. Electrical or potassium stimulation in the present experiments has, on the other hand, been found to increase not only the affinity of the IR but also the maximum number of binding sites. This rise in receptor affinity means that, for a given hormone concentration, more complexes can be formed and the muscle can extract more fuel from the blood. The increased availability of metabolic energy could permit more activity in the active ion-pumping systems and thus restore both membrane excitability and the contraction properties of the muscle. Similar increases in IR affinity with acute stimulation in tissues of experimental animals and man have been reported [18, 30] and, as in the present studies, the duration of these effects long outlasts the stimulus [18]. In the same way, the exercised diabetic may need less insulin for days. It is of interest that muscle contraction may not be responsible for stimulation effects. The ability of a steady potassium depolarization to affect a property of the IR is paralleled by similar effects of x-irradiation on the same molecule [17] and suggests that "field effects" in the membrane may be of importance in the response to stimulation. These local effects could alter a subunit of the receptor complex, leaving others unchanged [9, 10, 22].

The second argument against a neurotrophic substance acting jointly on AChR and IR arises in light of the effects of excess insulin in vitro on AChR. The question is why an increase in ligand for one receptor should reduce the availability of binding sites for an entirely different substance, in this case, the neuromuscular transmitter. One explanation might be that IR, through its metabotrophic capabilities, simply regulates the turnover and replacement of AChR in the membrane. Many experiments confirm that both synthesis and degradation of AChR require energy, and it is possible that loss of the "trophic" substance from the nerve differentially affects the two processes, allowing synthesis to run unchecked, while degradation is unchanged [15]. The inference to account for insulin's action on AChR in this case is that the energy provided by the hormone's reaction with its receptor works in the opposite way, limiting AChR synthesis and leaving breakdown as before. The similarity of the actions of electrical stimulation and insulin on denervated muscle could then be viewed as the net result of more IR complexing (increased affinity) or saturation (more ligand) of all available sites for energy transduction. Since both stimulation and insulin can increase the influx of Ca^{2+} ions and since Ca^{2+}, insulin, and the calcium ionophore A23187 normalize RMP and AChR in denervated muscle [4], it is possible that calcium is the metabolic "second messenger".

A post-denervation rise in the number of insulin receptor sites on muscle, theoretically adequate to enhance all energy-requiring activities, is still not enough to sustain the fibers, and the question is, why not. The density of transmitter receptors (AChR) and membrane electrophysiological properties can be moved toward normal by IR interactions, if enough complexes are formed, but functions deeper in the sarcoplasm are not fully maintained. It may be that the velocity of important intracellular biochemical processes is not directly susceptible to "trophic" influences having major effects on and within the membrane. Post-receptor reactions are known to be subject to rate-limiting steps which may not apply on the surface [27]. From the present experiments it is possible to infer another more direct explanation for the failure of excess IR to maintain a denervated muscle in normal condition. What seems most reasonable is that the large number of unmasked or newly synthesized insulin-binding fragments are not metabolically active. From structural analysis by other workers on the IR unit, it is known that the total receptor complex is made up of several discrete components [9, 10, 17, 22]. The results here suggest that, when newly inserted into the membrane, the various parts are not coupled for activity and the final arrangement for maximum sensitivity and efficiency requires some form of membrane depolarization, or muscle activity, or both. The increased binding affinity demonstrated here after stimulation may reflect this cooperative action, though a shift of the binding curve to the left does not necessarily confer greater metabolic effect.

The trophic effect of nerve can be considered, generally, as a combination of its ability to control in the resting state the fuel and supplies for basic repair, growth, turnover, and activity of all muscle components. In light of present results, this control is mediated through a direct effect on the number of insulin-reactive sites on the membrane. Activity in the form of impulses would affect mainly the affinity of the insulin receptor (inactivity – low binding), or the rate of whatever intrafiber insulin-sensitive systems were activity- or Ca^{2+}-dependent. The lack of critical balance between these functions in experimental situations may account for failure to prevent all denervation effects.

The experiments with botulinium toxin show that it affects something besides ACh release from the nerve terminal. Lack of this substance, or interference by colchicine with its delivery, clearly interferes with some nerve property that regulates insulin-binding sites. Taken together, these latter findings argue in favor of the theory that the nerve does liberate a controlling material affecting a number of muscle membrane properties. That the effects of a "trophic" element can be mimicked by a circulating hormone strongly suggests that control is mediated by that hormone's receptor.

By way of speculation, it might be supposed that insulin-like substances would behave the same way. It has already been shown, for example, that concanavalin A has insulinomimetic actions [8] and inhibits AChR on neurones [25]. Preliminary studies with 7-S NGF [12] show that, after a 48-h incubation period, the NGF-treated mouse diaphragm developed 43% fewer bungarotoxin binding sites than incubated controls and only 26% more than freshly excised muscle. It is suggested that future studies of neurotrophic relationships should include the search for neurally released substances which can alter the number, the affinity, the efficiency, or the post-combination metabolic effects of the insulin receptor. Such materials, though not directly controlling the intensively studied post-denervation chemosensitivity itself, may govern more subtle reactions, the failure of which leads both to the observed membrane changes and the ultimate fiber atrophy and death.

Summary

This study deals with the metabotrophic interaction between a motor nerve and the muscle it supplies. The data suggest that the metabolic and structural integrity of the muscle is maintained in part by a surface macromolecular complex, the insulin receptor (IR), and that the amount of the latter is regulated by some factor delivered by axoplasmic flow to the nerve terminals. After denervation there is, as with the transmitter receptor (AChR), a rise in IR, but the binding sites are not normally effective. The increase in IR may be considered compensatory, but it is either too brief or inefficient to preserve the muscle.

---

[12] Purified 7-S Nerve Growth Factor. Hy-Clone Tissue Culture Products, Logan, UT 84321, USA.

References

1. Albuquerque EX, Thesleff S (1968) A comparative study of membrane properties of innervated and chronically denervated fast and slow muscles of the rat. Acta Physiol Scand 73:471–480
2. Albuquerque EX, Warnick JE, Tasse JR, Sansone FM (1972) Effects of vinblastine and colchicine on neural regulation of the fast and slow skeletal muscles of the rat. Exp Neurol 37:607–634
3. Bray JJ, Harris AJ (1975) Dissociation between nerve-muscle transmission and nerve trophic effects on rat diaphragm using type D botulinum toxin. J Physiol (Lond) 253:53–77
4. Bray JJ, Forrest JW, Hubbard JI (1982) Evidence for the role of non-quantal acetylcholine in the maintenance of the membrane potential of rat skeletal muscle. J Physiol (Lond) 326:285–296
5. Bretag AH (1969) Synthetic interstitial fluid for isolated mammalian tissue. Life Sci 8:319–329
6. Cangiano A, Magherini PC, Pasino E, Pellegrino M, Risaliti R (1984) Interaction of inactivity and nerve breakdown products in the origin of acute denervation changes in rat skeletal muscle. J Physiol (Lond) 355:345–365
7. Chang CC, Chuang S-T, Huang MC (1975) Effects of chronic treatment with various neuromuscular blocking agents on the number and distribution of acetylcholine receptors in the rat diaphragm. J Physiol (Lond) 250:161–173
8. Cuatrecasas P, Tell GPE (1973) Insulin-like activity of concanavalin A and wheat germ agglutinin-direct interactions with insulin receptors. Proc Nat Acad Sci USA 70:485–489
9. Czech MP (1977) Molecular basis of insulin action. Ann Rev Biochem 46:359–384
10. Czech MP (1979) Insulin action and the regulation of hexose transport. Diabetes 29:399–409
11. Drachman DB (1967) Is acetylcholine the trophic neuromuscular transmitter? Arch Neurol 17:206–218
12. Drachman DB (1974) Trophic functions of the neuron. Ann NY Acad Sci 228:1–423
13. Drachman DB (1976) Trophic interactions between nerves and muscles: the role of cholinergic transmission (including usage) and other factors. In: Goldberg AM, Hanin I (eds) Biology of cholinergic function. Raven, New York, pp 161–186
14. Drachman DB, Johnston DM (1975) Neurotrophic regulation of dynamic properties of skeletal muscle: effects of botulinum toxin and denervation. J Physiol (Lond) 252:657–667
15. Fambrough DM (1979) Control of acetylcholine receptors in skeletal muscle. Physiol Rev 59:165–227
16. Gutmann E (1976) Neurotrophic relations. Physiol Rev 38:177–216
17. Harmon JT, Kahn CR, Kempner ES, Schlegel W (1980) Characterization of the insulin receptor in its membrane environment by radiation inactivation. J Biol Chem 255:3412–3419
18. Heat GW, Gavin JR III, Hinderliter JM, Hagberg JM, Bloomfield SA, Holloszy JO (1983) Effects of exercise and lack of exercise on glucose tolerance and insulin sensitivity. J Appl Physiol 55:512–517
19. Hofmann WW, Thesleff S (1972) Studies on the trophic influence of nerve on muscle. Eur J Pharmacol 20:256–260
20. Hofmann WW, Nakano D (1981) Control of muscle insulin receptors by the motor nerve. Experientia 37:286–287
21. Jones R, Vrbova G (1974) Two factors for the development of denervation hypersensitivity. J Physiol (Lond) 236:517–538
22. Kasuga M, Karlsson FA, Kahn CR (1982) Insulin stimulates the phosphorylation of the 95,000 Dalton unit of its own receptor. Science 215:185–186
23. Lømo T, Rosenthal J (1972) Control of ACh sensitivity by muscle activity in the rat. J Physiol (Lond) 221:493–513
24. Lømo T, Westgaard RH (1975) Further studies on the control of ACh sensitivity by muscle activity in the rat. J Physiol (Lond) 252:603–626
25. Messing A, Bizzini B, Gonatas NK (1984) Concanavalin A inhibits nicotinic acetylcholine receptor function in cultured chick ciliary ganglion cells. Brain Res 303:241–249

26. Nicholls JG (1956) The electrical properties of denervated skeletal muscle. J Physiol (Lond) 131:1–12
27. Olefsky JM (1981) Insulin resistance and insulin action. Diabetes 30:148–162
28. Pestronk A, Drachman DB, Stanley EF, Price DL, Griffin JW (1980) Cholinergic transmission regulates extrajunctional acetylcholine receptors. Exp Neurol 70:609–696
29. Purves D, Sakmann B (1974) The effect of contractile activity on fibrillation and extrajunctional acetylcholine sensitivity in rat muscle maintained in organ culture. J Physiol (Lond) 237:157–182
30. Richter EA, Garetto LP, Goodman NM, Ruderman NB (1972) Muscle glucose metabolism following exercise in the rat. J Clin Invest 69:785–793
31. Stanley EF, Drachman DB, Pestronk A, Price DL (1980) The role of ACh transmission in the neural regulation of muscle resting membrane potential. Neurosci Abstr 6:384
32. Thesleff S (1960) Effects of motor innervation on the chemical sensitivity of skeletal muscle. Physiol Rev 40:734–752
33. Uchitel O, Robbins N (1978) On the appearance of acetylcholine receptors in denervated rat diaphragm and its dependence on nerve stump length. Brain Res 153:539–548
34. Westgaard RH (1975) Influence of activity on the passive electrical properties of denervated soleus muscle fibers in the rat. J Physiol (Lond) 251:683–697
35. Zierler KL (1972) Insulin, ions, and membrane potentials. In: Handbook of physiology-endocrinology. American Physiol Soc, Washington D.C., vol 1, pp 347–408

Persistent Depolarization of Muscle Fibers: A Common Cause of Weakness in Muscle Disorders *

F. LEHMANN-HORN [1] and G. KÜTHER [1]

There exists a group of muscular diseases, called episodic paralyses, which is characterized by transient hyperexcitability followed by flaccid weakness. Included in these myopathies are the hypokalemic periodic paralysis (mostly without any signs of hyperexcitability), the hyperkalemic periodic paralysis and paramyotonia congenita. In all these diseases, a transient, fully reversible weakness can be triggered by specific provocative procedures [6].

In hypokalemic periodic paralysis, weakness is provoked by oral application of carbohydrates or prolonged inactivity after strenous work, or by exposure to cold environment. During the attacks, an abnormally low serum potassium concentration is found.

In hyperkalemic periodic paralysis, weakness is accompanied by an excessive increase of serum potassium. The attacks develop during rest after strenous work or by oral potassium administration.

In paramyotonia congenita, cooling of skeletal muscle leads to stiffness and subsequent weakness. The weakness lasts several hours even when the muscles have been rewarmed [5].

Experimental results have provided evidence that in all these diseases the underlying defect is myogenic in nature and not due to alterations of the motor nerves or defects in neuromuscular transmission mechanisms. The contractile apparatus also is intact. Therefore, the muscle fiber membrane is a likely location for the pathological process. To test this hypothesis we recorded membrane potential and the membrane conductances of muscle fibers excised intact from patients with these three diseases. The muscle bundles were placed in an experimental chamber. Thus we were able to change the concentration of the extracellular fluid. For example, we varied the external potassium concentration, decreased the temperature, or added tetrodotoxin (TTX) to block Na channels. Simultaneously we measured the isometric tetanic force elicited by supramaximal stimulation.

The biopsied intercostal muscle bundles showed the same phenomena as observed in vivo [3, 4, 7]:

1. Muscle bundles from *hypokalemia patients* had a decreased isometric force when exposed to a low potassium medium.
2. Muscle bundles from *hyperkalemia patients* became paralyzed when exposed to a high potassium solution. Under these conditions normal muscle fibers only showed a slight decrease of tetanic force.
3. Fibers from *paramyotonia patients* developed weakness after cooling.

---

* This study was conducted in collaboration with R. Rüdel, Abteilung für Allgemeine Physiologie der Universität Ulm, and K. Ricker, Neurologische Universitätsklinik, Würzburg.

[1] Neurologische Klinik und Poliklinik der Technischen Universität München, Möhlstr. 28, D-8000 München 80, FRG.

Clinical Aspects of Sensory Motor Integration
Ed. by A. Struppler and A. Weindl
© Springer-Verlag Berlin Heidelberg 1987

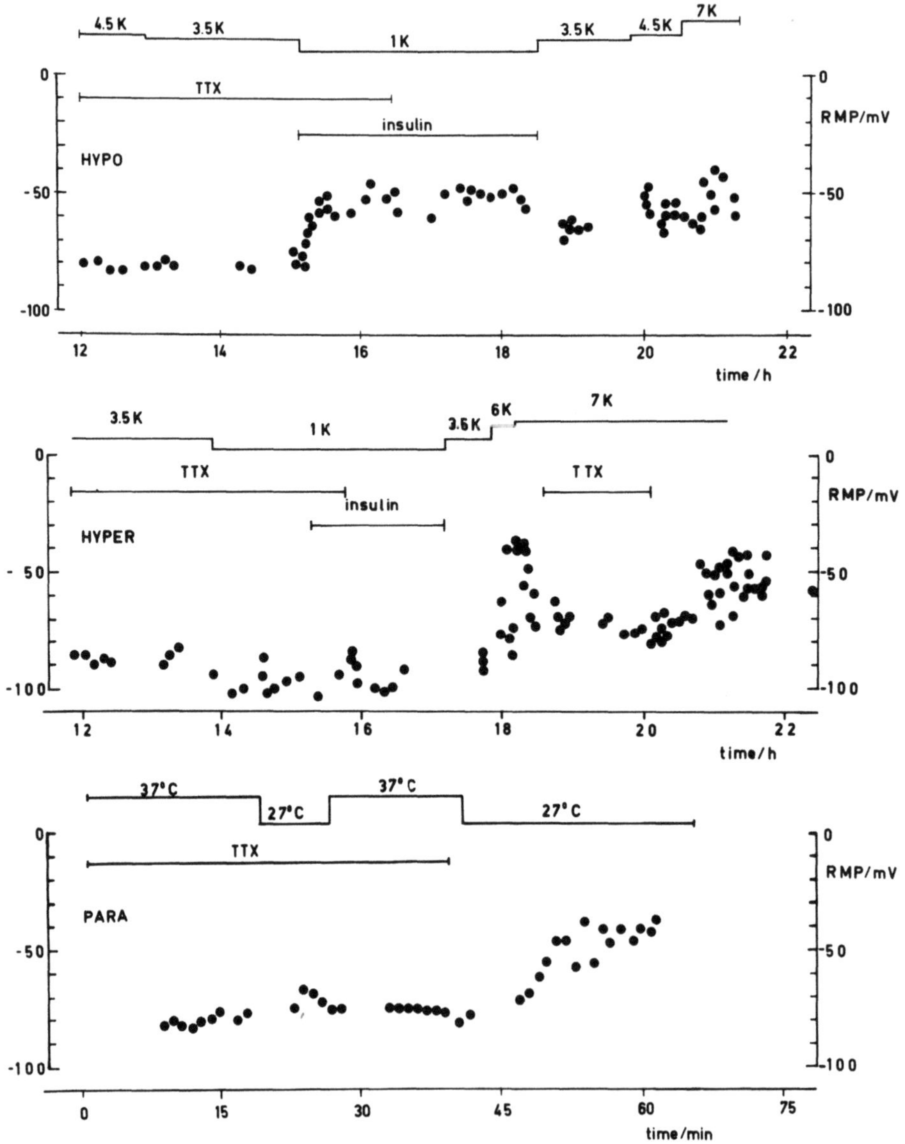

Fig. 1. Resting membrane potentials (*RMP*) in mV of muscle fibers from patients with hypokalemic periodic paralysis (*HYPO*), hyperkalemic periodic paralysis (*HYPER*) and paramyotonia congenita (*PARA*)

Figure 1 shows resting membrane potentials when the relevant parameters were changed. The upper panel is from a patient with hypokalemic periodic paralysis. At normal extracellular potassium concentration, almost normal resting potentials were recorded. A challenge with 1 m*M* potassium plus insulin led to a marked depolarization to approximately −55 mV. Increasing the extracellular potassium to normal (3.5–4.5 m*M*) or even higher values (7 m*M*) did not restore the resting potential.

In the second panel, resting potential of fibers from a patient with hyperkalemic periodic paralysis are plotted. The same treatment, 1 m*M* potassium plus insulin, caused a hyperpolarization as in normal fibers. However, when the potassium concentration was increased to 6 or even 7 m*M*, the fibers excessively depolarized to approximately −50 mV. Tetrodotoxin restored the membrane potential to values which can normally be expected for 7 m*M* potassium concentrations.

In paramyotonia congenita, the relevant challenge is a lowering of the temperature as shown in the third panel. At a temperature of 27 °C, all fibers were depolarized below −50 mV. This decrease could be prevented by tetrodotoxin. Under all these conditions, normal fibers show only slight alterations of the membrane potential (Table 1).

The question to be answered is: Can the persistent depolarization explain the paralysis? One possible answer can be suggested, if the extent of the depolarization is compared with the threshold potential for excitation of the fiber membranes and if the inactivation of ion currents is considered. The threshold potential for excitation in muscle fibers is about −60 mV [2]. Therefore the abnormal depolarization may initially lead to transient spontaneous activity. As the depolarization continues, muscle fibers become inexcitable and, as a consequence, weakness occurs.

The next question to be answered concerns the reason for the abnormally large depolarization seen in the diseased fibers. In principle, a depolarization could be explained by an increased sodium conductance, a decreased potassium conductance or an inhibition of the sodium-potassium pump. Furthermore, instabilities of the membrane potential can occur when the chloride conductance is decreased, as in myotonia congenita. By voltage clamping of individual excised muscle fibers the component conductance for the different ions has been determined. In none of the three disorders was there evidence for a pathological alteration of the potassium or chloride conductances or a disturbance of the sodium-potassium pump. These findings suggest that an abnormally large sodium current may be the cause of the depolarization. More direct evidence for such a sodium current was obtained in experiments in which the sodium conductance was pharmacologically blocked by tetrodotoxin or by local anesthetics.

In paramyotonic fibers, both TTX and tocainide, a local anesthetic, prevented the occurrence of membrane depolarization. In hyperkalemic fibers only TTX was able to prevent or restore the abnormal, potassium-induced depolarization.

Table 1. Resting membrane potentials (*mV*) of muscle fibers (*n*) from patients with hypokalemic periodic paralysis (*HYPO*), hyperkalemic periodic paralysis (*HYPER*), paramyotonia congenita (*PARA*), and healthy controls

Solutions	HYPO	HYPER	PARA	Controls
3.5 mM K, 37° C	−74.9 (5)	−82.1 (3)	−82.3 (7)	−83.3 (15)
3.5 mM K, 27° C	−49.8 (1)	–	−40.3 (7)	−77.9 (15)
1.0 mM K, 37° C	−54.9 (5)	−93.4 (1)	–	−99.6 (15)
7.0 mM K, 37° C	–	−51.9 (3)	−71.1 (1)	−69.0 (15)

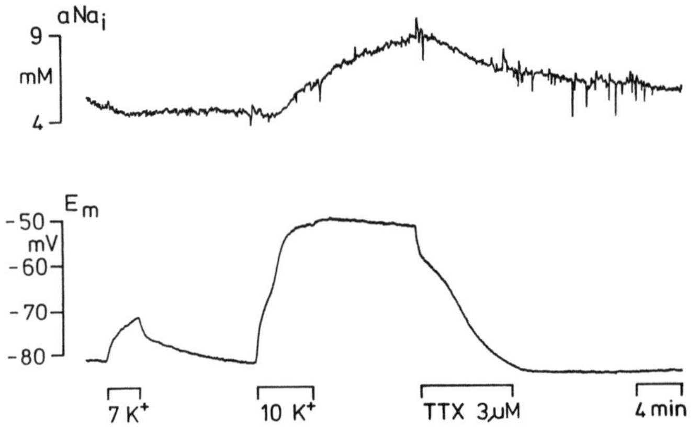

Fig. 2. Changes of both intracellular sodium activity (*upper trace*) and resting membrane potential (*lower trace*) elicited by an increase of the extracellular potassium concentration of muscle fibers obtained from a patient with hyperkalemic periodic paralysis

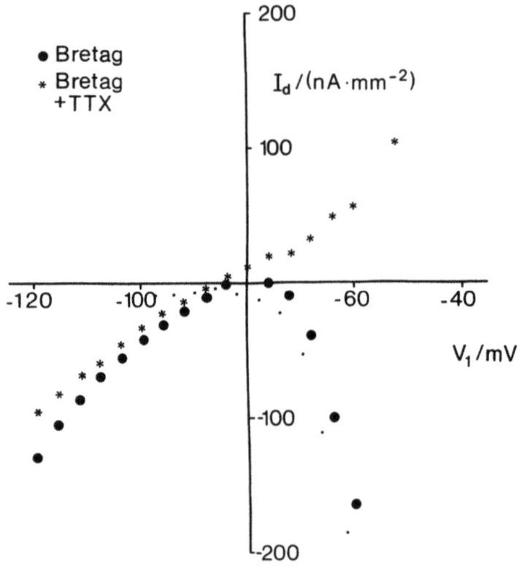

Fig. 3. Membrane current density – membrane potential relationship of muscle fibers from a patient with hyperkalemic periodic paralysis in normal solution with (*) and without TTX (●). The difference between the currents determined in these both solutions is also shown (·)

Therefore, at least in these two disorders a pathological sodium conductance can be assumed. No effect of TTX or tocainide was observed in hypokalemic fibers.

Recently, Grafe et al. [1] were able to measure the intracellular ion activities of fibers of a patient with hyperkalemic periodic paralysis. Exposure of these fibers to high potassium concentrations led to an abnormal increase of the intracellular sodium activity (Fig. 2). In normal fibers, no change of the sodium activity could be detected. However, the voltage dependence of this sodium current could be determined by voltage-clamping measurements of fibers from the same patient (Fig. 3). In this figure, the time-independent membrane currents are plotted against the membrane potential. Normal fibers or tetrodotoxin-treated hyperkalemic fibers show the upper characteristic curve. In contrast to solutions containing tetrodotoxin, an additional inward current can be seen in normal solutions starting at membrane potentials of −80 mV.

Conclusions

Good evidence was found for an alteration of the sodium channels at least in paramyotonia congenita and hyperkalemic periodic paralysis. In both diseases, a non-inactivation of some sodium channels keeps the fibers depolarized. Tetrodotoxin was able to prevent the depolarization or even to restore the membrane potential.

Non-inactivating sodium channels could belong either to an altered channel type resulting from a defect in the genetic code for the channels, or to a physiological type abnormally increased in number. Apparently the effect of the local anesthetics is linked to the presence of an inactivation gate. The result that the lidocaine derivative tocainide is without therapeutic effect in hyperkalemic periodic paralysis may suggest an absence of the inactivation gate in some Na channels. In contrast to paramyotonia congenita and hyperkalemic periodic paralysis, our concepts of the membrane defect in hypokalemic periodic paralysis are less concrete. But in principle, the applied electrophysiological technique should also help to elucidate the defects in this disease.

In summary, the results presented here show that long-lasting depolarization may be a common cause for transient weakness in different myopathies. It may be assumed that this concept can be applied to other muscle diseases.

Summary

There exists a group of muscle diseases (hypokalemic periodic paralysis, hyperkalemic periodic paralysis, and paramyotonia congenita) in which weakness can be triggered by various clinical tests. We investigated intact intercostal muscle biopsies from patients with these myopathies. In vitro, similar triggers as those used clinically (i.e. cooling, altered extracellular potassium concentrations) caused a weakness (decreased amplitudes of twitch and tetanic contractions) in these muscle bundles. In each myopathy, a membrane depolarization war correlated to the triggered weakness. Abnormal functioning of the sarcolemmal sodium channels appear clearly to be the cause of the depolarization in: hyperkalemic periodic paralysis muscle fibers, which showed a non-inactivating sodium current and an increase of the intracellular sodium activity; and in paramyotonic fibres, in which an abnormal sarcolemmal current-voltage relationship was observed after cooling. Abnormal sodium conductance is also suggested to be the underlying defect and the cause of the depolarization in hypokalemic periodic paralysis muscle fibers, due to the abnormal sensitivity of these fibers to low concentrations of extracellular potassium.

References

1. Grafe P, Ballanyi K, Küther G (1985) Intracellular Na^+-activity in normal intercostal muscle and in one case of hyperkalemic periodic paralysis. Pflügers Arch 403:R58
2. Kwiecinski H, Lehmann-Horn F, Rüdel R (1984) The resting parameters of human intercostal muscle at low, normal, and high extracellular potassium. Muscle Nerve 7:60–65
3. Lehmann-Horn F, Rüdel R, Dengler R, Lorkovic H, Haass A, Ricker K (1981) Membrane defects in paramyotonia congenita with and without myotonia in a warm environment. Muscle Nerve 4:396–406
4. Lehmann-Horn F, Rüdel R, Ricker K, Lorkovic H, Dengler R, Hopf HC (1983) Two cases of adynamia episodica hereditaria: in vitro investigation of muscle cell membrane and contraction parameters. Muscle Nerve 6:113–121
5. Lehmann-Horn F, Höpfel D, Rüdel R, Ricker K, Küther G (1985) In vivo P-NMR spectroscopy: muscle energy exchange in paramyotonia patients. Muscle Nerve 8:606–610
6. Rüdel R, Lehmann-Horn F (1985) Membrane changes in cells from myotonia patients. Physiol Rev 65:310–356
7. Rüdel R, Lehmann-Horn F, Ricker K, Küther G (1984) Hypokalemic periodic paralysis: in vitro investigations of muscle fiber membrane parameters. Muscle Nerve 7:110–120

IV. Convergence on the Final Common Path

Ultrastructural Analysis of Target-Dependent Properties of Mammalian Motoneurones

T. A. SEARS [1], I. P. JOHNSON [1], and A. H. PULLEN [1]

Introduction

Mature motoneurones demonstrate a major dependence on the periphery for normal maintenance, as revealed through their retrograde response to axotomy, interruption of axonal transport, or blockade of neuromuscular transmission [8, 12, 20]. Likewise the immature motoneurone is dependent on a maintained functional contact with the muscle fibres it innervates for its differentiation and survival [4]. Immature motoneurones die on losing contact with their muscle targets following motor nerve crush (i.e. axotomy; cf. [2]), while fully mature motoneurones survive under similar circumstances providing they regain functional contact with their targets following axonal regeneration [16]. To investigate further this target-dependence of motoneurones, we have used the paradigm of reversible axotomy (nerve crush) or chronic axotomy (nerve section with proximal ligation) of intercostal nerves in adult cats to study changes in Nissl-body ultrastructure as a measure of altered protein synthesis. This approach follows our recent experience with the topographically distinct Nissl body that is located postsynaptically and immediately subjacent to the C-type synapse and its subsynaptic cistern [14, 15]. With the chronic partial central deafferentation that occurs following spinal hemisection, the presynaptic axon terminal of the C-type synapse selectively hypertrophies, and this presynaptic response is accompanied by an increase in size and a change in the ribosomal organisation of the postsynaptic Nissl body [15]. Since the synthesis of particular classes of protein has been associated with particular forms of ribosomal organisation [11], functional correlates of altered protein synthesis can be inferred from changes in the ribosomal organisation of Nissl bodies. This approach has now been extended to the analysis of Nissl bodies sited in the general cytoplasm of normal and axotomised motoneurones.

Materials and Methods

Animals

Nine adult cats, including two normal animals (controls), were examined; anaesthesia was by intraperitoneal injection of sodium pentobarbitone (45 mg/kg). Control animals were not subjected to operative procedures and provided normal spinal cord tissue.

[1] Sobell Department of Neurophysiology, Institute of Neurology, National Hospital for Nervous Diseases, Queen Square, London WC1N 3BG, UK.

Clinical Aspects of Sensory Motor Integration
Ed. by A. Struppler and A. Weindl
© Springer-Verlag Berlin Heidelberg 1987

General Protocol

Intercostal nerves were exposed in two non-adjacent segments 20 mm from the dorsal mid-line and either (a) crushed with fine forceps, or (b) tightly ligated, sectioned, and with a 5 mm portion of the nerves removed distally. Axotomised motoneurones were examined 1, 4, 8, 33, 64, and 208 days later at both light and electron-microscope levels.

Labelling Axotomised Motoneurones

In order to provide independent proof that motoneurones had been axotomised, the retrograde transport of horseradish peroxidase (HRP) was used to label them. For the 1-day survival period, HRP (Sigma Type VI; 40% in saline) was applied directly to the site of the original axotomy; for longer survival periods, animals were allowed to recover and were anaesthetised 24 h prior to perfusion. The intercostal nerves were re-exposed and newly lesioned 2 mm proximal to the original lesion for application of HRP.

Perfusions

Animals were perfusion-fixed with 2.5% glutaraldehyde plus 2% paraformaldehyde in 0.2 *M* cacodylate buffer, pH 7.4 (1 "control") or 2% glutaraldehyde plus 1% paraformaldehyde in 0.1 *M* phosphate buffer, pH 7.4 (1 "control", all axotomised-HRP labelled).

Processing (Non-Labelled Tissues)

Non-labelled spinal cord was cut into 1 mm transverse slices, post-fixed in 1% osmium tetroxide, dehydrated in ascending ethanols, equilibrated with Araldite using propylene oxide and embedded in pure Araldite. Slices were subsequently used to prepare both semi-thin (0.5 μm) plastic sections for light microscopy (LM), and ultrathin (silver) sections stained with uranyl acetate and lead citrate for electron-microscopy (EM).

Processing (HRP-Labelled Tissues)

Transverse (70 μm) slices of spinal cord cut on a Vibroslice vibrating-blade microtome were processed to demonstrate HRP activity using 3-3′diaminobenzidine [1] with cobalt intensification of reaction product. Reacted slices were osmicated, dehydrated, equilibrated with Araldite and flat-embedded in Araldite between two PTFE-coated microscope slides [17].

Identification of HRP-Labelled Motoneurones

HRP-labelled motoneurones were identified by LM in transilluminated 70 μm slices of spinal cord, and isolated within a trimmed mesa. Labelled cells reidentified in 0.5 μm toluidine-blue stained sections of the "mesa" and thin-sectioned for EM were subsequently analysed. Data were obtained from complete photomontages of each motoneurone with magnifications between 10,000 and 20,000 X, and selected higher magnifications provided details of ribosomal organisation.

In both control and axotomised motoneurones, representative Nissl bodies were selected by LM, and the *same* Nissl bodies were reidentified by EM.

In addition, specific morphometric analyses were performed; the results are reported elsewhere [6, 7] and will not be discussed here.

Results

Normal Nissl Bodies

Using classical criteria, normal Nissl bodies were identified in 0.5 μm plastic LM sections as irregular patches of cytoplasmic basophilia (Fig. 1 a). Adjacent ultrathin EM sections revealed that the ultrastructure of the same Nissl bodies com-

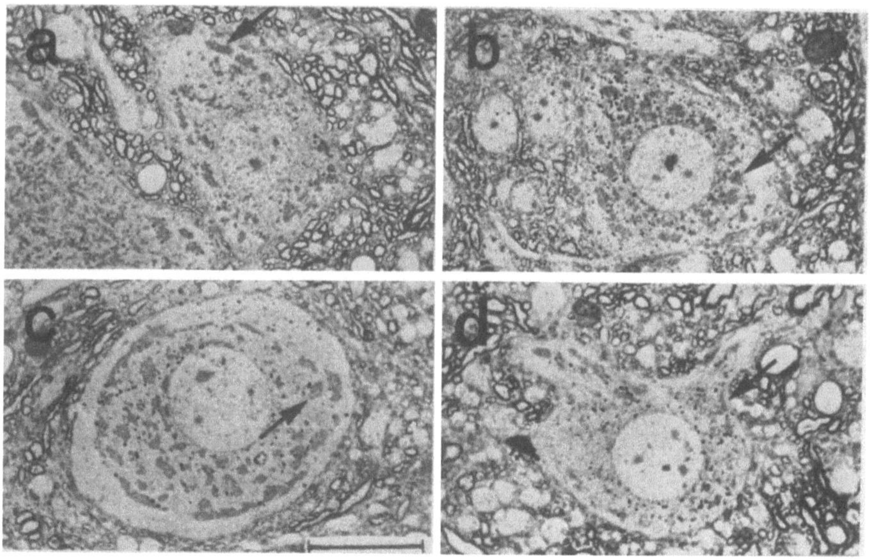

Fig. 1 a–d. 0.5 μm toluidine blue-stained sections of cat thoracic motoneuronal cell bodies. **a** Normal motoneurone with prominent Nissl bodies (*arrow*); **b** motoneurone 4 days following nerve crush, Nissl bodies (*arrow*) can still be seen; **c** Nissl bodies (*arrow*) within a motoneurone 64 days following nerve transection; **d** Nissl bodies (*arrow*) within a motoneurone 64 days following nerve crush (scale bar = 25 μm)

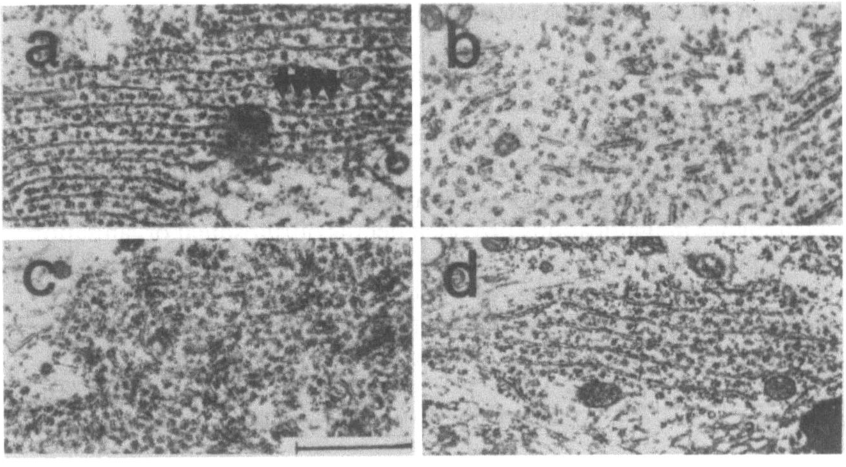

Fig. 2 a–d. Electron micrographs of the Nissl bodies arrowed in Fig. 1. **a** Normal Nissl body exhibiting highly organised rER lamellae and numerous interposed "lamellae-associated" polyribosomes (*arrows*); **b** abnormal Nissl body 4 days following nerve crush comprising an aggregate of polyribosomes interspersed with short rER fragments; **c** abnormal Nissl body 64 days following nerve transection; **d** normal Nissl body 64 days following nerve crush (scale bar = 1 μm)

prised highly ordered stacks of rough endoplasmic reticulum (rER) and linear arrays of polyribosomes between individual lamellae (Fig. 2a). These polyribosomes are described as 'lamellae-associated polyribosomes' to distinguish them from those polyribosomes found elsewhere and not necessarily in the vicinity of rER, and also the single ribosomes bound to the rER itself. This nomenclature we introduced previously for the C-type synapse [15].

Earlier Stages Following Axotomy (1–33 Days)

Nissl bodies with normal LM and EM structure were seen 1 day following either chronic or reversible axotomy. Apparently 'normal' Nissl bodies were still present by LM 4 days following reversible axotomy (Fig. 1 b), but by EM the same Nissl bodies had lost their normal, orderly ultrastructure. They completely lacked the stacks of parallel lamellae of rER and were composed, instead, simply of large aggregates of polyribosomes within which short fragments of rER were present (Fig. 2 b). By 4 days following chronic axotomy, LM showed early chromatolysis in which motoneurones exhibited diffuse cytoplasmic basophilia due to Nissl fragmentation and dispersion. By EM, these Nissl fragments consisted simply of *small* aggregates of polyribosomes; no ordered rER lamellae were present. Thus the diffuse basophilia seen by LM corresponded to a reduction in size of the polyribosomal aggregates close to the limits of resolution for LM, but not to Nissl body disorganisation per se. By LM, chromatolysis was fully developed between 8 and 33 days following chronic or reversible axotomy, the ultrastructure of the two types of axotomised motoneurones being indistinguishable.

Later Stages Following Axotomy (64–208 Days)

By LM, many motoneurones were still chromatolytic at 64–208 days, while others, irrespective of whether axotomy was chronic or reversible, had apparently reformed normal Nissl bodies (cf. [18]). However, by EM a striking difference in Nissl body ultrastructure was apparent. In chronically axotomised motoneurones, Nissl bodies were composed of dense aggregates of randomly sited polyribosomes, interspersed with short fragments of rER (Figs. 1c and 2c). In contrast, with reversible axotomy, Nissl bodies now displayed their normal ordered appearance due to the regeneration of long lengths of rER arranged in stacks and with individual lamellae separated by ordered linear arrays of lamellae-associated polyribosomes (Figs. 1 d and 2 d). From morphometric analyses we estimate that, for chronically axotomised motoneurones, the total length of rER within ribosomal aggregates is only 30%–40% of that organised as parallel lamellae in motoneurones subjected to reversible axotomy.

Examination of distal nerve stumps at 64 and 208 days confirmed that axonal regeneration had occurred following reversible axotomy but not with chronic axotomy.

Further Test of Target-Dependence of Nissl Bodies

To demonstrate further this dependence of the normal Nissl ultrastructure on the periphery, chronic axotomies of the external intercostal nerves were performed in two non-adjacent thoracic segments of the same animal. Animals were allowed to survive for an initial period of 49 days. At 49 days, one segment was selected for further experimentation (test segment); the other served as a control. Axons terminating in the neuroma of the external intercostal nerve of the test segment were re-sectioned and allowed to regenerate into the distal stump of a newly sectioned internal intercostal nerve of the same segment. When examined 64 days later, normal Nissl bodies had reformed within the HRP-identified external intercostal motoneurones (Fig. 3 a), but not in the still chronically axotomised motoneurones of the adjacent-but-one control segment (Fig. 3 b).

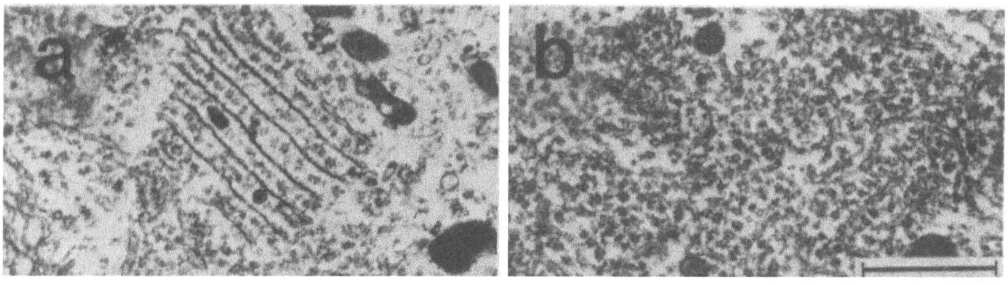

Fig. 3. a Normal Nissl ultrastructure with muscle reinnervation. **b** Abnormal Nissl ultrastructure with chronic axotomy (scale bar = 1 μm)

Discussion

Previous investigators of Nissl responses to axotomy have relied on LM evidence of chromatolysis or on ultrastructural evidence of Nissl body fragmentation to formulate hypotheses [8, 9], and as a result would have missed the mismatch described here between the results based on correlative LM and EM. In contrast, by utilising an independent label of axotomised motoneurones (HRP) we have identified two distinct forms of Nissl body ultrastructure: a "normal" one, comprising highly ordered stacks of rER lamellae with linear arrays of lamellae-associated polyribosomes between individual rER cisternae, and an "abnormal" one, which is disordered and comprises aggregates of free polyribosomes interspersed with short rER fragments. "Normal" Nissl ultrastructure characterises both normal motoneurones and motoneurones from 64 days post-axotomy when axonal regeneration is successful (reversible axotomy; crush lesions). "Abnormal" Nissl ultrastructure characterises axotomised motoneurones ending in a neuroma (chronic axotomy; nerve section with ligation). These differences in Nissl ultrastructure are unlikely to depend on the synaptic loss associated with axotomy (cf. [3]), since normal Nissl ultrastructure, including that of the C-type synapse, persists in the face of an extensive synaptic loss, due to central deafferentation [14], comparable in extent to the synaptic depletion following axotomy of intercostal motoneurones [6]. Nor is axonal elongation per se a sufficient condition for regeneration of "normal" Nissl ultrastructure since it was actively proceeding at 33 days when Nissl body regeneration was absent. This also discounts Schwann cells or the peripheral nerve sheath as the essential source of a signal for Nissl body regeneration, notwithstanding the role such signals play in axonal elongation [10, 13]. Thus "normal" Nissl ultrastructure appears to depend on either the presynaptic motor nerve terminal or the muscle fibre itself. In the former case, an inactive protein derived from the soma and transported orthogradely to the terminal might there be converted post-translationally (e.g. by transfer-RNA mediated aminoacylation; cf. [5, 21]) to serve, after retrograde transport, as a signal governing gene expression. The effect might be to induce rER synthesis de novo, or to act as a repressor of growth-associated proteins (GAPS) that characterise the regenerating state (see [22]), and thus signal a transition of metabolic state. Alternatively, a muscle-derived signal, either acting independently or through the presynaptic mechanism discussed above, would represent a further postsynaptic example of a target-dependent neurotrophic signal [19]. Whatever its source, the signal/s is/are likely to operate through transcriptional controls that codetermine, on the one hand, synthesis of additional rER, with its various membrane receptors (ribophorins, signal recognition proteins, etc.) that allow the co-translational insertion and vectorial discharge of membrane and secretory proteins; and on the other, the formation of cytoskeletal elements that create the highly ordered structure of the normal Nissl body. Such maturation might well depend on the appearance of a cross-linking neurofilament protein maintaining the orderly configuration of the rER analagous to the ^{H}polypeptide, whose appearance in the axon during development marks the transition from a state of cytoskeletal growth to one of cytoskeletal stability [22].

Summary

The effects on Nissl body (NB) ultrastructure of muscle reinnervation or neuroma formation were determined in HRP identified cat thoracic motoneurones subjected to axotomy by either nerve crush or nerve section with proximal ligation. Normal NB ultrastructure comprised highly ordered lamellae of rough endoplasmic reticulum (rER) with associated linear arrays of unbound polyribosomes. This ultrastructural orderliness was lost following axotomy, with or without evidence of chromatolysis in the light microscope. While NBs were present at late stages following both nerve crush and ligation, normal NB ultrastructure was only observed following nerve crush. An inductive effect of the periphery on NB ultrastructure is suggested and the implications of NB ultrastructure are discussed in relation to protein synthesis.

Acknowledgments. This work was supported by grants to T.A.S. from the International Spinal Research Trust and the National Fund for Research into Crippling Diseases.

References

1. Adams JC (1977) Technical considerations on the use of horseradish peroxidase as a neuronal marker. Neuroscience 2:141–145
2. Bearcroft CP, Lowrie MB, Peterson DC (1983) Selective effect of postnatal nerve crush on "fast" motor units in the rat. J Physiol (Lond) 338:12P
3. Blinzinger K, Kreutzberg G (1968) Displacement of synaptic terminals from regenerating motoneurones by microglial cells. Z Zellforsch Mikrosk Anat 85:145–157
4. Hamburger V (1977) The developmental history of the motor neuron. Neurosci Prog Res Bull 15:1–37
5. Ingoglia NA, Zanakis MF, Chakraborty G (1984) Transfer-RNA-mediated post-translational aminoacylation of proteins in axons. In: Elam JS, Cancalon P (eds) Axonal transport in neuronal growth and regeneration. Plenum, London (Advances in neurochemistry, vol 6, pp 119–136
6. Johnson IP (1983) Morphological correlates of altered protein synthesis: an ultrastructural analysis of axotomy and diphtheritic intoxication. PhD thesis, University of London, London
7. Johnson IP, Pullen AH, Sears TA (1986) Correlative optical and electronmicroscope analysis of Nissl body structure in cat thoracic motoneurones (to be published)
8. Lieberman AR (1971) The axon reaction: a review of the principle features of perikaryal responses to axon injury. Int Rev Neurobiol 14:49–124
9. Lieberman AR (1974) Some factors affecting the retrograde responses to axonal lesions. In: Bellairs R, Gray EG (eds) Essays on the nervous system. Clarendon, Oxford, pp 71–105
10. Lundberg G, Longo FM, Varon S (1982) Nerve regeneration model and trophic factors in vivo. Brain Res 232:157–161
11. Palade G (1975) Intracellular aspects of protein synthesis. Science 189:347–358
12. Pilar G, Landmesser L (1972) Axotomy mimicked by localised colchicine application. Science 177:1116–1118
13. Politis MJ, Ederle K, Spencer PS (1982) Tropism in nerve regeneration in vivo. Attraction of regenerating axons by diffusible factors derived from cells in distal stumps of transected peripheral nerves. Brain Res 253:1–12
14. Pullen AH, Sears TA (1978) Modification of C-synapse following partial central deafferentation of thoracic motoneurones. Brain Res 145:141–146

15. Pullen AH, Sears TA (1983) Trophism between C-type axon terminals and thoracic motoneurones in the cat. J Physiol (Lond) 337:373–388
16. Romanes GL (1946) Motor localisation and the effect of nerve injury on the ventral horn cells of the spinal cord. J Anat 80:117–131
17. Romanovicz DK, Hanker JS (1977) Wafer embedding: specimen selection in electronmicroscopic cytochemistry with osmiophilic polymers. Histochem J 9:317–327
18. Søreide AJ (1981) Variations in the axon reaction after different types of nerve lesion. Light and electron-microscopic studies on the facial nucleus of the rat. Acta Anat (Basel) 110:173–188
19. Varon S, Bunge RP (1978) Trophic mechanisms in the peripheral nervous system. Ann Rev Neurosci 1:327–361
20. Watson WE (1969) The response of motor neurones to intramuscular injection of botulinum toxin. J Physiol (Lond) 202:611–630
21. Willard M, Skene JHP (1982) Molecular events in axonal regeneration. In: Nicholls JG (ed) Repair and regeneration of the nervous system. Springer, Berlin Heidelberg New York (Dahlemer Konferenzen, pp 71–89)
22. Willard M, Skene JHP, Simon C, Meiri K, Hirokawa N, Glicksman M (1984) Regulation of axon growth and cytoskeletal development. In: Elam JS, Cancalon P (eds) Axonal transport in neuronal growth and regeneration. Plenum, London (Advances in neurochemistry, vol 6, pp 171–183)

Ultrastructural Analysis of C-Type Synapses in Thoracic Motoneurones of the Cat

A. H. PULLEN [1]

Introduction

The C-type synapse is one of five main morphological 'types' contacting cat thoracic alpha motoneurones. Its name derives from a morphologically distinctive 10–15 nm subsynaptic cistern which underlies its major length [5], and this is topographically associated with a subjacent contiguous postsynaptic Nissl body. Indirect evidence attributes the C-type to intrasegmental short-axon propriospinal pathways [6, 21, 22], a source directly proven in cat cervical and opossum trigeminal nuclei [11, 20]. C-type synapses characterise alpha motoneurones of the oculomotor [31], hypoglossal [4], phrenic [9], cervical [20], thoracic [24] and lumbosacral nuclei [5]. They do not occur on cat lumbosacral Renshaw cells [17] or thoracic gamma motoneurones [12], and recent preliminary studies of HRP-labelled cat alpha motoneurones innervating cat sphincter ani muscle have also failed to identify C-types despite their presence on neighbouring unlabelled large diameter motoneurones (Pullen, unpublished observations).

On the basis of a coupled hypertrophy of the presynaptic terminal, and an enlargment of its postsynaptic Nissl body in C-type synapses surviving partial central deafferentation by spinal hemisection, a mutual trophic interaction has been proposed between the C-type synapse and its postsynaptic motoneurone [25]. Arising from this, further morphological questions concerning the nature of the synaptic site(s) and synaptic vesicles have been examined [23] and are briefly presented here with a discussion of their implications.

Ultrastructural Analysis of the Synaptic Site

Definitive ultrastructural features of chemically mediated synapses include *focal* densities of the pre- and postsynaptic membranes which together are the presumed active synaptic site [29]. In aldehyde-fixed, osmicated cat spinal cord, C-type synapses display focal densities only in peripheral regions where no subsynaptic cistern occurs [5]. In contrast, aldehyde-fixed but non-osmicated C-type synapses exhibit discrete densities along the entire presynaptic membrane, but these are opposed postsynaptically, by a *continuous* density of uncertain relationship to focal densities [28]. This relationship has been examined in cat thoracic motoneurones stained selectively for synaptic sites using ethanolic phosphotungstic acid (EPTA) [3].

---

[1] Sobell Department of Neurophysiology, Institute of Neurology, National Hospital for Nervous Diseases, Queen Square, London, W.C.1N 3BG, UK.

Clinical Aspects of Sensory Motor Integration
Ed. by A. Struppler and A. Weindl
© Springer-Verlag Berlin Heidelberg 1987

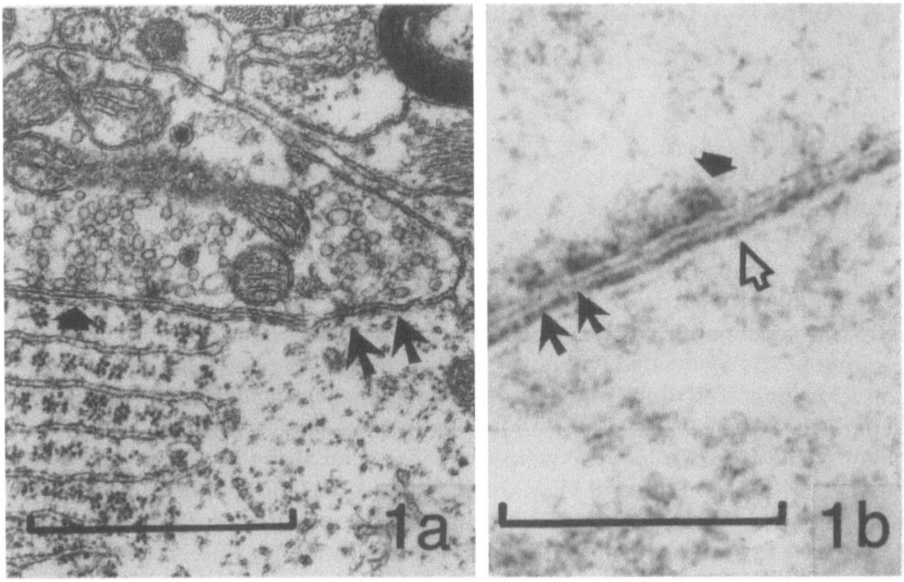

Fig. 1. a Osmicated C-type showing the subsynaptic cistern (*single arrow*), postsynaptic Nissl body (*asterisk*) and peripheral focal synaptic site (*double arrow*); **b** EPTA stained C-type showing the focal presynaptic densities (*single arrow*), common postsynaptic density (*double arrows*) and unstained cistern (*open arrow*) [scale bars = 1 μm (1A), 0.5 μm (1b)]

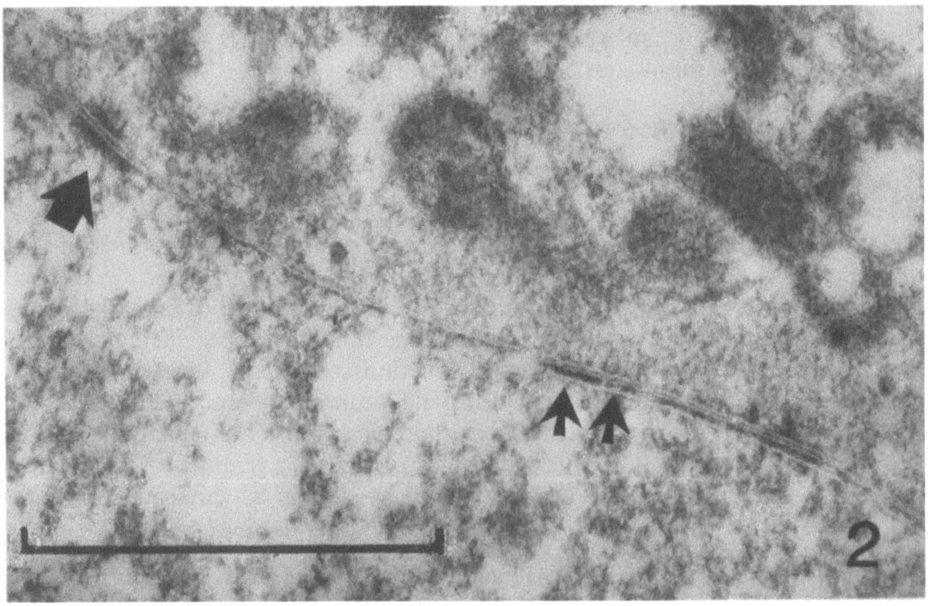

Fig. 2. EPTA stained C-type showing the spatial separation between the peripheral focal synaptic site (*single arrow*) and cisternal common postsynaptic site (*double arrows*) (scale bar = 1 μm)

Results

The ultrastructure of the C-type synapse in glutaraldehyde-fixed, osmicated cat thoracic spinal cord (Fig. 1 a) contrasts with that seen in EPTA-stained material (Fig. 1 b). In osmicated C-types, directly opposing focal densities of the pre- and postsynaptic membranes only occurred in peripheral regions which also lacked a subsynaptic cistern. No focal densities were found where a cistern was present (i.e. in "cisternal" regions). EPTA-stained C-type synapses displayed two distinct synaptic regions: first, directly opposed focal densities of the pre- and postsynaptic membranes in peripheral regions, as observed in osmicated material, and second, in cisternal regions, a series of discrete focal presynaptic densities which were opposed postsynaptically, by a common, continuous electron-dense layer. This possessed neither intrinsic focal densities or inclusions (e.g. microvesicles) nor contiguity with the focal postsynaptic density of peripheral regions (Fig. 2), and it was positioned between the unstained postsynaptic membrane and unstained cistern.

Conclusions

EPTA staining of C-type synapses thus

1. confirmed the existence of focal presynaptic membrane densities in cisternal regions and, in particular, their opposition to a single, common postsynaptic site;
2. newly demonstrated that synaptic sites of the cisternal region are both structurally and spatially distinct from focal densities in peripheral regions lacking a subsynaptic cistern.

Ultrastructural Analysis of the Synaptic Vesicles

C-type axon terminals exhibit two morphological classes of synaptic vesicle: first, the dominant 50 nm electronlucent spherical class [5, 24], and second, the sparse 100 nm diameter class containing a 75 nm electron-dense core (Fig. 1 a). No data exist for the cat thoracic C-type concerning the functional contents of either class. The only insight derives from an illustrated axon terminal resembling a C-type, complete with its postsynaptic Nissl body, which is seen to stain for acetylcholinesterase [18]. On this basis, a possible cholinergic function has been associated with C-types [7]. This possibility was further examined in cat thoracic C-type terminals stained for choline acetyltransferase (CAT), using the method by Ogawa et al. [13], in which the reaction product appears intravesicular with minimal "spread" of the vesicle contour. However, this method does have the disadvantage that only a small proportion of vesicles stain, even in proven cholinergic pathways. The larger dense-core-containing vesicles were distinguished histochemically from the others by their reaction to chromaffin stains, which at pH 6 are claimed to stain catecholamines [30].

Table 1. Results of ultrastructural analysis of synaptic vesicles

Preparative method	Property	Class (a)	Class (b)
Aldehyde-fixed, osmicated	Size	50 nm	100 nm
	Shape	Spherical	Spherical
	Appearance	e-lucent	e-lucent rim
	Inclusions	None	75 nm dense core
CAT stain	Size	60–65 nm	120 nm
	Shape	Spherical	Spherical
	Appearance	e-dense	e-lucent
	Reactivity	+	–
Chromaffin stain	Size	55 nm	125 nm
	Shape	Spherical	Spherical
	Appearance	e-lucent	e-lucent rim, e-dense core (90 nm)
	Reactivity	–	+

Results

For brevity, these are tabulated (Table 1). Generally, smaller vesicles were CAT (+) but chromaffin (–), with the larger cored vesicles exhibiting the converse reactivities. Size relationships between the two vesicle classes were maintained with both histochemical techniques.

Conclusions

1. Two distinct classes of synaptic vesicle exist in cat thoracic C-type terminals. They differ both in relative size and reaction to CAT and chromaffin stains.
2. Results obtained do not necessarily imply cholinergic or aminergic function until supported by definitive proof from tests using transmitter-specific immunocytochemical and/or pharmacological probes.

General Discussion

The major result for discussion must be the demonstration of two distinct synaptic sites for the cat thoracic C-type synapse, since it enables functional considerations of its morphology further to those already introduced from this laboratory [24, 25].

Ethanolic phosphotungstic acid (EPTA) reveals two structurally and spatially distinct synaptic sites: one in peripheral regions lacking a subsynaptic cistern, the other in cisternal regions. The peripheral site comprises directly opposed focal densities of the pre- and postsynaptic membranes, and its structure thus appears identical to the directly opposed focal densities of presumed sites of chemically mediated synaptic transmission [29]. This function, therefore, may be tentatively attributed to the peripheral focal sites of the C-type terminal. Such an exclusive function, however, would be difficult also to associate with cisternal regions which have an altogether different ultrastructure. While cisternal regions exhibit a series of discrete focal densities presynaptically, these are opposed postsynaptically, by a common density which extends parallel to, and along the entire length of, the subsynaptic cistern. This postsynaptic site traverses the neuronal cytoplasm between cistern and postsynaptic membrane and contains no intrinsic focalities. A more likely co-function association for cisternal regions would be with the postulated mutual trophic interaction between the C-type axon terminal and its postsynaptic motoneurone [25].

Briefly, C-type axon terminals, in response to partial central deafferentation of their postsynaptic intercostal motoneurones, increased their presynaptic territory by increases in both absolute number (i.e. axonal sprouting) and individual size (paraterminal extension) (see [24]). Both processes of reinnervation are analagous to those occurring following deafferentation of septal nuclei [26]. Furthermore, increase in size of individual C-type terminals was accompanied postsynaptically, by extension of the subsynaptic cistern with enlargement of the subjacent Nissl body and proliferation of single ribosomes bound to Nissl body lamellae and the lamellae-associated polyribosomes [25]. This local postsynaptic increase in capacity for protein synthesis was attributed to a trophic effect of enhanced synaptic activation of the C-type pathway [25], analagous to the increased postsynaptic incorporation of labelled nucleotide into RNA evoked in molluscan motoneurones by enhanced presynaptic activity [16]. In C-types, increased protein synthesis in the motoneurone may provide suitable "receptors" which, after insertion into the postsynaptic membrane, enable paraterminal growth of existing terminals (e.g. [14, 26]). On the other hand, a mutual trophism between presynaptic terminal and postsynaptic motoneurone could provide overall "feedback" control of reinnervation, possibly by balancing the number and rate of de novo postsynaptic "receptors" synthesised, against a retrograde control of presynaptic terminal growth via regulation of gene expression in the parent interneurone (see [33] for pertinent mechanisms).

The EPTA-stained cisternal postsynaptic site may therefore represent the final pathway for "receptor", proteins, synthesised in the Nissl body and post-translationally conveyed to the agranular subsynaptic cistern for discharge, prior to their insertion into the postsynaptic membrane. The process of synthesising nascent polypeptides for insertion into, and through, membranes of Nissl body lamellae to the luminal side for post-translational processing and eventual utilisation as "secretory" protein has a secure biochemical basis [1, 2, 19, 27]. Other mechanisms however, must complete the discharge of "secretory" protein from subsynaptic cisternae to an immediate postsynaptic position since no "secretory" vesicles [8, 27] occur in the immediate subsynaptic cytosol of the C-type synapse, nor are they apparent in the EPTA-stained postsynaptic site. Secretory molecules, if able to gain access to the cisternal postsynaptic site, must incorporate into the postsynaptic membrane with their active groups on the outer postsynaptic membrane surface as required for surface-active "recognition" glycoproteins [14, 15, 27].

Since postsynaptic sites persist following deafferentation [26], paraterminal extension must involve incorporation of vacated postsynaptic sites formerly occupied by other synaptic "types", into the C-type territory. Separate processes of reinnervation are therefore required for peripheral and cisternal postsynaptic sites. Focal postsynaptic sites, in addition to their association with synaptic transmission, also act as "recognition" sites for corresponding presynaptic elements [14, 15]. Although generally of universal composition in the CNS, the glycoproteins of these sites also include specific factors signalling for given classes of terminal [14]. Thus, vacated postsynaptic sites destined to become within C-type territory need first to lose, or modify, this special factor by transcriptional controls first repressing its gene expression, and then initiating de novo synthesis of new factors appropriate to the C-type synapse. For cisternal regions, however, extension of the existing elongated postsynaptic site would merely require de novo synthesis of surface glycoproteins coding for a cisternal position and function. In either case, modifications to, and synthesis of, postsynaptic glycoprotein is likely to be accompanied by restructuring associated cytoskeletal elements of the postsynaptic density itself [10, 32].

Enhanced synaptic activation of the C-type pathways thus evokes in the postsynaptic cell the synthesis of different classes of protein: first, a class associated with nascent polypeptides that become permanently resident in the Nissl membranes; second, a class associated with the synthesis of "secretory" proteins; and third, a class responsible for the synthesis of cytoskeletal elements ("structural" proteins). Similarly, three classes would also be required presynaptically (i.e. in the parent interneurone) to promote (a) increased synthesis of "transmitter" (trophic agent?), and growth-associated ("secretory") proteins, (b) increased axonal terminal membrane with its associated increased numbers of presynaptic focal densities in cisternal regions, and (c) the extra cytoskeleton of the enlarged axon terminal, including that linked with the presynaptic densities ("structural" proteins).

Summary

Histochemical procedures applied to C-type synapses on cat thoracic motoneurones revealed two morphologically and spatially distinct post-synaptic sites, and two classes of vesicles in the presynaptic terminal. The implications of the two post-synaptic sites are presented in relation to previous findings concerning the response of C-type synapses to partial central deafferentation of their postsynaptic motoneurones following spinal hemisection.

Acknowledgments. These studies form part of a wider investigation in collaboration with Prof. T. A. Sears and Dr. I. P. Johnson. I am indebted to both for provocative and helpful discussions and encouragement with this manuscript.

References

1. Blobel G, Dobberstein B (1975 a) Transfer of proteins across membranes. I. Presence of proteolytically processed and unprocessed nascent immunoglobulin light chains on membrane-bound ribosomes of murine myeloma. J Cell Biol 67:835–851
2. Blobel G, Dobberstein B (1975 b) Transfer of proteins across membranes. II. Reconstitution of functional rough microsomes from heterologous components. J Cell Biol 67:852–862
3. Bloom FE, Aghajanian GK (1968) Fine structural and cytochemical analysis of the staining of synaptic junctions with phosphotungstic acid. J Ultrastruct Res 22:361–375
4. Boone TB, Aldes LD (1984) Synaptology of the hypoglossal nucleus in the rat. Exp Brain Res 57:22–32
5. Conradi S (1969 a) Ultrastructure and distribution of neuronal and glial elements of the motoneurone surface in the lumbosacral spinal cord of the adult cat. Acta Physiol Scand [Suppl 332]:5–48
6. Conradi S (1969 b) Ultrastructure of dorsal root boutons on lumbosacral motoneurones of the adult cat as revealed by dorsal root section. Acta Physiol Scand [Suppl 332]:85–111
7. Conradi S (1976) Functional anatomy of the anterior horn motor neuron. In: Landon DN (ed) The peripheral nerve. Chapman-Hall, London, pp 279–340
8. Dautry-Varsat A, Lodish HF (1983) The Golgi complex and the sorting of membrane and secreted proteins. Trends Neurosci 6:484–489
9. Goshgarian HG, Rafols JA (1984) The ultrastructure and synaptic architecture of phrenic motor neurons in the spinal cord of the adult rat. J Neurocytol 13:85–109
10. Gray EG, Westrum LE (1976) Microtubules associated with nuclear pore complexes and coated pits in the CNS. Cell Tiss Res 168:445–453
11. Hamos JE, King JS (1980) The synaptic organisation of the motor nucleus of the trigeminal nerve in the opossum. J Comp Neurol 194:441–463
12. Johnson IP (1985) Ultrastructure of gamma motoneurones in the cat thoracic spinal cord. J Physiol (Lond) 360:46P
13. Kaiya H, Iwata T, Namba M, Ago Y, Mayahara H, Ogawa K (1979) Ultracytochemistry of cholineacetyltransferase in the neostriatum of the rat brain. Acta Histochem Cytochem 12:391–404
14. Kelly PT, Cotman CW (1977) Identification of glycoproteins and proteins at synapses in the central nervous system. J Cell Biol 252:786–793
15. Kelly PT, Cotman CW, Gentry C, Nicolson GL (1976) Distribution and mobility of lectin receptors on synaptic membranes of identified CNS neurons. J Cell Biol 71:487–496
16. Kernell D, Price-Peterson R (1970) The effect of spike activity versus synaptic activation on the metabolism of ribonucleic acid in a molluscan giant neurone. J Neurochem 17:1087–1094
17. Lagerback PA, Ronnevi LO (1982) An ultrastructural study of serially sectioned Renshaw cells. II. Synaptic types. Brain Res 246:181–192
18. Lewis PR, Shute CCD (1966) The distribution of cholinesterase in cholinergic neurons demonstrated with the electron microscope. J Cell Sci 1:381–390
19. Loh YP, Brownstein MJ, Gainer H (1984) Proteolysis in neuropeptide processing and other neural functions. Ann Rev Neurosci 7:189–222
20. Matsushita M, Ikeda M (1973) Propriospinal fibre connections of the cervical motor nuclei in the cat: a light- and electron-microscope study. J Comp Neurol 150:1–32
21. McLaughlin BJ (1972 a) Dorsal root projections to the motor nuclei in the cat spinal cord. J Comp Neurol 144:461–474
22. McLaughlin J (1972 b) Propriospinal and supraspinal projections to the motor nuclei in the cat spinal cord. J Comp Neurol 144:475–500
23. Pullen AH (1985) Two postsynaptic regions revealed in cat C-type terminals using ethanolic phosphotungstic acid. J Physiol (Lond) 360:47P
24. Pullen AH, Sears TA (1978) Modification of "C" synapses following partial central deafferentation of thoracic motoneurones. Brain Res 145:141–146
25. Pullen AH, Sears TA (1983) Trophism between C-type axon terminals and thoracic motoneurones in the cat. J Physiol (Lond) 337:373–388

26. Raisman G, Field PM (1973) A quantitative investigation of the development of collateral reinnervation after partial deafferentation of the septal nuclei. Brain Res 50:241–264
27. Sabatini DD, Kreibich G, Morimoto T, Adensik M (1982) Mechanisms for the incorporation of proteins in membranes and organelles. J Cell Biol 92:1–22
28. Schrøder HD (1979) Paramembraneous densities of "C" terminal motoneurone synapse in the spinal cord of the rat. J Neurocytol 8:47–52
29. Szenthagothai J (1970) The morphological identification of the active synaptic regions: aspects of general arrangement, of geometry and topography. In: Andersen P, Jansen JKS (eds) Excitatory synaptic mechanisms. Universitetsforfloget, Oslo, pp 9–26
30. Tranzer JP, Richards JG (1976) Ultrastructural cytochemistry of biogenic amines in nervous tissue; methodological improvements. J Histochem Cytochem 24:1178–1193
31. Tredici G, Pizzini G, Milanesi S (1976) The ultrastructure of the nucleus of the oculomotor nerve (somatic efferent portion) of the cat. Anat Embryol 149:323–346
32. Westrum LE, Gray EG (1977) Microtubules associated with postsynaptic "thickenings". J Neurocytol 6:505–518
33. Willard M, Skene JHP, Simon C, Meiri K, Hirokawa N, Glicksman M (1984) Regulation of axon growth and cytoskeletal development. In: Elam JS, Cancalon P (eds) Axonal transport in neuronal growth and regeneration. Adv Neurochem 6:171–183

Physiology and Pathophysiology of Reciprocal Inhibition in the Human Forearm

B. L. Day [1], J. C. Rothwell [1], J. A. Obeso [1], P. Thompson [1], A. Cozens [1], and C. D. Marsden [1]

Active inhibition of antagonist muscles during voluntary movement is accomplished largely from two sources: by central descending commands from the brain and by peripheral input from agonist muscle spindle Ia afferents. Work on animals has demonstrated that both the central and peripheral sources of inhibition act through the same system of spinal Ia inhibitory interneurones (see [1]).

Peripheral Reciprocal Inhibition

It is possible to study the system of Ia inhibitory interneurones which act between extensor and flexor muscles in the forearm of intact, relaxed human subjects [2]. A single submotor threshold shock to the radial nerve, which supplies the extensor muscles in the forearm, produces an abrupt and short-lasting inhibition of the flexor monosynaptic H-reflex (Fig. 1 a). The inhibition is present with radial nerve stimulus intensity as low as 0.7 times the motor threshold and has a central delay of 0.9 ms in excess of that for the H-reflex. The pathway is likely, therefore, to involve a single inhibitory interneurone which is excited by extensor muscle Ia afferents.

In addition to disynaptic inhibition (first phase), the radial nerve shock produces two further phases of inhibition of the flexor H-reflex (second and third phase) shown in Fig. 1 b. The depth of inhibition of these later phases is comparable to the first phase but is much longer-lasting.

A clue to the mechanism underlying the second phase of inhibition may be obtained by testing the excitability of the flexor motoneurone pool with a different input to the H-reflex. A synchronous muscle discharge can be evoked by stimulating the corticospinal tract with a shock applied to the scalp [4]. As with an H-reflex, changes in flexor motoneurone excitability which result from a radial nerve conditioning stimulus are reflected in the size of the muscle potential evoked by cortical stimulation. This method of testing the flexor pool shows interesting similarities and differences when compared to H-reflex testing. The disynaptic inhibition is seen using both methods of testing. This demonstrates that it is a true postsynaptic inhibition of the flexor pool. However, the second phase of inhibition, which is seen using H-reflex testing, is not apparent when tested with cortical stimulation. It seems as if the second phase of inhibition acts exclusively on some site of the H-reflex pathway other than postsynaptically on the flexor motoneurones. A reasonable explanation is that the second phase of inhibition acts presynaptically on the Ia afferent terminals, a view which is compatible with its long-lasting time course.

[1] Institute of Psychiatry, Department of Neurology, De Crespigny park, Denmark Hill, London SE5 8AF, UK.

Clinical Aspects of Sensory Motor Integration
Ed. by A. Struppler and A. Weindl
© Springer-Verlag Berlin Heidelberg 1987

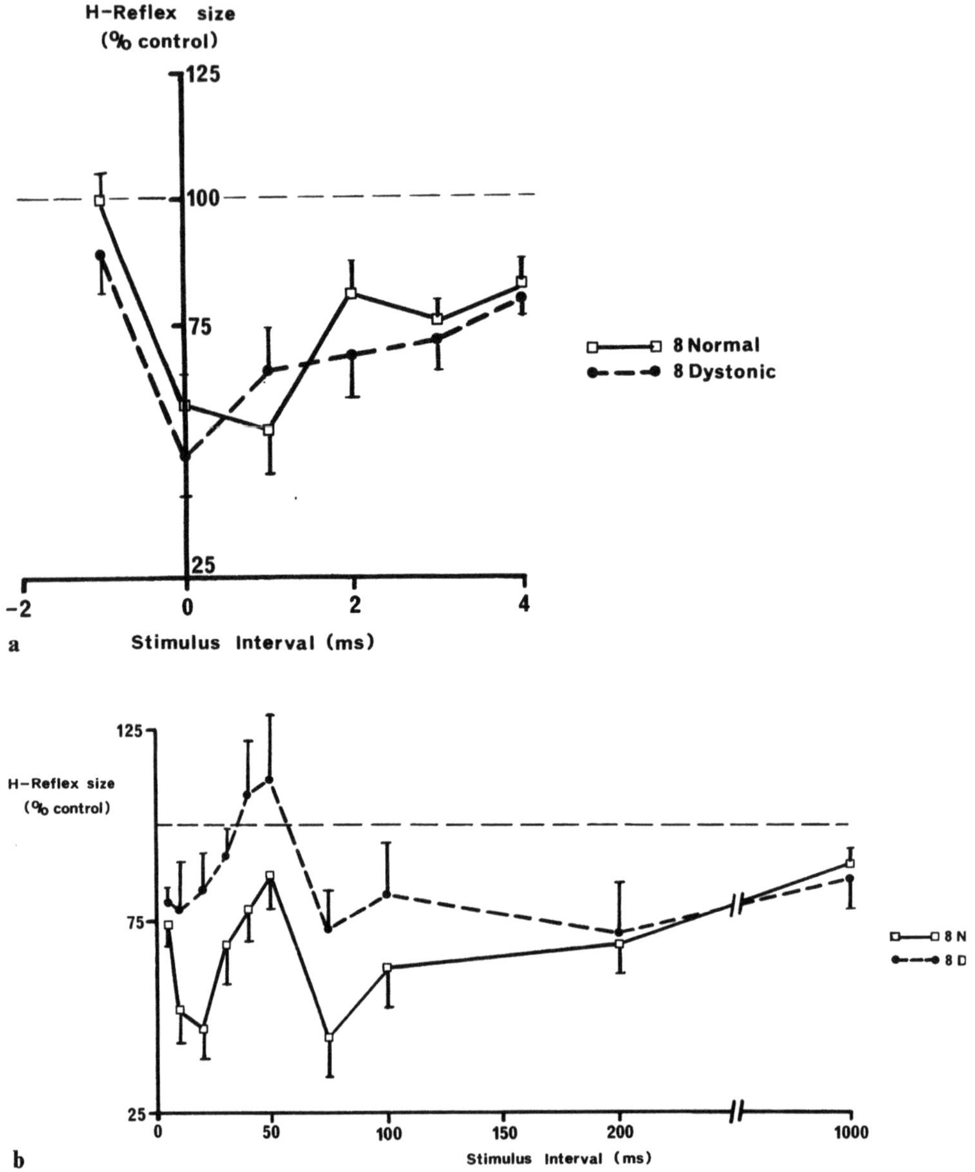

Fig. 1 a, b. H-reflexes were evoked in flexor muscles in the forearm at different times relative to a radial nerve conditioning shock given at time zero. The conditioned H-reflex size was expressed as a percentage of the control size. **a** First phase (disynaptic) inhibition. **b** The longer-lasting second and third phases of inhibition. Both graphs illustrate the mean data obtained from 8 normal subjects (*solid line*) and the mean data obtained from 8 patients with idiopathic torsion dystonia (*dashed line*)

Descending Influences on Reciprocal Inhibition

Transmission in the Ia inhibitory interneurones is modulated by descending input from the brain. This can be seen by measuring changes in disynaptic inhibition during the interval leading up to a voluntary movement of the wrist ([3]; see also [7]). Prior to a wrist extension movement the depth of inhibition is increased whilst it is decreased prior to a wrist flexion movement. This is consistent with the notion that the command to the agonist motoneurones acts also on a specific population of Ia inhibitory interneurones. Descending input onto the Ia inhibitory interneurones from extensors to flexors can also be studied using the technique of cortical stimulation [6]. A single subthreshold shock to the scalp produces a corticospinal volley which inhibits this population of interneurones. The method allows accurate timing of the effect. The timing is compatible with either a monosynaptic or disynaptic corticospinal projection onto the spinal inhibitory interneurones. Whether or not the later phases of inhibition can be modulated by descending control has yet to be established.

Pathophysiology of Reciprocal Inhibition

We have studied the reciprocal inhibitory pathways in two groups of patients. One group of patients exhibited some degree of hemiparesis following a stroke. The other group suffered from idiopathic torsion dystonia [5]. The aim of the study was to explore the possibility that the cocontraction of antagonist muscles seen in both groups of patients when attempting voluntary movement may be due to an abnormality of some aspect of the inhibitory mechanisms described above.

The 8 patients with dystonia had no known brain lesion. 4 patients were classified as having a focal dystonia, 2 patients had segmental dystonia and 2 patients had generalised dystonia. As a group they showed an abnormality of later phases of inhibition whilst the first-phase (disynaptic) inhibition was normal (Fig. 1 a, b). The abnormality was characterised by an excitation of the flexor H-reflex some 40–50 ms after the radial nerve conditioning shock. In normal subjects this is around the timing that a partial or complete recovery separates the second and third phases of inhibition. It is possible that the recovery arises from an excitatory process superimposed on top of a single long-lasting inhibition. The later phase abnormality can then be viewed as either an exaggerated excitation or a reduced inhibition.

Of the 13 stroke patients studied, 6 showed a reduction or absence of the first phase (disynaptic) of inhibition. Interestingly, 5 of these 6 patients exhibited increased muscle tone when tested clinically, whereas only 1 of the remaining 7 patients showed any increase. 6 patients showed an abnormality of the later phases of inhibition. The abnormality was very similar to that seen in the dystonic patients. The normal recovery seen between the second and third phase of inhibition was replaced by a frank excitation of the flexor H-reflex.

In 11 of the 13 stroke patients it was possible to obtain some idea of the extent and site of the brain lesion from CT scan evidence. The evidence leads us to believe that a significant interruption of corticospinal pathways which lead to spasticity may result in a reduction in the amount of disynaptic inhibition that is normally present in the relaxed subject. The implication is that such a lesion either removes tonic descending excitatory input or releases a descending tonic inhibition of the spinal Ia inhibitory interneurones. The second-phase abnormality appeared to be present in those stroke patients in whom the basal ganglia were affected by the lesion. This idea is attractive since the dystonic patients showed a similar abnormality and dystonia is thought to result from basal ganglia malfunction. Presumably these abnormalities in reciprocal inhibition contribute, at least partially, to the movement difficulties experienced by these patients.

Summary

A single shock delivered to the radial nerve at motor threshold intensity produces a complex pattern of inhibition of the H-reflex evoked in relaxed wrist and finger flexor muscles. The first phase of inhibition is ascribed to action of the spinal Ia inhibitory interneurones. Presynaptic inhibition of the flexor Ia afferent terminals is thought to contribute to the subsequent phases of inhibition. Descending control of the Ia inhibitory interneurones can be observed by measuring changes in the depth of the first phase of inhibition either prior to voluntary movement or in response to electrical stimulation of the brain. Brain lesions also may affect the depth of reciprocal inhibition at rest. Thus, stroke patients with spasticity show reduced or absent disynaptic inhibition. Patients with idiopathic torsion dystonia and some stroke patients have normal disynaptic inhibition but abnormal later phases of inhibition.

References

1. Baldissera F, Hultborn M, Illert M (1981) Integration in spinal neuronal systems. In: Brooks VB (ed) Handbook of physiology, sect 1: The nervous system, part 2, vol 2. American Physiological Society, Bethesda, pp 509–593
2. Day BL, Marsden CD, Obeso JA, Rothwell JC (1984) Reciprocal inhibition between the muscles of the human forearm. J Physiol 349:519–534
3. Day BL, Rothwell JC, Marsden CD (1983) Transmission in the spinal reciprocal Ia inhibitory pathway preceding willed movements of the human wrist. Neurosci Lett 37:245–250
4. Merton PA, Morton HB (1980) Stimulation of the cerebral cortex in the intact human subject. Nature 285:227
5. Rothwell JC, Day BL, Berardelli A, Marsden CD (1984) Effects of motor cortex stimulation on spinal interneurones in intact man. Exp Brain Res 54:382–384
6. Rothwell JC, Obeso JA, Day BL, Marsden CD (1983) Pathophysiology of dystonias. In: Desmedt JE (ed) Motor control mechanisms in health and disease. Raven, New York, pp 851–863
7. Simoyama M, Tanaka R (1974) Reciprocal Ia inhibition at the onset of voluntary movements in man. Brain Res 82:529–540
8. Yanagisawa N, Tanaka R, Ito Z (1976) Reciprocal Ia inhibition in spastic hemiplegia of man. Brain 99:555–574

V. Long-Loop Reflexes: Concepts and Consequences

The Use of Short- and Long-Latency Reflex Testing in Leg Muscles of Neurological Patients

J. DICHGANS [1] and H. C. DIENER [1]

Hammond [27, 28] originally described electromyographic (EMG) responses of long latency, which were evoked 50 ms later than the stretch reflex, when the biceps brachii muscle was suddenly extended and the subject was instructed to resist the displacement. These EMG responses were considered to be "automatic" in nature, for their latency was 40 ms shorter than the fastest voluntary reaction to a mechanical stimulus. There has been a long controversy as to the possible pathway and generators as well as the function of "long-loop reflexes" (for reviews see [8, 12, 13, 57]). Despite the fact that different types of EMG responses can be evoked from finger, hand, arm, and leg muscles labelled with a varying terminology, there are some common features of these long-latency responses. The best way to generate them is a sudden angular displacement of the joint with a certain minimal velocity and duration [5, 8, 34]. Prestretching or loading of the stretched muscle is required but can be replaced by an instruction given to the subject to resist the imposed displacement. Functional requirements obviously determine strength and temporal pattern of these responses [15, 30].

The present paper tries to add some arguments to the ongoing controversy about the generation of medium- and long-latency responses, but its main aim is to review our own results concerning the extent to which recordings of medium- and long-latency responses from leg muscles can provide important information for the neurologist. For the latter, more pragmatic approach it is less significant whether these responses are explained by segmented afferent activity [22, 26], by a late contribution from slowly conducting group II afferents [40], by spinal mechanisms [24, 31, 42, 54], or, finally, by a transcortical loop [38, 52].

We evoke short-, medium-, and long-latency responses (SL, ML, and LL, respectively) from distal leg muscles by suddenly tilting a supporting platform toe-up or -down around the ankle joint. This procedure, performed in quietly standing humans, has several advantages over the experimental paradigms most frequently used. In the arm and the leg of a sitting human, ML and LL occur in the same muscle and are often fused and therefore difficult to evaluate independently. With our experimental setup, SL and ML can be recorded from the stretched muscle, whereas the LL response is generated in the shortened antagonist [15]. The analogy of this antagonistic response with LL in a stretched muscle (M 3; nomenclature from [52]) has been debated. Indeed, the LL response has some features in common with the well-known shortening response occuring at very similar latency [56], but it also differs in three important points: the size of the shortening reaction is independent of the velocity of stretch; the elimination of joint af-

[1] Department of Neurology, University of Tübingen, Liebermeisterstr. 18–20, D-7400 Tübingen, FRG.

Clinical Aspects of Sensory Motor Integration
Ed. by A. Struppler and A. Weindl
© Springer-Verlag Berlin Heidelberg 1987

ferents abolishes the shortening reaction; and there exists a striking asymmetry in the frequency of occurence of the shortening reaction in the anterior tibial (56%) and triceps surae muscle (6%) [4, 32]. Although these differences may possibly be explained by the fact that shortening reactions so far have invariably been tested in subjects lying supine or sitting down, it may be pointed out that LL in a standing subject is clearly dependent on the velocity of stretch [18], ischemia of the foot and ankle joint leaves the latency and size of LL unchanged [19], LL in triceps surae can be recorded in 100% of normals by tilting the platform toe-down. Finally, LL is observed in the stretched triceps surae muscle and not in the shortened antagonist when the standing person is subjected to linear displacements of the platform [43].

Some additional features qualify our experimental setup:

1. It does not require an active preshortening of the stretched muscle.
2. No special instruction has to be given to the subject. Standing comes with a subconscious setting of the system controlling posture.
3. The stereotyped responses do not adapt over consecutive trials and thereby allow for averaging of as few as eight platform tilts.

Patients

Several groups of patients with lesions of various localization within the motor system were tested to examine whether characteristic alterations of the reflex pattern exist. The *cerebellar disease* group included 122 patients. These were divided into the following subgroups: (a) anterior lobe disease (late cortical cerebellar atrophy in alcoholics), (b) lower vermis (tumor), (c) cerebellar hemisphere diseases, (d) diffuse cerebellar atrophy, and (e) Friedreich's disease. Basal ganglia disorders included 33 patients with *Parkinson's disease* and 3 patients with *Huntington's chorea.* 53 patients had *spinal lesions,* and 54 had *central lesions,* tumors, or ischemic deficits involving the internal capsule, or sensorimotor cortex. Finally, in order to specify the influence of afferent information for the generation of the three EMG responses, we additionally investigated 18 patients with an acquired axonal or demyelinating *polyneuropathy* and 7 patients with *hereditary motor-sensory neuropathy.* 7 additional patients had complete bilateral *vestibular inexcitability.*

Short-Latency Response (SL)

The short-latency (SL) response is most likely generated on a segmental spinal level via mono- and polysynaptic pathways. Its mean latency in the triceps surae after platform tilt toe-up in normals is 48.9 ± 6 ms. There exists a striking asymmetry, since platform tilts toe-down do not evoke a stretch reflex in the anterior tibial muscle (see also [25]). Patients with peripheral neuropathy consistently presented delayed (in the case of demyelination) or abolished SL responses. The cor-

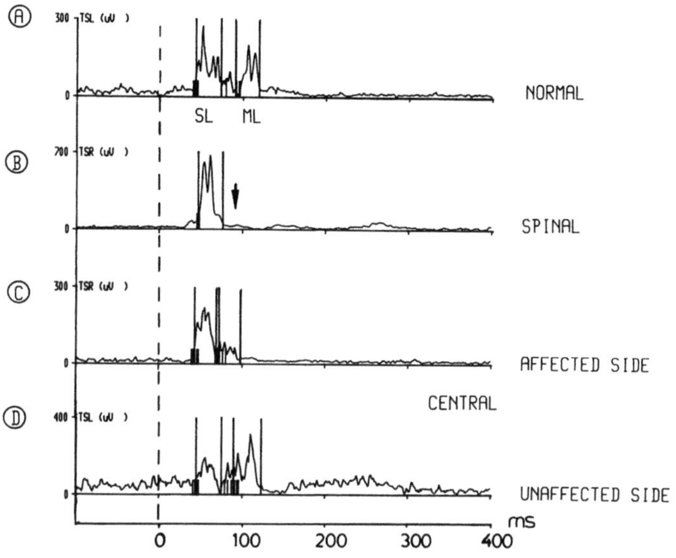

Fig. 1. Rectified averaged EMG responses from the triceps surae muscle after platform tilt toe-up (50°/s; 4°). *SL*, short-latency response; *ML*, medium-latency response. The *dashed line* indicates the beginning of platform tilt. The prominent *vertical line* indicates the beginning and end of each EMG response, and the *small lines* indicate the latency of single trials. Note the missing ML response in patients with a spinal tumor (*spinal*) and the increase of SL and decrease of ML in a patient with a hemiparesis due to ischemia in the region of the medial cerebral artery (*central*). Also note the different scalings of the ordinate in A–D

relation between the latency of SL and the H-reflex latency was very high ($r=0.94$). The latency of SL was normal in all patients with central motor disorders who did not show clinical signs of polyneuropathy. Quite expectedly, the mean integral of the rectified SL response (4.0 ± 2.8 arbitrary units) was significantly increased in patients with spasticity due to spinal lesions (Fig. 1). Patients with central lesions showed slightly greater SL responses on the affected side (2.5 ± 1.2 arbitrary units) than in the normal leg (2.3 ± 1.8 arbitrary units) (Fig. 1; see [1]). These findings correspond to the clinical observation of more exaggerated tendon reflexes in spinal than in central lesions. Similar increases in the size of SL in upper motor neuron syndromes were reported in the leg [7] and hand muscles [35, 55] of humans and monkeys [37].

The equivalence of SL and the stretch reflex obtained by clinicians is further supported by the fact that its size increases with velocity [18] and acceleration [5] but not with the amplitude of stretch [18]. SL, in contrast to ML and LL, is suppressed by vibration which saturates Ia afferents [5, 29]. One may consequently assume that the afference starting ML and LL is different from that arousing SL. The mechanical impact of the SL response is small, since normal subjects without stretch reflexes exhibit exactly the same body displacement after platform tilt as those with intact stretch reflexes.

Medium-Latency Response (ML)

The mean latency of ML is 89.5 ± 10 ms in triceps surae of normals. The latency of ML was normal in every group of patients except for patients with a decrease in peripheral nerve conduction velocity due to polyneuropathy. The normal latency of ML even in cases with spinal and central lesions affecting the dorsal columns and/or the pyramidal tracts (including patients with multiple slcerosis) argues against a transcortical loop for this response. ML in the leg may very well be a segmental reflex response whose origin could be explained by a late contribution from slowly conducting group II afferents according to Matthews [40]. According to Noth et al. [45], this interpretation does not hold for M2 in distal muscles of the arm.

The size of ML, however, is under supraspinal control. Patients with Parkinson's disease present increased ML responses [49], whereas this EMG response was invariably abolished in our patients with Huntington's chorea (Fig. 2). An increase in size of the M2/M3 response but normal latencies in parkinsonians have also been observed by others in hand, arm [33, 48, 50, 51, 53], and distal leg muscles [6, 9]. The increase of ML is not due to a high level of tonic background activity in the stretched muscle, since the size of SL reflecting the excitability of the alpha motoneuron pool is normal or quite frequently decreased in our patients [49]. Furthermore, Tatton et al. [53] found normal or even lower resting background EMG levels in parkinsonians and, thus, no indication of an increased descending drive to motoneurons. This result favors the assumption that the modulating influence of the basal ganglia is exerted via polysynaptic pathways on spinal interneurons. The question of whether the increase of M2/M3 complexes or the ML response could account for the amount of rigidity is controversial. We and others [48, 49] have found no positive correlation, but Tatton and Lee [51]

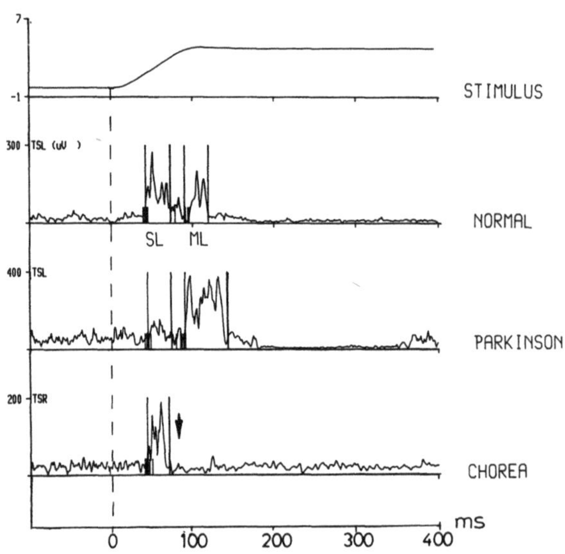

Fig. 2. Rectified averaged EMG responses from the triceps surae muscle after platform tilt toe-up. The medium-latency response (*ML*) is increased in a patient with Parkinson's disease and abolished in Huntington's chorea (*arrow*). Note the different scalings of the ordinate

and Berardelli et al. [6] have. With respect to Huntington's chorea, there is also an analogous modulation of M2 or ML in arm and in leg. The second EMG response of the first interosseus muscle, evoked by short displacements of the index finger (latency 56.5 ms), was absent in nearly all patients studied by Noth et al. [45]. The fact that these patients had normal short-latency responses, among other arguments, may be taken as evidence against the hypothesis of Hagbarth et al. [26] that ML responses are due to repetitive bursts of primary muscle-spindle endings caused by oscillations during the nonlinear stretch of a muscle. Noth et al. [44] assumed that the lack of long-loop reflexes in Huntington's chorea might be a consequence of decreased somatosensory input to the cerebral cortex as revealed by the recording of sensory-evoked cortical potentials [44, 46]. Whereas this explanation may be valid for M2 in finger muscles, it cannot account for our results. SL is normal in our patients with chorea, and both latency and amplitude of the LL response, which clearly depends on intact afferent information to sensorimotor cortex [20], are also normal. In analogy with the changes of ML observed in Parkinson's disease, we would suggest in chorea a modulatory, inhibiting influence of the extrapyramidal system on spinal interneurons which transmit the ML response.

ML is decreased and often abolished in patients with spinal lesions [21]. The size of ML is significantly decreased in the affected leg as compared with the normal leg in patients with central lesions. We found no correlation between the increased size of SL in these patients and the decreased size of ML, disproving the hypothesis of Beradelli et al. [7], who assume that if the motoneuron pool produces a large monosynaptic response which includes all motor units, it cannot be excited by subsequent afferent stimuli within a short time interval.

Long-Latency Responses (LL)

Our experimental evidence votes against the possible assumption that the LL response in the antagonist of the stretched muscle is a voluntary one. Its latency is significantly shorter than the fastest voluntary reaction to a somatosensory stimulus [18], and the standard deviation of succeeding trials is five times smaller in LL responses than with the reaction-time paradigm. Finally, progressive training shortens the latency of voluntary reactions whereas LL responses occur after a fixed interval. We conclude that LL is an automatic response triggered by the platform displacement.

The results from the different patient groups indicate that LL travels transcortically and that its duration and amplitude, but not its onset, are modulated by the cerebellum. The increase in latency of LL (132 ± 15 ms in normals) is characteristic in lesions of the spinal cord (170 ± 34 ms), internal capsule, and sensorimotor cortex (156 ± 27 ms on the affected side) in cases with purely sensory, purely motor, and cases with combined dysfunctions of the two as assessed by clinical examination (Fig. 3; see [21]). Cerebellar dysfunctions predominantly of the anterior lobe prolong the duration and increase the amplitude of LL [14, 16, 17], whereas the M2/M3 complex in hand muscles is increased in patients with

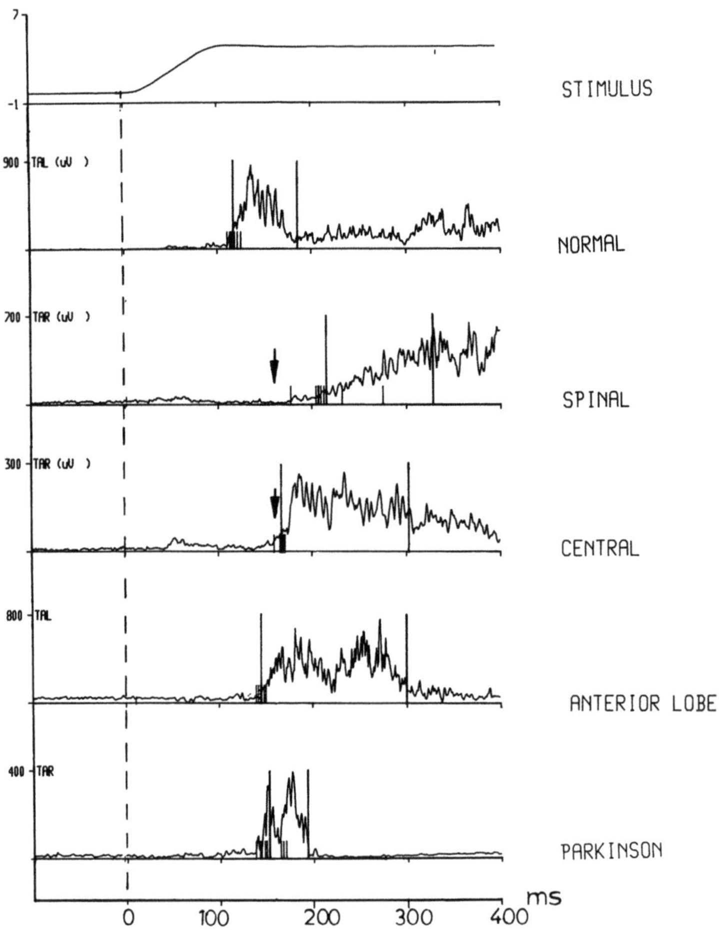

Fig. 3. Rectified averaged long-latency (*LL*) EMG responses from the anterior tibial muscle after platform tilt toe-up (stimulus). The *black arrow* indicates the upper limit of normality for the latency of LL (AM + 2 SD). LL latencies are increased in a patient with a spinal angioma (*spinal*) and a patient with a tumor of the sensorimotor cortex (*central*). Note the increased duration and integral of LL in a patient with atrophy of the anterior lobe of the cerebellum due to chronic alcoholism. Normal LL response in Parkinson's disease. Note the different scalings of the ordinate

lesions of the cerebellar hemisphere [23]. Patients with Friedreich's disease mainly involving the posterior columns and spino-cerebellar pathways invariably show prolonged latencies of LL [16] and either missing tibialis-evoked sensory cortical potentials or increased latencies of the latter.

The close correlation between the latency of LL and the latency of sensory-evoked potentials in response to stimulation of the tibial nerve in multiple sclerosis [20] and spinal compressive or vascular lesions [1] supports our hypothesis of a transcortical reflex. The fact that a few patients with purely sensory deficits show normal sensory-evoked potentials but delayed LL reflexes (and vice versa) may indicate different pathways for the transmission of afferent sensory information to consciousness and the sensory input relevant to elicitation of sensory-evoked potentials as well as transcortical reflexes. Increased LL latencies may be caused by damage of efferent cortico- or subcorticospinal pathways. Indeed, LL is delayed in patients who present only motor symptoms without sensory involvement. The increase in latency and the decrease in amplitude of LL, the latter of which may indicate desynchronization, are correlated with the severity of paresis in these patients. The recording of LL responses apparently offers a way to objectively measure the function of efferent pathways, for instance, especially in patients who complain of weakness in one leg but do not yet present the neurological signs of pathology.

The question as to whether LL is simply abolished or reduced in spinal and central lesions [10, 35, 39, 55] and replaced by activity travelling along alternative pathways, or whether it is simply delayed, is difficult to answer. Our follow-up study in patients with spinal lesions who underwent successful surgery with clinical improvement and a progressive normalization of LL latencies indicates a continuous decrease in latency and not a clustering of latencies along a discrete temporal scale.

Latencies and integrals of LL were normal in our patients with disorders of the basal ganglia while standing. While sitting, however, LL is not decreased in Parkinsonians as it is the case in normals (decreased functional adaptation). Correspondingly, the shortening reaction has been reported to increase in amplitude in patients with Parkinson's disease [3, 47].

The afferent input for the generation of the LL response in some respects must be different from the one which triggers SL and ML. Our patients with hereditary sensorimotor neuropathy (HSMN I), who presented extremely slowed peripheral nerve conduction velocities, invariably had abolished ML and SL responses but normal latencies of the LL response (when present). Furthermore, application of bilateral ischemia in normals at the level of the thigh to suppress group I afferents under preservation of the efferent innervation of leg muscles did not change the latency of LL [19].

A strong influence of the vestibular system on medium- and long-latency responses is unlikely. Seven patients with a bilateral peripheral vestibular loss exhibited normal SL, ML, and LL responses. Vestibulospinal influences, however, seem to be important under conditions where the center of gravity is displaced prior to platform tilt or when the eyes are closed [2].

Conclusions

Recordings of short-, medium-, and long-latency responses offer a fast and atraumatic method to evaluate the function of: (a) the segmental stretch reflex arc (SL) in peripheral neuropathy and spasticity, (b) the descending inhibitory or excitatory influences of the basal ganglia on spinal interneurons (ML), and (c) the conduction velocity of the transcortical reflex (LL). The combination of LL response measurements and recordings of scalp potentials, both evoked by platform displacement toe-up and thus stretching triceps surae, allows for a selective evalua-

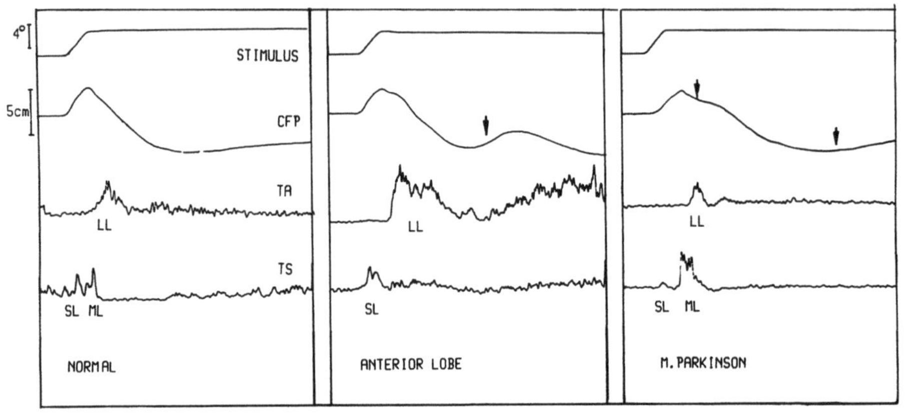

Fig. 4. Averaged rectified EMG responses from the anterior tibial (*TA*) and the triceps surae muscle (*TS*) after platform tilt toe-up (50°/s; 4°; stimulus). The recording of center-of-foot pressure (*CFP*) indicates an overshoot in the correction of the passively induced body movement backwards leading to foreward body movement in anterior lobe atrophy (*arrow*) and insufficient correction of body displacement in Parkinson's disease due to an increased ML response but normal LL

tion of the afferent and efferent limb of the transcortical reflex loop in a small number of patients with lesions limited to one or the other system.

Pathological changes of medium- and long-latency responses are not only of topodiagnostic significance, but by way of their obvious biomechanical consequences they also help in understanding the abnormal compensation of passive displacements of the body (by platform tilt) in some patients. The increased ML response in the stretched muscle of a parkinsonian for example, by way of adding to the destabilization of posture, may explain the retropulsion in these patients (Fig. 4). Another example: patients with cerebellar anterior lobe lesions presenting increased and prolonged LL responses show an overcompensation of the corrective body movement, which in turn elicits another stretch, and so on. This results in an anterior-posterior body sway of 2–3 Hz (Fig. 4; see [17, 41]).

The results from our study in leg muscles cannot be generalized. There is increasing evidence that M2 in distal upper limb muscles is dependent on the function of the motor cortex [11, 37], whereas M2 in proximal muscles of the arm and ML in leg muscles are likely to be local, segmental polysynaptic spinal responses to the stimulus [21, 36, 42].

Summary

Short (SL), medium (ML), and long latency (LL) EMG responses to stretch of lower leg muscles were recorded in patients suffering from a variety of localized neurological lesions. The study was performed in order to elucidate the possible contribution of different parts of the motor system in the generation or transmission of these responses. Stretch was applied while standing on a tiltable platform. SL is delayed or absent with polyneuropathy and is increased in amplitude with spasticity. It corresponds to the spinal stretch reflex. ML is modulated in amplitude with basal ganglia diseases; it is increased with Parkinson's disease and abolished with Huntington's chorea. It is a polysynaptic response of the spinal cord. LL depends in latency on the integrity of posterior columns, sensory motor cortex, and pyramidal tracts, and therefore must travel transcortically. Its size and duration are under cerebellar control.

Acknowledgment. This research was supported by the Deutsche Forschungsgemeinschaft, Grant No. SFB 307-A3.

References

1. Ackermann H, Diener HC, Dichgans J (1985) Diagnostische Bedeutung von Long loop-Reflexen bei spinalen Läsionen. EEG EMG 16:35
2. Allum JHJ, Pfaltz CR (1985) Visual and vestibular contributions to pitch sway stabilization in the ankle muscles of normals and patients with bilateral peripheral vestibular deficits. Exp Brain Res 58:82–94
3. Angel RW (1982) Shortening reaction in normal and parkinsonian subjects. Neurology (Ny) 32:246–251
4. Bathien N, Toma S, Rondot P (1981) Etude de la réaction de raccourcissement présente chez l'homme dans diverses affections neurologiques. Rôle des afférences articulaires. Electroencephalogr Clin Neurophysiol 51:156–164
5. Berardelli A, Hallett M, Kaufman C, Fine E, Berenberg W, Simon SR (1982) Stretch reflexes of triceps surae in normal man. J Neurol Neurosurg Psychiatry 45:513–525
6. Berardelli A, Sabra AF, Hallett M (1983a) Physiological mechanisms of rigidity in Parkinson's disease. J Neurol Neurosurg Psychiatry 46:45–53
7. Berardelli A, Sabra AF, Hallett M, Berenberg W, Simon SR (1983b) Stretch reflexes of triceps surae in patients with upper motor neuron syndromes. J Neurol Neurosurg Psychiatry 46:54–60
8. Chan CWY (1983) Segmental versus suprasegmental contributions to long-latency stretch responses in man. In: Desmedt JE (ed) Motor control mechanisms in health and disease. Raven, New York, pp 467–487
9. Chan CWY, Kearney RE, Melvill Jones G (1979a) Tibialis anterior responses to sudden ankle displacements in normal and Parkinsonian subjects. Brain Res 173:303–314
10. Chan CWY, Melvill Jones G, Kearney RE, Watt DGD (1979b) The "late" electromyographic response to limb displacement in man. Electroencephalogr Clin Neurophysiol 46:173–181
11. Cheney PD, Fetz EE (1984) Corticomotoneural cells contribute to long-latency stretch reflexes in the rhesus monkey. J Physiol (Lond) 349:249–272
12. Desmedt JE (ed) (1978) Cerebral motor control in man: long loop mechanisms. Karger, Basel
13. Desmedt JE (ed) (1983) Motor control mechanisms in health and disease. Raven, New York
14. Dichgans J, Diener HC (1985) Clinical evidence for functional compartmentalization of the cerebellum. In: Bloedel JR, Dichgans J, Precht W (eds) Cerebellar functions. Springer, Berlin Heidelberg New York, pp 126–147
15. Diener HC, Bootz F, Dichgans J, Bruzek W (1983) Variability of postural "reflexes" in humans. Exp Brain Res 52:423–428
16. Diener HC, Dichgans J, Bacher M, Guschlbauer B (1984a) Characteristic alterations of long loop "reflexes" in patients with Friedreich's ataxia and late atrophy of the anterior cerebellar lobe. J Neurol Neurosurg Psychiatry 47:679–685

17. Diener HC, Dichgans J, Bacher M, Guschlbauer B (1984b) Improvement of ataxia in late cortical cerebellar atrophy through alcohol abstinence. J Neurol 231:258–262
18. Diener HC, Dichgans J, Bootz F, Bacher M (1984c) Early stabilization of human posture after a sudden disturbance: influence of rate and amplitude of displacement. Exp Brain Res 56:126–134
19. Diener HC, Dichgans J, Guschlbauer B, Mau H (1984d) The significance of proprioception on postural stabilization as assessed by ischemia. Brain Res 296:103–109
20. Diener HC, Dichgans J, Hülser PJ, Buettner UW, Bacher M, Guschlbauer B (1984e) The significance of long loop "reflexes" for the diagnosis of multiple sclerosis. Electroencephalogr Clin Neurophysiol 57:336–342
21. Diener HC, Ackermann H, Dichgans J, Guschlbauer B (1985) Medium- and long-latency responses to displacements of the ankle joint in patients with spinal and central lesions. Electroencephalogr Clin Neurophysiol 60:407–416
22. Eklund G, Hagbarth KE, Hagglund JU, Wallin EU (1982) Mechanical oscillations contributing to the segmentation of the reflex electromyogram response to stretching human muscles. J Physiol 326:65–77
23. Friedemann HH, Noth J, Diener HC, Bacher M (1987) Long-latency EMG responses in hand and leg muscles. I. Cerebellar disorders. J Neurol Neurosurg Psychiatry 50:71–77
24. Ghez C, Shinoda Y (1978) Spinal mechanisms of the functional stretch reflex. Exp Brain Res 32:55–68
25. Gottlieb GL, Agarwal CG (1979) Response to sudden torques about the ankle in man: Myotatic reflex. J Neurophysiol 42:91–105
26. Hagbarth KE, Hägglund JV, Wallin EU, Young KK (1981) Grouped spindle and electromyographic responses to abrupt wrist extension movements in man. J Physiol 312:81–96
27. Hammond PH (1956) The influence of prior instruction to the subject on an apparently involuntary neuromuscular response. J Physiol (Lond) 122:17P–18P
28. Hammond PH (1960) An experimental study of servo action in human muscular control. Proceedings: IIIrd Int Conference on Medical Electronics:190–199. Institution of Electrical Engineers, London
29. Hendrie A, Lee RG (1978) Selective effects of vibration on human spinal and long loop reflexes. Brain Res 157:369–375
30. Hore J, Vilis T (1985) A cerebellar-dependent efference copy mechanism for generating appropriate muscle responses to limb perturbations. In: Bloedel JR, Dichgans J, Precht W (eds) Cerebellar functions. Springer, Berlin Heidelberg New York, pp 24–35
31. Hultborn H, Wigström H (1980) Motor response with long latency and maintained duration evoked by activity in Ia afferents. In: Desmedt JE (ed) Spinal and supra-spinal mechanisms of voluntary motor control and locomotion. Karger, Basel, pp 99–116
32. Katz R, Rondot P (1978) Muscle reaction to passive shortening in normal man. Electroencephalogr Clin Neurophysiol 45:90–99
33. Lee RG, Tatton WG (1975) Motor responses to sudden limb displacements in primates with specific CNS lesions and in human patients with motor system disorders. Can J Neurol Sci 2:285–293
34. Lee RG, Tatton WG (1982) Long-latency reflexes to imposed displacements of the human wrist: dependence on duration of movement. Exp Brain Res 45:207–216
35. Lee RG, Murphy JT, Tatton WG (1983) Long-latency myotatic reflexes in man: mechanisms, functional significance and changes in patients with Parkinson's disease or hemiplegia. In: Desmedt JE (ed) Motor control mechanism in health and disease. Raven, New York, pp 489–508
36. Lenz FA, Tatton WG, Tasker RR (1983a) Electromyographic response to displacement of different forelimb joints in the squirrel monkey. J Neurosci 3:783–794
37. Lenz FA, Tatton WG, Tasker RR (1983b) The effect of cortical lesions on the electromyographic response to joint displacement in the squirrel monkey forelimb. J Neurosci 3:795–805
38. Marsden CD, Merton PA, Morton HB (1973) Latency measurements compatible with a cortical pathway for the stretch reflex in man. J Physiol 230:58–59
39. Marsden CD, Merton PA, Morton HB, Adam J (1978) The effect of lesions of the central nervous system on long-latency stretch reflexes in the human thumb. In: Desmedt JE (ed) Cerebral motor control in man: long loop mechanisms. Karger, Basel, pp 334–341

40. Matthews PBC (1984) Evidence from the use of vibration that the human long-latency stretch reflex depends upon spindle secondary afferents. J Physiol 348:383–415
41. Mauritz KH, Schmitt C, Dichgans J (1981) Delayed and enhanced long-latency reflexes as the possible cause of postural tremor in late cerebellar atrophy. Brain 104:97–116
42. Miller AD, Brooks VB (1981) Late muscular responses to arm perturbations persist during supraspinal dysfunctions in monkeys. Exp Brain Res 41:146–158
43. Nashner LM (1983) Analysis of movement control in man using the movable platform. In: Desmedt JE (ed) Motor control mechanisms in health and disease. Raven, New York, pp 607–619
44. Noth J, Engel L, Friedemann HH, Lange HW (1984) Evoked potentials in patients with Huntington's disease and their offspring. I. Somatosensory evoked potentials. Electroencephalogr Clin Neurophysiol 59:134–141
45. Noth J, Podoll K, Friedemann HH (1985) Long-loop reflexes in small hand muscles studied in normal subjects and in patients with Huntington's disease. Brain 108:65–80
46. Oepen G, Doerr M, Thoden U (1981) Visual (VEP) and somatosensory (SSEP) evoked potentials in Huntington's chorea. Electroencephalogr Clin Neurophysiol 51:666–670
47. Rondot P, Metral S (1973) Analysis of the shortening reaction in man. In: Desmedt JE (ed) New developments in electromyography and clinical neurophysiology. Karger, Basel, pp 629–634
48. Rothwell JC, Obeso JA, Traub MM, Marsden CD (1983) The behaviour of the long-latency stretch reflex in patients with Parkinson's disease. Electroencephalogr Clin Neurophysiol 46:35–44
49. Scholz E, Diener HC, Noth J, Friedemann H, Dichgans J, Bacher M (1986) Long-latency EMG responses in leg muscles. II. Parkinson's disease. J Neurol Neurosurg Psychiatry 50:66–70
50. Struppler F, Lehmann-Horn F, Klein W, Lücking CH, Deuschl G (1984) Effect of stereoencephalotomy on long-latency EMG responses and motor control of arm movements in Parkinson's syndrome. Adv Neurol 40:437–445
51. Tatton WG, Lee RG (1975) Evidence for abnormal long-loop reflexes in rigid parkinsonian patients. Brain Res 100:671–676
52. Tatton WG, Forner SD, Gerstein GL, Chambers WW, Lin CM (1975) The effect of postural lesions on motor responses to sudden upper limb displacement in monkeys. Brain Res 96:108–113
53. Tatton WG, Bedingham W, Verrier MC, Blair RDG (1984) Characteristic alterations in responses to imposed wrist displacements in parkinsonian rigidity and dystonia musculorum deformans. Can J Neurol Sci 11:281–287
54. Tracey DJ, Walmsley B, Brinkman J (1980) Long-loop reflexes can be obtained in spinal monkeys. Neurosci Lett 18:59–65
55. Verrier MC, Tatton WG, Blair RDG (1984) Characteristics of EMG responses to imposed limb displacement in patients with vascular hemiplegia. Can J Neurol Sci 11:288–296
56. Westphal C (1880) Über eine Art paradoxer Muskelkontraktion. Arch Psychiatr Nervenkr 10:243–248
57. Wiesendanger M, Miles TS (1982) Ascending pathway to low-threshold muscle afferents to the cerebral cortex and its possible role in motor control. Physiol Rev 62:1234–1270

Long-Latency Stretch Responses in Man – Segmental versus Suprasegmental Hypothesis

L. GERILOVKSY [1], H. RIESCHER [1], and A. STRUPPLER [1]

The existence of long-latency responses to a sudden stretch of steadily activated muscles in humans raised two main questions:

1. To what extent can such responses account for segmental or suprasegmental mechanisms, involving the higher centers of the CNS?
2. What is their functional significance?

These two questions raised a number of investigations and hypotheses (for review see [2, 11]). The electromyographic (EMG) stretch response consists of three (sometimes even four) consecutive bursts: M_1, M_2, and M_3, according to Tatton et al. [9]. The typical segmented EMG pattern shows a variability of latencies increasing in the separate burst, while the degree of synchronization decreases from M_1 to M_3.

What is the process of segmentation itself? We do not deal with the wide range of modulations, resulting from the object of investigation and the experimental paradigm.

Animal experiments showed clearly that the mechanisms of the segmentation are built into the "hardware" programs and structures of the spinal segmental apparatus [4, 10]. We also performed a series of studies to prove these findings in healthy human subjects.

Stretch Response in the Relaxed Human Muscle

Using single motor unit (MU) recordings, the segmentation in the stretch response of brachial biceps was investigated. In order to avoid voluntary descending facilitatory drive, the experiments were done in absence of (a) initial voluntary activity in the muscle under stretch, and (b) prior instruction "to resist" against the perturbation. Therefore, the muscle investigated was fully relaxed and the instruction was "do not intervene against perturbation". Sudden torque pulses of various intensities, from 3.0 to 6.0 Nm, were applied.

The stretch-response latency gradually decreased with increasing perturbation torque and reached monosynaptic values only at high intensities of the torque pulse, from 5.0 to 6.0 Nm (Fig. 1).

At relatively weak intensities the sole maximum in the poststimulus histogram consisted almost entirely of the first MU spikes and only on rare occasions of an

[1] Neurologische Klinik und Poliklinik der Technischen Universität München, Möhlstr. 28, D-8000 München 80, FRG.

Clinical Aspects of Sensory Motor Integration
Ed. by A. Struppler and A. Weindl
© Springer-Verlag Berlin Heidelberg 1987

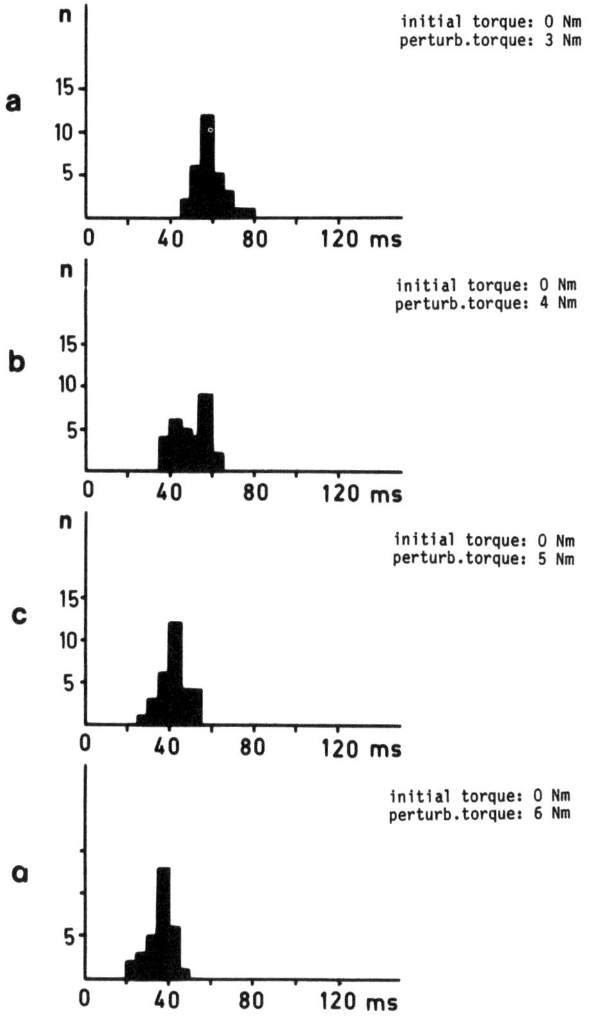

Fig. 1. Latency histograms of stretch responses evoked from initially inactive M. biceps brachii upon increasing the perturbation torque

insignificant number of second spikes at its end. An increase in the intensity of the torque pulse led to a segmentation of the stretch response into two or even three components. The poststimulus histogram showed that the first maximum, with highest amplitude, was formed predominantly from the first MU spikes, following the onset of the perturbation, while the later low-amplitude maxima are formed exclusively from the second MU spikes (Fig. 2).

The results showed that the segmentation of the stretch response, although not well expressed, could be observed in relaxed muscle without any instruction "to resist".

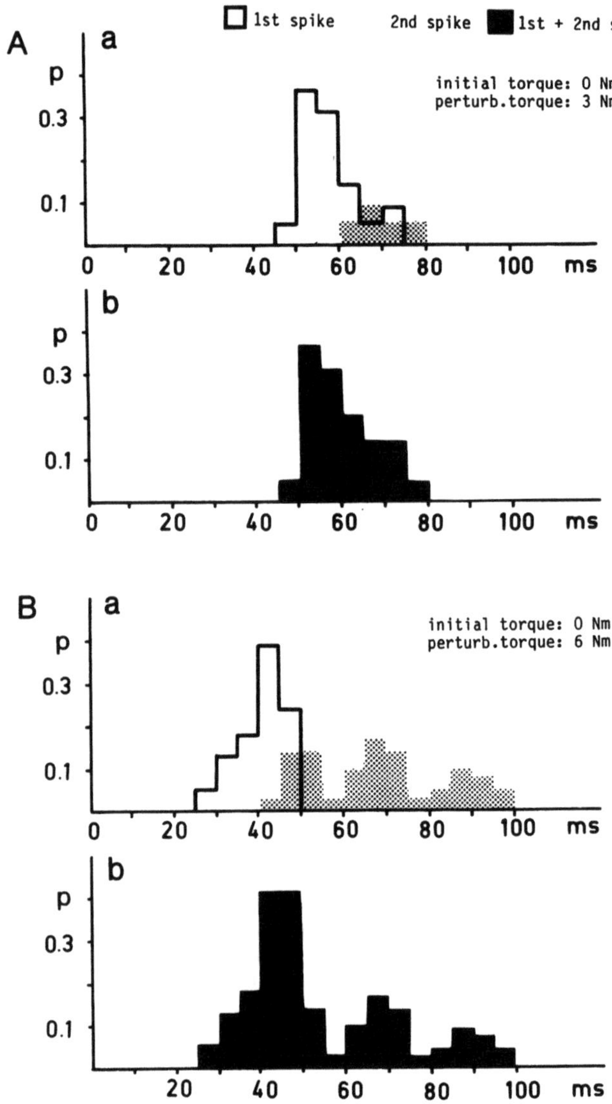

Fig. 2 A, B. Single motor unit poststimulus histograms of initially inactive M. biceps brachii upon increasing the perturbation torque. The histograms of the distribution of the 1st and 2nd MU spikes are shown after the onset of perturbation. The index *p* (on the *y-axis*) is calculated for each 5-ms time interval (on the *x-axis*) as a sum of the spikes in the interval divided by the number of trials (*n*). **A** (*a*) Histogram of 1st MU spike (*white bars*) and 2nd MU spike (*dotted bars*). (*b*) Histogram of the sum of the 1st and 2nd spikes (*black bars*). **B** Histograms of the same MU spikes as in **A** at increased perturbation torque

Stretch Response of Steadily Voluntary Activated Muscle

Under this condition the peristimulus histograms of the initially active and inactive MUs were different from the resting state.

Recruitment of Initially Active MUs in the Stretch Response

1. Two or three maxima could be seen in the peristimulus histogram, depending on the stimulus intensity; at higher intensities a third maximum appeared between the two already existing ones, the latter shifting slightly back in time.
2. The three maxima were better expressed than in a resting state.
3. In comparison with relaxed muscle, the latency of the earliest maximum at low intensity of the perturbation torque reached monosynaptic values sooner (Fig. 3).

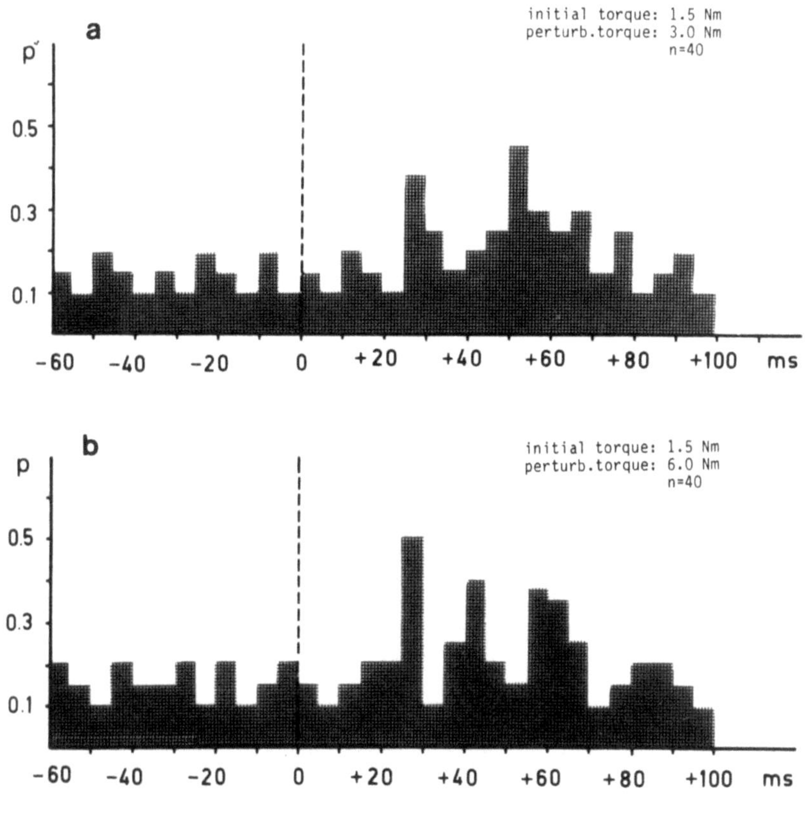

Fig. 3 a, b. Peristimulus histograms of initially active single motor unit of M. biceps brachii upon increasing the perturbation torque. Normalized histograms of the spike distribution before and after torque perturbation **a** *O*, onset of perturbation. **b** Results from the same MU at constant initial and different perturbation torque

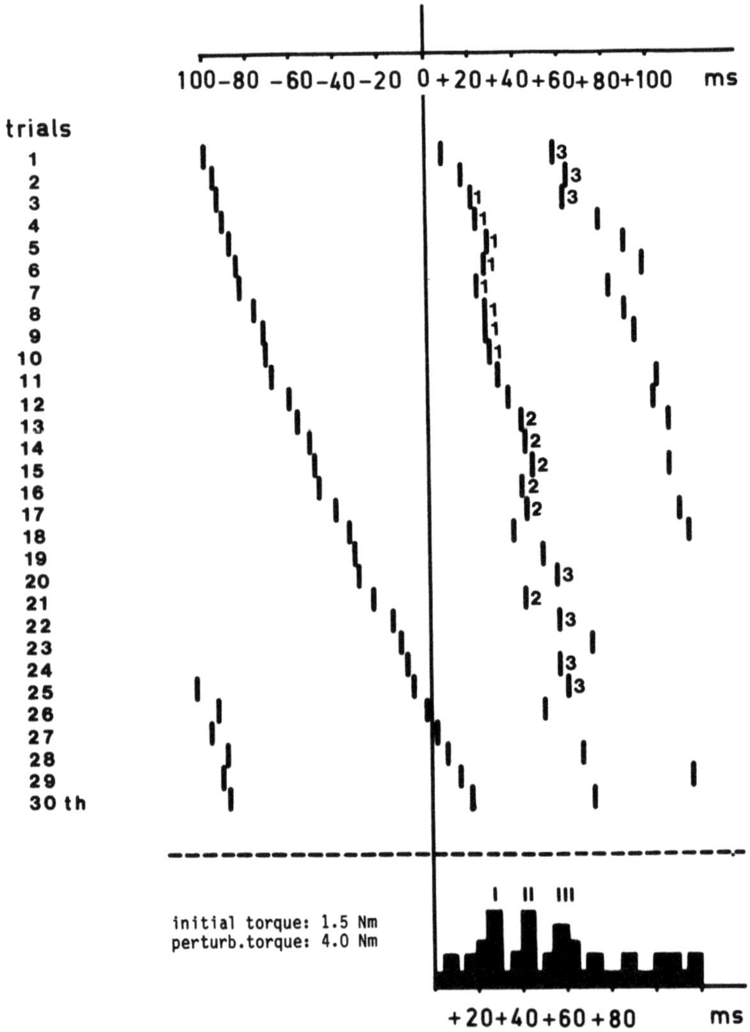

Fig. 4. Motor unit recruitment in stretch response of M. biceps brachii depending on the time interval between the onset of perturbation and the last motor unit spike preceeding it. Thirty trials with the same motor unit. *O*, onset of perturbation (MU spikes marked by *vertical bars*). *1*, *2*, and *3* denote the spikes participating in the 1st, 2nd and 3rd maxima, respectively (see the poststimulus histogram at the *bottom right corner*)

When the interval between the onset of the perturbation and the last preceding MU spike was sufficiently long, the first MU spike after the onset of the perturbation participated in the first maximum of the peristimulus histogram (from 3rd to 10th trials in Fig. 4). Shortening this interval, the MU spike participated in the second maximum, jumping over the first one (trials no. 13, 14, 15, 16, 17, and 21). Having appeared shortly before the first maximum, this very MU spike in a similar way jumped over the first and second maxima and sometimes participated in the third one (trials no. 1, 2, 20, 22, 24, and 25). Rarely did MU spikes recruited in the first maximum take part in the third one (3rd trial). These results confirmed previous findings of Gydikov et al. [5] and Noth et al. [8].

Recruitment of Initially Inactive MUs in the Stretch Response

1. The latency of the stretch response of the initially inactive MUs was longer than that of the active ones; its minimal values at high stimulus intensities rarely reached monosynaptic values.
2. Increasing the stimulus intensity the reflex volley was segmented into components.
3. The maxima in the peristimulus histogram (most frequently two) coincided in time with the corresponding second and third maxima of the initially active MUs (Fig. 5).

It is obvious that upon isometric voluntary activity, the earliest, short-latency response (M_1) is due to the spikes of the initially active MUs, while the two successive long-latency responses (M_2 and M_3) are formed of both initially active and inactive MUs.

Most of these results could be explained satisfactorily only by typical spinal mechanisms, i.e., after hyperpolarization, recurrent inhibition, influences via primary and secondary spindle afferents, Golgi tendon organs, etc.

It would, however, be an oversimplification to assume that the above-mentioned basic segmental mechanisms are the only ones responsible both for the segmentation of the stretch response into its components and for the determination of the parameters of the separate responses (especially the long-latency ones). The

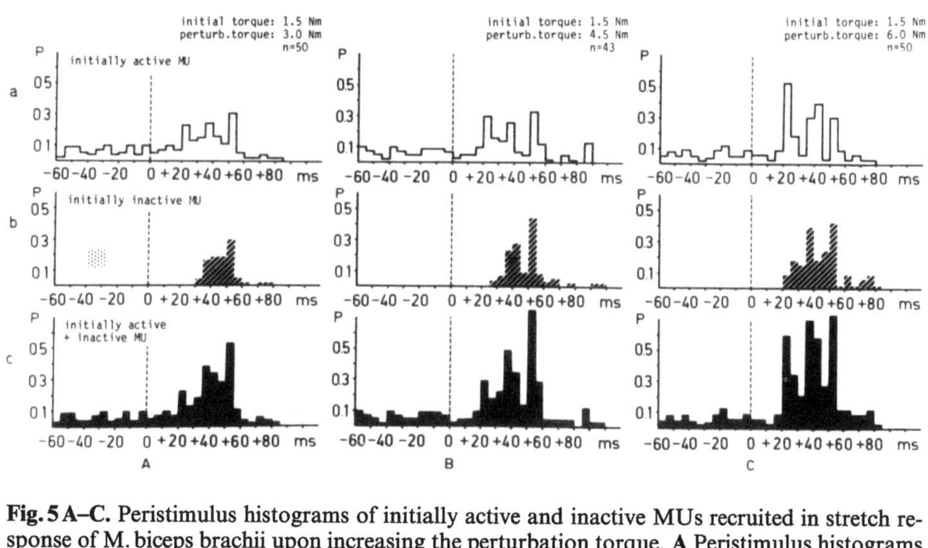

Fig. 5 A–C. Peristimulus histograms of initially active and inactive MUs recruited in stretch response of M. biceps brachii upon increasing the perturbation torque. **A** Peristimulus histograms of initially active MU at constant initial and different perturbation torque. **B** The same with initially inactive MU. **C** Summated histograms of initially active and inactive MUs

great number of remarkable investigations in this field (recently reviewed by Bonnet [1], Chan [2], and Wiesendanger and Miles [11]) undoubtedly show the coexistence of both spinal and supraspinal mechanisms in the stretch response; influences via supraspinal, transcortical or transcerebellar loops counteracting with the "basic" stretch response pattern would introduce essential modifications in the long-latency responses, especially in the condition of intended movement.

Voluntary Response to a Sudden Stretch: The Dual Nature of the Kinesthetic Reaction Time

The early EMG response to kinesthetic stimuli showed surprisingly low values – less than 70 ms, comparable with the latencies of the transcortical loops. An EMG response with such a latency interpreted as a reflex or automatic response [3] or a voluntary one [7] could play a crucial role in the interrelation between the spinal and supraspinal mechanisms.

In order to distinguish the early "kinesthetic" changes in the EMG from the "basic" stretch response pattern, we have used unilateral perturbation with a voluntary response to be performed with both hands.

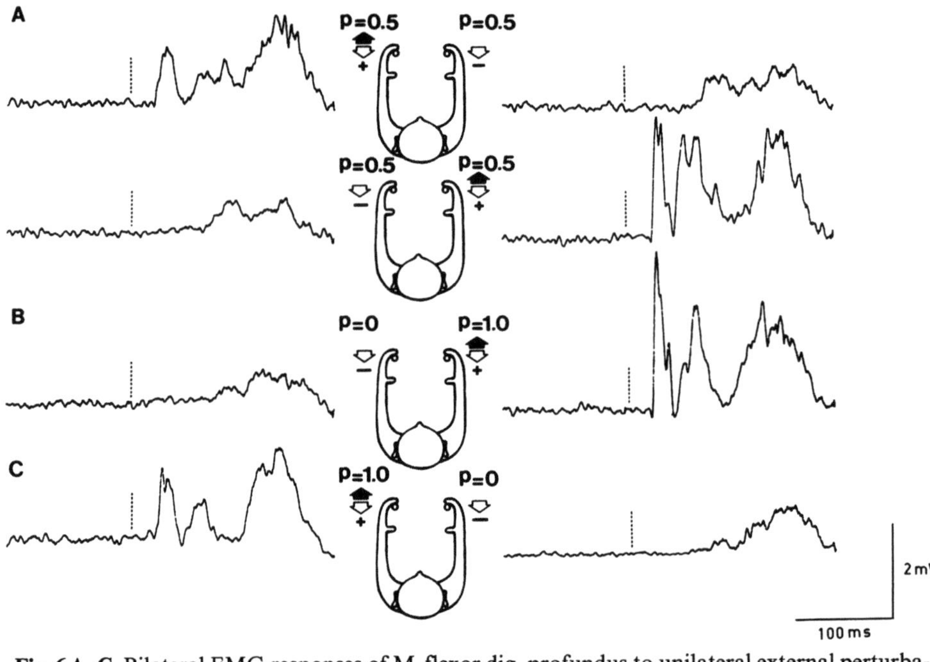

Fig. 6 A–C. Bilateral EMG responses of M. flexor dig. profundus to unilateral external perturbation. Rectified and averaged EMG responses (128 averagings on each side). **A** Randomly and equiprobably applied perturbation on both sides ($p = 0.05$). *Upper trace:* left side (+) is perturbated. **B** Only the right side (+) is perturbated ($p = 1.0$). **C** The left side (+) is perturbated ($p = 1.0$). Note the differences in EMG responses on the non-perturbated side between **A**, **B** and **C**, respectively

Bilateral Responses to Randomly and Equiprobably Distributed Unilateral Torque Perturbation

The subject resisted steadily with the fingers of both hands against an external force of 1.0 Nm, applied to the top phalangues by means of the levers of two precise, independently operating torque motors. According to the prior instruction, the subject flexed simultaneously the fingers of both hands, immediately after he felt the stretch. The torque perturbation (1.5 Nm) was equiprobably and randomly distributed to the left and right side, so that the subject did not know in advance which hand would be perturbated; such an experimental situation is optimal for activation of the hypothetical long loops of the deep finger flexors of both hands. During the experiments, the EMG recorded from the deep finger flexors with the use of wire electrodes, the mechanogram and dynamogram of both sides had been registrated (Fig. 6). Triggering with the onset of perturbation (Fig. 6 A) we found:

1. The EMG response of the perturbated side consisted of a short-latency component (23–26 ms) and two successive long-latency components – the first with a latency of 55–80 ms and the second with a latency of 115–135 ms.
2. The EMG response of the non-perturbated side consisted of two successive components – the first with a latency of 60–80 ms and a component with a latency of 115–135 ms, coinciding with the reaction time values to kinesthetic stimuli [3].
3. As a rule the second response on the perturbated side coincided in time with the first component of the non-perturbated side, with the only one exception (see B in Fig. 7).
4. The latency of the second component on the non-perturbated side coincided in time with the third response on the perturbated side, i.e., with the onset of the voluntary response to the kinesthetic stimulus.

While the second long-latency response on the non-perturbated side is a voluntary one, the first component could be (a) reflex or automatic, (b) a result of preset due to the forthcoming voluntary response, or (c) voluntary response.

Triggering with the onset of the dynamogram of the non-perturbated hand, the amplitude of the second long-latency EMG response on this (non-perturbated) side increased, while the first response disappeared (Fig. 8, right).

The results show the dual nature of the kinesthetic EMG response. While the second component is a voluntary one, the first component is time locked with the onset of perturbation and could not be a voluntary response or result of preset due to the forthcoming movement.

Triggering with the onset of perturbation, the first and second EMG components on the non-perturbated side had comparable amplitudes. The amplitude of the first component was originally low, whereas the amplitude of the second one was low after averaging due to the variability in the latency of the voluntary response. Triggering with the onset of the mechanical response, the amplitude of the second component increased and its values became comparable to that on the perturbated side (Fig. 8, right).

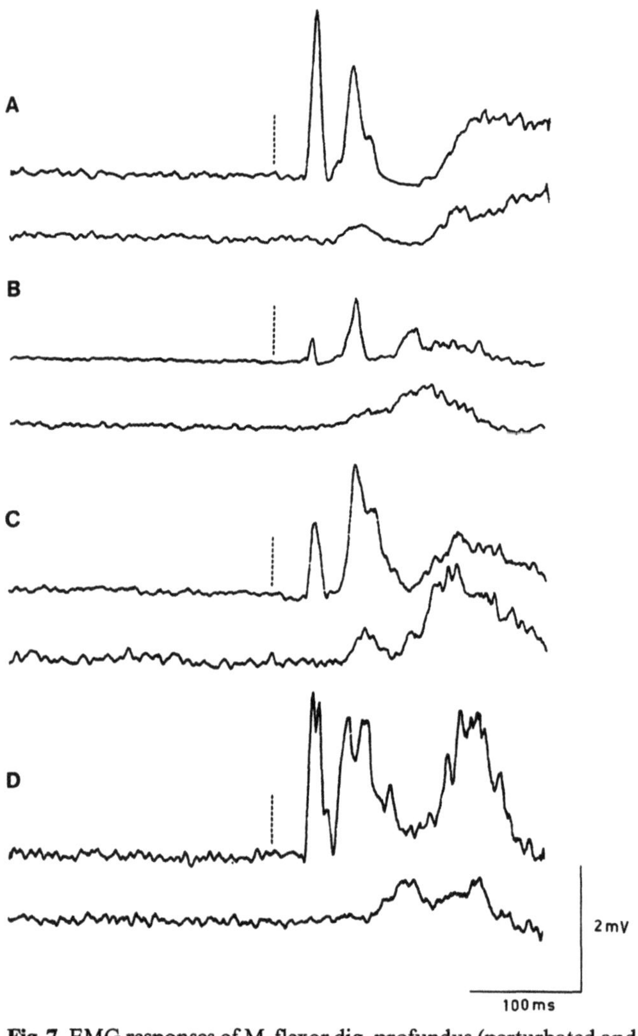

Fig. 7. EMG responses of M. flexor dig. profundus (perturbated and non-perturbated side). Rectified and averaged EMG responses (128 averagings on each side) in four subjects (**A, B, C,** and **D**). *Upper trace:* perturbated side. *Lower trace:* non-perturbated side

Bilateral Response to Expected Unilateral Perturbation

When the subject knew in advance which of the two sides would be perturbated (right hand in Fig. 6B and left hand in Fig. 6C), the first EMG component of the non-perturbated muscle tended to be lower, while the second one remained unchanged in comparison to Fig. 6A, when the perturbation was randomly and equiprobably applied on both sides; so the subject did not know in advance which hand would be perturbated.

The first component of the kinesthetic response could be spinal or supraspinal in origin.

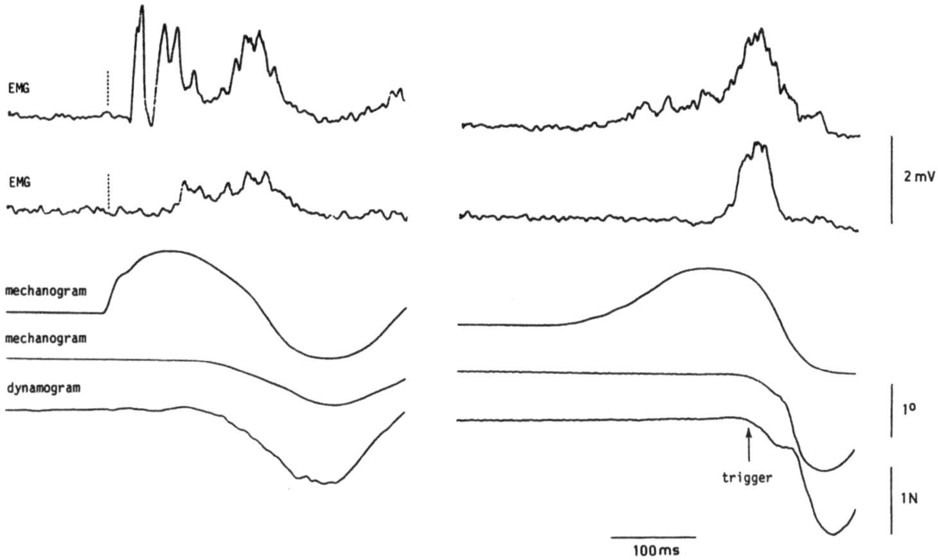

Fig. 8. Unilateral EMG responses depending on the mode of the triggering. *Left:* the averager is triggered by the onset of the perturbation (*vertical dashed line*). 128 averaged responses. *Right:* the same responses, when the averager is triggered by the onset of the voluntary response (*vertical arrow*). From *top* to *bottom:* EMG response after perturbation, EMG response on the same side, non-perturbated, mechanogram after perturbation, mechanogram without perturbation, dynamogram without perturbation

The *spinal hypothesis* includes two possibilities:

1. Excitation of the homonymous Ia endings of the non-perturbated finger flexors, due to the "jaw" from the torque motor on the contralateral side at the onset of the perturbation. This did not seem very likely because the latency of the first component was too long. Such an effect would have produced a response comparable with that of a monosynaptic one.
2. Influences from the contralateral perturbated side mediated via oligo- or polysynaptic pathways. Such a suggestion cannot be excluded.

The *supraspinal hypothesis,* although speculative, is very attractive for the following reasons:

1. It could explain one essential side of the interactions between spinal and supraspinal mechanisms during torque perturbation.
   If the first component, clearly expressed in the EMG response on the non-perturbated side, were a result of influences mediated via supraspinal pathways, one could expect similar influences on the perturbated side. These influences would be "masked" by the spinal stretch response pattern, however they could modify the long-latency responses.
2. Investigations dealing with the influence of the prior instruction on the long-latency (and, to a lesser extent, the short-latency responses had been started with Hammond's [6] experiments. Emphasis was placed mainly on the instruc-

tions (a) to "resist" or "not resist" a forthcoming displacement, and (b) to perform a voluntary movement with a given direction in response to perturbation, whether it lengthens or shortens the muscle, etc.

Assuming a supraspinal origin of the first component on the non-perturbated side, the present experiments suggest a new aspect in the possible influences of the prior instruction on the long-latency responses. In kinesthetic reaction time task (bimanual voluntary response to one-side torque perturbation) with two identical kinesthetic inputs from the agonists of the forthcoming movement, the amplitude of the first component depends on whether or not the subject knows in advance through which of both inputs the perturbation will come. In these cases, the prior instruction, determining the probability of the appearance of the perturbation for each input, could influence the activity of the relevant supraspinal, transcortical or transcerebellar loops. The mechanism underlying these hypothetical processes as well as the levels in the CNS at which they develop remain unclear. They probably belong to the category which psychophysiologists usually relate to the so-called "local attention".

Summary

The amount of segmental or suprasegmental mechanisms and their functional significance were investigated in forearm flexor muscles of normal subjects. The segmentation of the stretch response was observed in relaxed muscles (no instruction to resist). Under isometric activity M1 is composed entirely of the spikes of the initially active MUs while M2 and M3 consisted of both initially active and inactive MUs.

To discriminate in the EMG the early kinesthetic changes from the basic stretch response, bilateral responses to randomly and equiprobably distributed unilateral torque perturbations were investigated. The non-perturbated side shows a first component, with a latency of ca. 70 ms, independent of the latency of the reactive component. The amplitude of the first component varies depending on whether or not the subject knows in advance through which of both inputs the perturbation will come. In these cases the prior instruction, determining the probability of the appearance of the perturbation for each input, could influence the activity of the relevant supraspinal, transcortical or transcerebellar loops.

Acknowledgment. Part of this work was carried out while L. G. from the Central Laboratory of Biophysics, Bulgarian Academy of Sciences, Sofia, Bulgaria, was working as an A. v. Humboldt fellow at the Department of Neurology, Technical University, Munich.

References

1. Bonnet M (1984) Étude réfléxologique chez l'homme de la préparation au movement. Thèse de Doctorat. Dept Psychobiol Exp I.N.P. – C.N.R.S., Marseille
2. Chan CWY (1983) Segmental versus suprasegmental contributions to long-latency stretch responses in man. In: Desmedt JE (ed) Motor control mechanisms in health and disease. Raven, New York, pp 467–487

3. Desmedt JE, Godaux E (1978) Ballistic skilled movements: load compensation and patterning of the motor commands. Cerebral motor control in man: long-loop mechanisms. Prog Clin Neurophysiol 4:21–56
4. Ghez C, Shinoda Y (1978) Spinal mechanisms of the functional stretch reflex. Exp Brain Res 32:55–68
5. Gydikov A, Kossev A, Radicheva N, Tankov N (1981) Interaction between reflexes and voluntary motor activity in man revealed by discharges of separate motor units. Exp Neurol 73:331–344
6. Hammond PH (1956) The influence of prior instruction to the subject on an apparently involuntary neuromuscular response. J Physiol 132:17
7. Hufschmidt JH, Killimov N, Linke D (1977) A very short reaction time in man. Electroencephalogr Clin Neurophysiol 43:622
8. Noth J, Podoll K, Friedemann H-H (1985) Long-loop reflexes in small hand muscles studied in normal subjects and in patients with Huntington's disease. Brain 108:65–80
9. Tatton WG, Bawa P, Bruce IC, Lee RG (1978) Long-loop reflexes in monkeys: an interpretative base for human reflexes. Cerebral motor control in man: long-loop mechanisms. Prog Clin Neurophysiol 4:229–245
10. Tracey DJ, Walmsley B, Brinkman J (1980) "Long-loop" reflexes can be obtained in spinal monkeys. Neurosci Lett 18:59–65
11. Wiesendanger M, Miles TS (1982) Ascending pathway of low-threshold muscle afferents to the cerebral cortex and its possible role in motor control. Physiol Rev 62:1234–1270

Habituation of the Human Long-Latency Stretch Reflex and Its Cerebral Correlates

J. C. Rothwell [1], B. L. Day [1], A. Berardelli [2], G. Abbruzzese [3], and C. D. Marsden [1]

Many human reflexes are reported to show the phenomenon of habituation, by which is meant the gradual decrease in size of a response to a repetitive series of identical stimuli. Among cutaneous reflexes, for example, the electrically induced blink reflex [6] and the flexor reflex after electrical stimulation of the foot [12] habituate to a series of stimuli. In both instances it is the late components of the response which are particularly affected [4, 11, 12]. However, habituation of the mechanically evoked stretch reflex has not been noted, and the frequency of stretches has not been considered as a possible influence on its size. Part of this work has been reported in full elsewhere [1, 2].

The effect of stretch presentation rate on the size of the stretch reflex electromyographic (EMG) response in the flexor muscles of the forearm was investigated in 7 normal individuals. Subjects were instructed to maintain a constant position of their wrist (160° flexion) against a steady background force offered by a torque motor, and stretches were given by increasing the torque applied by the motor by a factor of six for 50 or 200 ms (see figure legends for details). This size of wrist stretch produced submaximal stretch reflexes (both short- and long-latency components) in all subjects studied. Stretches were given in batches of 16 trials at regular intervals from 0.1 Hz to 1 Hz. Data from two sets of 16 trials were then averaged. Subjects were instructed not to react to the stretch in any way (i.e., by opposing or assisting the motor), but to maintain a constant level of muscle activation throughout the experiment. The size of the stretch reflexes was measured from average records of surface, rectified EMG from the flexor muscles of the forearm. First, the duration of the spinal and long-latency components of the EMG responses were estimated visually. Then the size of the reflex at every repetition rate was taken as the integral of the EMG activity within that interval and normalised to the background level of EMG in the 50-ms period prior to stretch onset.

Figure 1 shows that the size of the stretch reflex EMG response was highly dependent upon the rate at which wrist stretches were applied. The size and peak velocity of the stretch was unaffected by presentation rate. The EMG response consisted of the usual short- and long-latency components with onset about 25 and 50 ms after stretch. The size of the short-latency component was the same at all presentation rates (see Figs. 1 and 2). However, there was a dramatic decrease

[1] Institute of Psychiatry, Department of Neurology, De Crespigny park, Denmark Hill, London, SE 5 8AF, UK

[2] Quinta Clinica Neurologica, Dipartimento di Scienzi Neurologiche, Università di Roma, I-Roma, Italy.

[3] Clinica Neurologica, Università di Genova, I-Genova, Italy

Clinical Aspects of Sensory Motor Integration
Ed. by A. Struppler and A. Weindl
© Springer-Verlag Berlin Heidelberg 1987

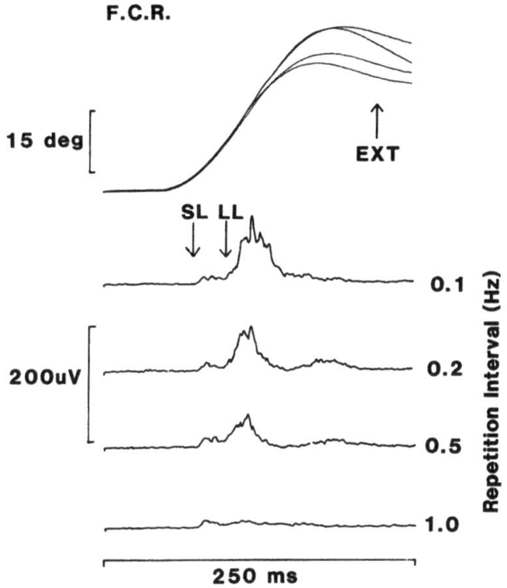

Fig. 1. Effect of changing stretch presentation rate on the size of the reflex EMG responses (*bottom 4 channels*) recorded in the flexor muscles of the forearm (*FCR*, flexor carpi radialis) following forcible extension of the wrist in one subject. Wrist position records are superimposed in the *top trace*. Background torque of 0.16 Nm rising to 1.1 Nm for 200 ms, 50 ms after start of sweep. The onset of the short-latency (*SL*) and long-latency (*LL*) components of the EMG response are marked by the *vertical arrows*. The size of the SL component remains the same at all four presentation rates, whilst the size of the LL component decreases in size considerably at higher frequencies. Mechanical consequences of the change in reflex size can be seen in the position records above. The records which show the most rapid return to the start position belong to the traces with the largest EMG responses. All traces are the grand average of 32 single trials

in the size of the long-latency response at short interstimulus intervals. Its size with repetition rates of 0.15 Hz was some 5–10 times larger than at 1.0 Hz.

We have shown previously that wrist stretch can evoke reproducible cerebral-evoked responses localised over the contralateral central area (see [1]). These potentials probably are produced by activity in muscle afferent fibres, since they persist during cutaneous anaesthesia of the hand. They consist of a primary negative/positive complex with onset latency of 25 ms and a secondary, positive/negative complex beginning about 40 ms after stretch. The size of these cerebral potentials, like that of the muscle stretch reflexes, was greatly affected by stimulus presentation rate (Fig. 2). The secondary component was particularly affected, decreasing in size progressively as the repetition rate increased from 0.15 to 0.5 to 1 Hz ($P<0.05$). The primary component was only reduced at the maximum repetition rate of 1 Hz.

Habituation of the human long-latency stretch reflex has not been noted before. It is a large effect which could be produced by changes at the level of alpha motoneurones, interneurones or sensory receptors within the reflex arc. Alpha motoneurones are unlikely to be the site of habituation in the present study in

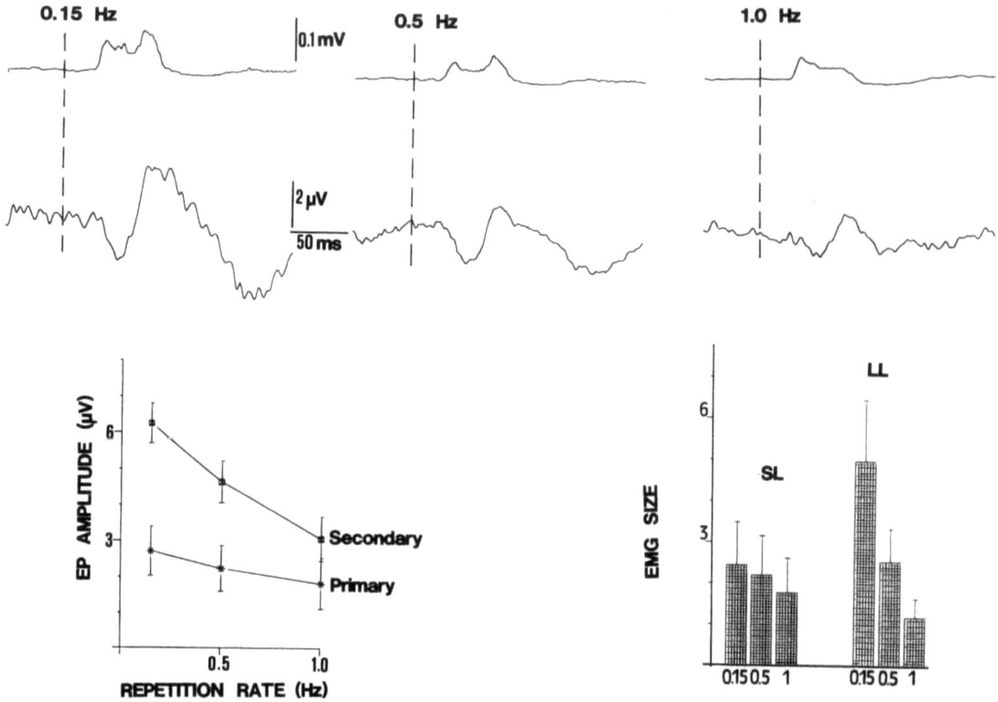

Fig. 2. Effect of repetition rate on the amplitude of the cerebral-evoked potentials and on the size of the short- (*SL*) and long-latency (*LL*) components of the muscle stretch reflex. At the *top* are shown the rectified EMG activities from the flexor muscles in the forearm (*upper trace*) and the evoked potentials from an electrode over C3 (referred to linked mastoid; filtered 0.8 Hz to 2.5 kHz) (*lower trace*) in a single normal subject at three different repetition rates (0.15, 0.5, and 1.0 Hz). The primary complex in the EEG is the first negative/positive deflection; the secondary complex is the following positive/negative wave (upwards deflection is negative). The *vertical dotted line* indicates the time of stretch onset. Background torque was 0.24 Nm, rising to 1.4 Nm for 50 ms. Each trace is the average of 128 trials. At the *bottom* is shown the effect of repetition rate on the peak-peak amplitude of primary and secondary complex of the cerebral potentials (*left*) and on the size of the SL and LL stretch reflexes (*right*). The results are the mean data ± 1 SE from 7 subjects. (Paired *t*-test in individual subjects on the logarithmically-transformed data: $P<0.05$ for the secondary complex of the EEG and $P<0.001$ for the long latency stretch reflexes at all repetition rats. Primary component of the cerebral potential only reduced at 1 Hz; $P<0.05$). (Abbruzzese [1])

view of the relatively long (1 s or more) interstimulus intervals used. Similarly, there is no indication from muscle spindle recordings made in man that spindle discharge is affected by the rate at which stretches are applied. We would favour the possibility that habituation occurred at the level of interneurones involved in the reflex arc. Differential habituation of short- and long-latency components of the reflex would then be consistent with the proposal that the two components of the reflex are mediated via two separate pathways, since different sets of interneurones would be involved in each pathway.

It has been suggested by many authors (e.g., [7, 8]) that the long-latency component of the stretch reflex traverses a transcortical pathway. If this is so, then the changes seen in the cerebral-evoked potentials may reflect changes in cortical motor output in response to sensory input. There is time for this to happen, since onset of the secondary complex (35 ms) is some 15 ms prior to onset of the long-latency muscle reflex, which is the time required to activate forearm muscles by direct stimulation of human motor cortex [9].

Adaptation of long-latency stretch reflexes is of importance when comparing reflex sizes seen in different pathological states in man. It may be that changes in reflex size are not due to any specific effect on the stretch reflex mechanism. Non-specific effects on the rate of reflex habituation also must be considered. For example, habituation of the blink and cutaneous reflexes is known to be decreased in patients with Parkinson's disease and to be increased in patients with Huntington's disease [3, 5]. If this applies to the long-latency stretch reflex it may explain why some authors have found these reflexes to be enhanced in Parkinson's disease [7] and decreased in Huntington's disease [10].

Summary

The effect of stretch presentation rate on the size of the stretch reflex EMG response in the flexor muscles of the forearm was investigated. When stretches were given every 1 s, the size of the long latency component was much smaller than if stretches were given every 10 s. The size of the short-latency component was unaffected. Stretch also produced scalp-recordable somatosensory-evoked potentials. The secondary component of the SEP (beginning about 40 ms after stretch) decreased in size as the stretch repetition interval was decreased. The importance of stretch repetition interval in testing for abnormalities of the stretch reflex or SEP in disease states is discussed.

Acknowledgments. We should like to thank Mr. H.C. Bertoya and Mr. R. Miller for building and maintaining the equipment used in these experiments. The work was funded by the Medical Research Council of Great Britain. A. B. was supported by the British Council, G. A. by the European Training Programme in Brain and Behaviour Research, and J.C.R. by The Royal Society.

References

1. Abbruzzese G, Berardelli A, Rothwell JC, Day BL, Marsden CD (1986) Cerebral potentials and electromyographic responses evoked by stretch of wrist muscles in man. Exp Br Res 58:544–551
2. Berardelli A, Day BL, Marsden CD, Rothwell JC (1986) Habituation of the human long latency stretch reflex. J Physiol 362:27P
3. Caraceni T, Avanzini G, Spreafico R, Negri S, Broggi G, Girotti F (1976) Study of the excitability cycle of the blink reflex in Huntington's chorea. Eur Neurol 14:465–472
4. Gregoric M (1973) Habituation of the blink reflex. In: Desmedt JE (ed) New developments in electromyography and clinical neurophysiology. Karger, Basel, pp 673–677
5. Kimura J (1973) Disorders of interneurons in parkinsonism. The orbicularis oculi reflex to paired stimuli. Brain 96:87–96

6. Kugelberg E (1952) Facial reflexes. Brain 75:385–396
7. Lee RG, Tatton WG (1975) Motor responses to sudden limb displacements in primates with specific CNS lesions and in human patients with motor system disorders. Can J Neurol Sci 2:285–293
8. Marsden CD, Merton PA, Morton HB (1973) Is the human stretch reflex cortical rather than spinal? Lancet I:759–761
9. Merton PA, Morton HB (1980) Stimulation of the cerebral cortex in the intact human subject. Nature 285:227
10. Noth J, Friedemann K, Podoll K, Lange HW (1983) Absence of long latency reflexes to imposed finger displacements in patients with Huntington's disease. Neurosci Lett 35:97–100
11. Penders CA, Delwaide PJ (1971) Blink reflexes studies in patients with Parkinsonism before and during therapy. J Neurol Neurosurg Psychiatry 36:674–678
12. Shahani BT, Young RR (1971) Human flexor reflexes. J Neurol Neurosurg Psychiatry 34:666–677

Torque-Induced Stretch Responses – Changes Due to Hypotonia

A. Struppler [1], H. Riescher [1], and L. Gerilovksy [1]

How can sudden external perturbation be compensated in hypotonic states? How will the CNS react to sudden stretch under isometric conditions?

In order to investigate this problem, a special type of hypotonia was chosen, i.e., the hypotonia following small stereotaxic lesions for treatment of parkinsonian rigidity and tremor.

Discrete stereotaxic lesions in the nucleus ventrolateralis of the thalamus (VL) and in the subthalamus can abolish resting, postural, and intention tremor. Concomitantly, rigidity as well as tonic reflex excitability are decreased or even abolished. Reinforcement (Jendrassik's maneuvre or mental drive) no longer evokes tremor or rigidity. The patient shows a postural hypotonia that can be compensated for by visual control. The rebound phenomenon (Stewart-Holmes) following sudden unloading of the muscle (isometrically activated by holding a light weight) is diminished. Motor force, rapid movement, and phasic stretch reflexes, however, remain unchanged.

If the lesion is small enough, there is no ataxia and no deficit of the so-called epicritic sensitivity, e.g., perception of touch or joint movement. Evaluation of muscle force, however, seems to be disturbed.

The flexor muscles of the forearm were investigated because they contribute to posture, particularly in the upright position. Moreover, these muscles are used predominantly for developing force and for adaptation to external perturbations in comparison with our finger muscles.

Table 1. Clinical findings following thalamotomy (*VL*) and subthalamotomy

Rigor	↓
Tremor – resting / postural / intentional	↓
Tonic stretch reflexes	↓
Postural tone	↓
Load compensation	↓
Phasic stretch reflexes	=
Maximal force	=
Muscle sense (force and movement perception)	?

[1] Neurologische Klinik und Poliklinik der Technischen Universität München, Möhlstr. 28, D-8000 München 80, FRG.

Clinical Aspects of Sensory Motor Integration
Ed. by A. Struppler and A. Weindl
© Springer-Verlag Berlin Heidelberg 1987

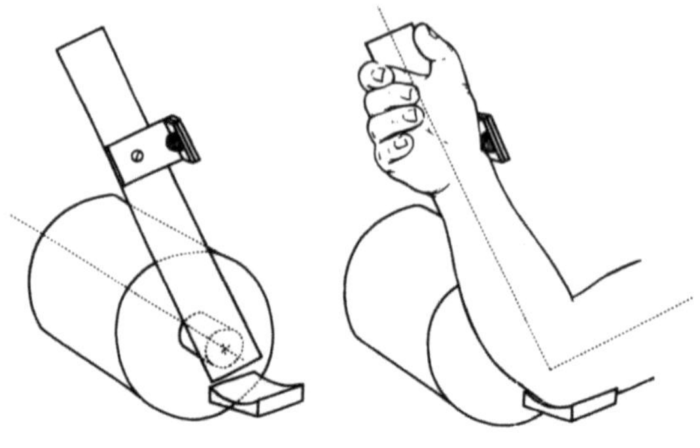

Fig. 1. Experimental setup

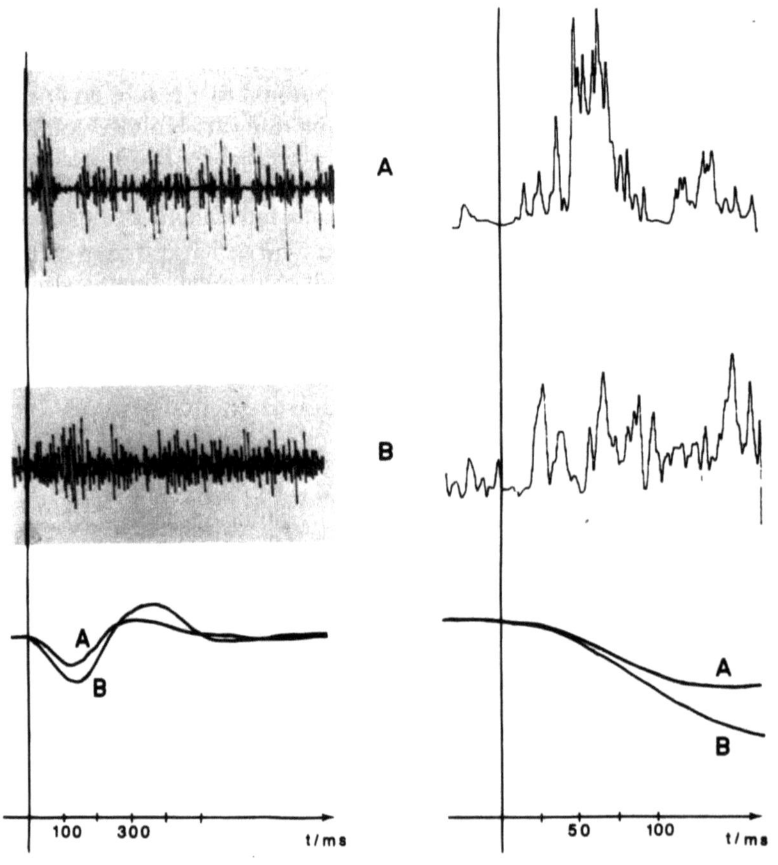

Fig. 2 A, B. Electromyographic (*upper* and *middle trace*) and mechanographic (*lower traces*) recordings of stretch-induced responses **A** before and **B** after operation. The perturbation begins at 0 ms. EMG recordings from brachial muscle, mechanical deviation of the forearm recorded at the elbow joints. Note the larger mechanical deflection following operation. *Left:* single sweep. *Right:* average of 8 sweeps. EMG activity rectified. (Lehmann-Horn et al. [1])

The following technique was used: both forearms of a subject were fixed to 2 levers moved by a pair of precision torque motors, which add a torque of 6 Nm to the gravity acting on the forearm. Additionally, sudden torque pulses (0–6 Nm, rise time 5 ms, duration 1 s) could be applied. EMG, joint angie (position), acceleration, and force were recorded.

In parkinsonian patients there is an increased M2 component, as shown by Tatton and Lee [4]. Following stereotaxic lesions, M2 was clearly normalized, associated with a pronounced overshoot and an elongation of the elbow flexors [1].

To find out whether the overshoot, for instance, following thalamotomy or even polyneuropathies, was only due to a decreased antagonistic triceps stretch reflex or caused by different types of innervation in the forearm flexor muscles during the holding of a weight, we tried to produce hypotonia only in the flexor muscle of normal subjects by applying a differential block of gamma axons con-

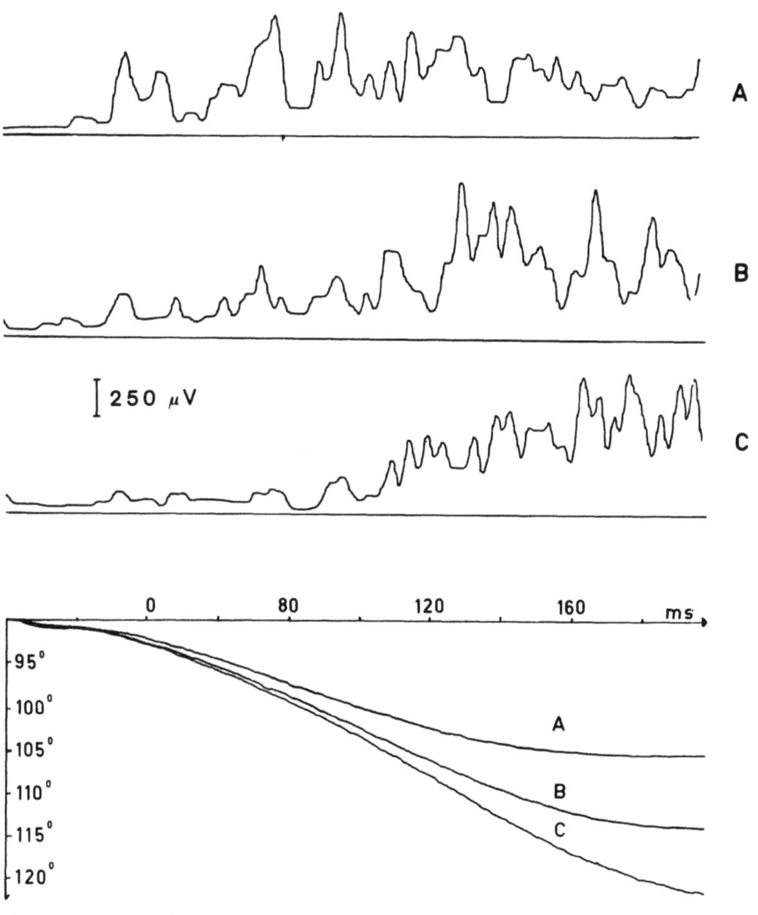

Fig. 3 A–C. The effect of differential block of the musculocutaneous nerve (Xylocaine). **A** before, **B** after 8 min, **C** after 18 min. Note that all M-compc ıents are abolished even though the initial velocity is higher, when the block is effective as shown in **C.** (Struppler [3])

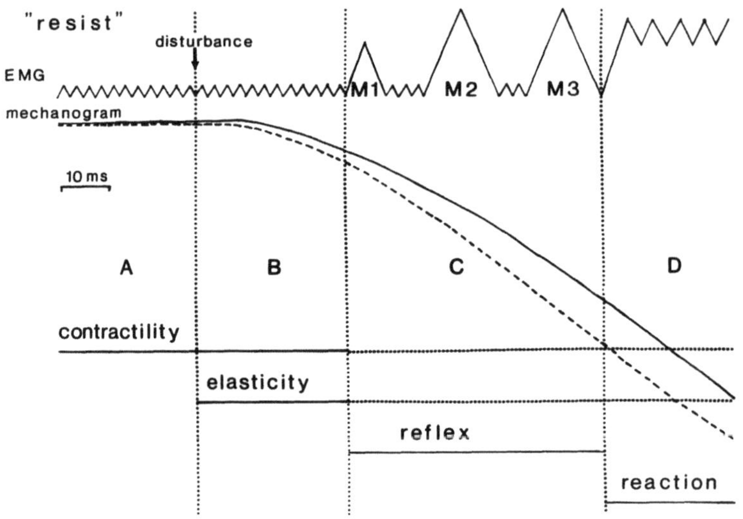

Fig. 4. Factors producing muscle stiffness during "resist" (**A–D**). The *dotted line* represents the slope of a poststereotaxic patient. (Struppler [3])

comitantly with other small nerve fibers in the musculocutaneous nerve. This procedure appears to be appropriate for studying the role of the gamma system during active innervation.

With this method we could restrict hypotonia to the agonistic muscle and avoid hypotonia in the antagonistic triceps. During differential block the torque-induced elongation of the flexor muscles was increased and the following compensation revealed on overshoot. The effect pointed to changes in the control of the flexor muscles and not to hypotonia of the antagonistic triceps muscles [2].

The compensation of the external perturbation is an adaptive function of our motor system. Here we may theoretically discriminate 3 main functions:

1. Contractility and elasticity of activated muscle, which counteract external disturbance.
2. Proprioceptive reflexes, whether spinally or supraspinally mediated.
3. Voluntary compensation.

Follwing the stereotaxic lesion we have observed two phenomena: (a) a normalization of the former increased amplitude of M2, and (b) changes in the mechanogram preceding the mechanical effect of the reflex activity.

Therefore the question arises: Is this kind of hypotonia primarily caused by a decrease of muscle stiffness?

In order to differentiate whether hypotonia is due to decreased reflex activity and additionally reduced non-reflex muscle stiffness, we tried to quantify the latter. With this methodology one can distinguish two modes of innervation serving muscle tone during isometric innervation: (a) early compensation of sudden external disturbances before reflex onset, and (b) compensation via reflex arc.

In order to evaluate indirectly muscle stiffness, we have to measure the mechanogram and the dynamogram, using a force transducer placed between the lever

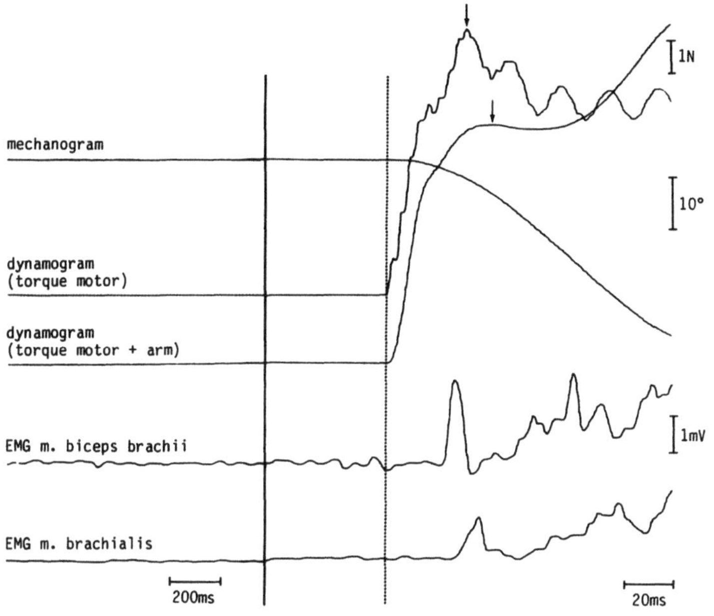

Fig. 5. Modification of mechanical parameters during torque disturbance

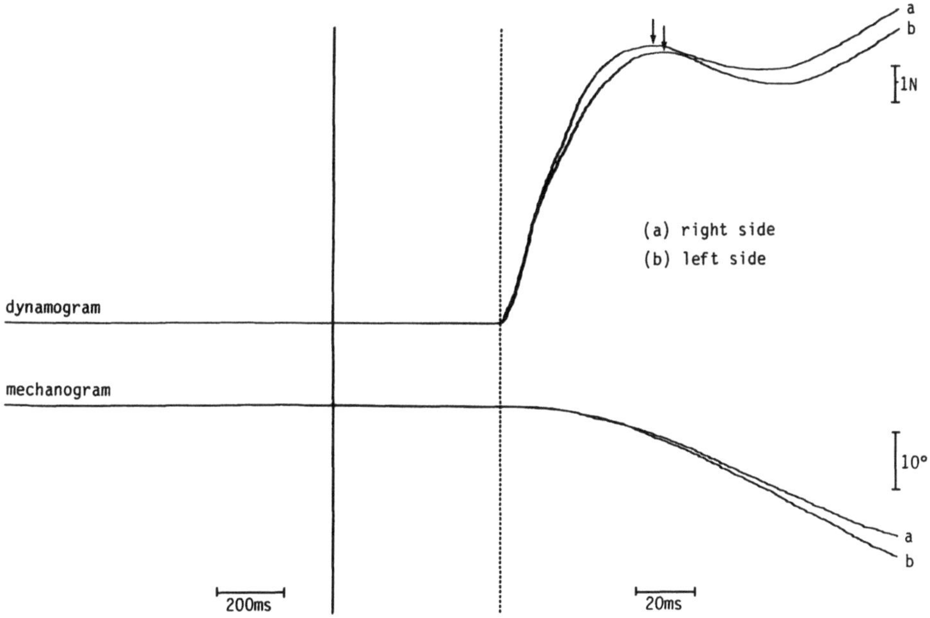

Fig. 6. Left-side hypotonia following stereotaxic lesion. The latencies of the local maxima are longer at the hypotonic side

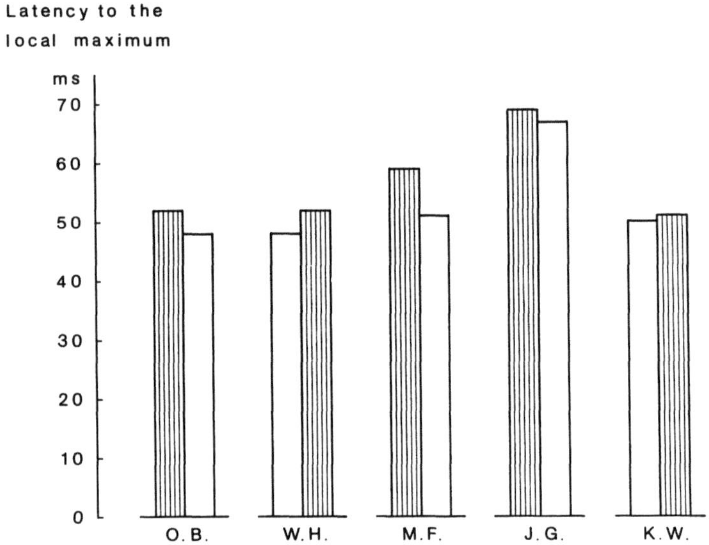

Fig. 7. Comparison of latencies between left (*left column*) and right side (*right column*) in 5 patients. The latency on the operated side (*filled column*) is longer

of the machine and the arm of the patient (see Fig. 3). Recording the dynamogram developed only by the torque motor under fixed conditions, we will find a typical curve with well-expressed local maxima and mechanical oscillations, superimposed on the curve.

Applying the force of the human arm, however, local maximum shifts backward in the time and the oscillations are abolished.

We investigated patients showing slight contralateral one-side hypotonia following small stereotaxic lesions, measuring the latency of the local maxima of the dynamogram and the changes in the mechanogram.

The latencies of the local maxima are longer on the hypotonic side. The changes following perturbation are better expressed in the dynamogram than in the mechanogram.

The absolute values of the latencies of the local maxima of both the operated and the non-operated side are shown in Fig. 9.

How can we interprete these findings?

Assuming that the results of decreased muscle stiffness following stereotaxic lesions can further be confirmed, two mechanisms could be considered:

1. The original population of motor units generating the controlled movement is changed to a new one with different intrinsic mechanical properties.
2. The distribution of activity into the synergistic muscles is changed, in this case into the various forearm flexor muscles (brachialis, biceps, and brachioradialis).

For immediate compensation of external perturbations muscle stiffness seems to be important. For this purpose we probably need a special feedback originating

Table 2. Different causes of hypotonia and their influence on the phasic stretch reflex

Hypotonia		
Can arise following:		Phasic stretch reflex
1. Dorsal root lesions	(Spindle afferents ↓)	↓
2. Differential block	(Gamma efferents ↓)	↓
3. Dorsal column lesions	(Ascending spinal afferents ↓)	Normal
4. Cerebellar lesions	(Alpha-gamma-linkage ↓)	Normal
5. Thalamotomy or subthalamotomy	(Ascending muscle afferents ↓?)	Normal

in the skeletal muscle for continuous holding of an adequate ratio between muscle force and length.

Hypotonia can be caused by lesions at various levels in the somatosensory afferent systems. In hypotonia of peripheral origin, e.g., the polyneuropathy syndrome, usually there are deficiencies in many somatosensory modalities such as touch and vibration, and the tendon reflexes are abolished. In the hypotonia of subthalamic origin, however, the so-called epicritic sensibility is unchanged. Here we could discuss the influence of muscle afferents, according to neuroanatomical and electrophysiological findings.

When hypotonia of subthalamic-thalamic origin is more pronounced than that of peripheral origin, as in polyneuropathy, one could assume a higher density in the representation of this afferent system at the thalamic level.

Summary

To differentiate whether hypotonia is due to decreased reflex activity and additionally reduced non-reflex muscle stiffness, torque-induced stretch responses in the forearm flexor muscles were investigated. In hypotonia following stereotaxic lesions on the subthalamic-thalamic level, the early compensation of sudden external disturbances is diminished *before* reflex onset. A special feedback system is discussed, originating in the skeletal muscle serving continuously the adequate muscle stiffness during sustained movements. This facilitatory system may have the highest density at thalamic level, in contrast to the representation in the peripheral nerve.

Acknowledgment. Part of this work was performed while L. G. from the Central Laboratory of Biophysics, Bulgarian Academy of Sciences, Sofia, Bulgaria, was working as an A. v. Humboldt fellow at the Department of Neurology, Technical University, Munich.

The excellent assistance of Ms. Kramp and Ms. Pahlke is gratefully acknowledged. This work was supported by a grant of the Deutsche Forschungsgemeinschaft (SFB 220).

References

1. Lehmann-Horn F, Struppler A, Klein W, Lücking CH, Burgmayer B, Deuschl G (1982) Veränderungen der motorischen Kontrolle bei Parkinson-Patienten. In: Struppler A (Hrsg) Elektrophysiologische Diagnostik in der Neurologie. Thieme, Stuttgart, S 236–237
2. Struppler A (1979) „Untersuchungen zur Kontrolle der Motorik am Menschen". Wissenschaftl Z Ernst-Moritz-Arndt-Universität, Greifswald, Jahrg XXVIII, Med Reihe, Heft 2, 181–187
3. Struppler A (1985) Stereoencephalotomy and motor control. In: Struppler A, Weindl A (eds) Electromyography and evoked potentials. Springer, Berlin Heidelberg New York Tokyo, pp 37–44
4. Tatton WG, Lee RG (1975) Motor responses to sudden upper limb displacements of the wrist in normal humans and parkinsonian patients. Clin Res 22:755a

VI. Motor Functions of Basal Ganglia

The Basal Ganglia and Sensorimotor Integration

M.R. De Long [1] and G.E. Alexander [1]

Functional Organization

Although the basal ganglia were once viewed as lacking in specificity of connection and function, evidence accumulated over the past two decades has revealed a level of organization and functional specificity paralleling that of the cerebral cortex itself. The purpose of this review is to outline our current view of the functional organization of the basal ganglia in primates, with emphasis on the basis of sensorimotor integration. For a more detailed discussion of this and related topics, see [5, 10, 12].

The putamen and caudate nucleus (which together constitute the primate neostriatum) are generally considered "receptive" portions of the basal ganglia, since in combination they receive topographic inputs from nearly the entire neocortex. Both the neostriatum and the subthalamic nucleus (STN) send topographically organized projections to both the internal segment of the globus pallidus (GPi) and the pars reticulata of the substantia nigra (SNr), which, in combination, give rise to most of the efferent projections and thus constitute the major "output" portion of the basal ganglia. The principal targets of GPi and SNr projections are portions of the ventrolateral and ventroanterior nuclei of the thalamus. Through these projections, basal ganglia influences are returned to discrete portions of premotor and prefrontal cortex. Through descending projections, directed principally to the superior colliculus and the pedunculopontine nucleus, the basal ganglia also exert some influence at upper brainstem levels. The basal ganglia do not appear to receive direct sensory input from the periphery, nor do they give rise to any direct projections to the segmental motor apparatus.

It had been suggested earlier that the basal ganglia might serve to integrate input from widespread cortical areas and to funnel these influences, via the ventrolateral thalamus, to the motor cortex [6, 16, 23]. The basal ganglia were thus viewed as providing a route whereby influences from the cortical association areas might be transmitted to the motor cortex and thereby participate in the initiation and control of movement. On the basis of further anatomical and physiological data, however, it was proposed recently that influences projected to the basal ganglia from cortical sensorimotor areas remain segregated from influences projected to the basal ganglia from cortical "association" areas [10]. Two functionally distinct loops through the basal ganglia were proposed: (a) a "motor" loop passing through the putamen, which receives inputs from premotor and sensorimotor cortex and whose influences are ultimately directed to certain premotor

[1] Departments of Neurology and Neuroscience, Johns Hopkins University School of Medicine, 5–185 Meyer Bldg., 600 N. Wolfe Street, Baltimore, Maryland, USA.

Clinical Aspects of Sensory Motor Integration
Ed. by A. Struppler and A. Weindl
© Springer-Verlag Berlin Heidelberg 1987

areas, and (b) an "association" (or "complex") loop passing through the caudate nucleus, which receives input from widespread association areas and whose influences are ultimately returned to portions of the prefrontal cortex [10, 11].

Recent anatomical and physiological findings have strengthened the concept of segregated basal ganglia-thalamocortical circuits and provided evidence for further anatomical and functional differentiation within these pathways [5]. The proposed circuits are shown schematically in Fig. 1. The basic organization is thought to be similar for each as indicated in Fig. 2. Each circuit appears to receive multiple, partially overlapping corticostriate inputs, which are progressively integrated in their subsequent passage throught the palladium and substantia nigra to a restricted portion of the thalamus, and from there back to a single cortical area. Each circuit is, therefore, partially closed by the restricted thalamocortical projection returned to one of that circuit's multiple sources of corticostriate input. It appears that integration within each of the proposed basal ganglia-thalamocortical circuits is carried out at striatal, pallidal/nigral, and thalamic levels. In this limited sense, this organizational scheme retains the concept of "funneling" that

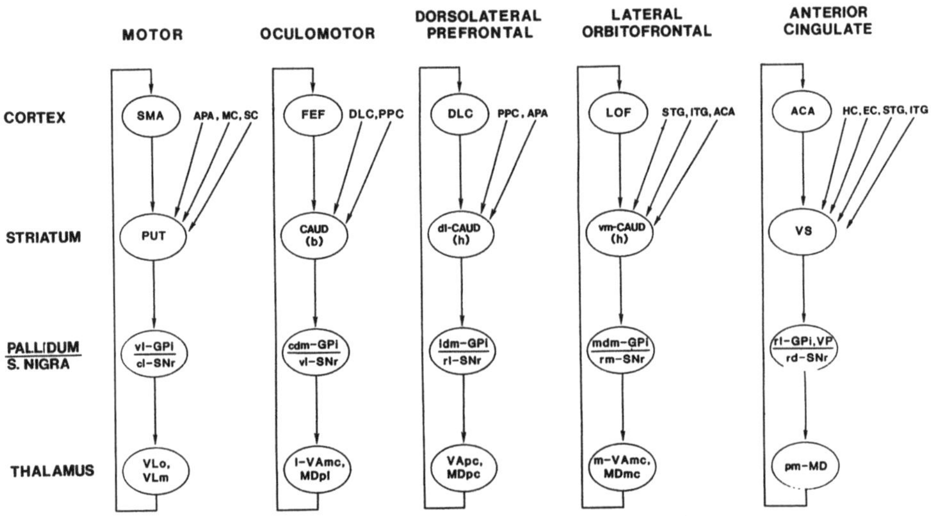

Fig. 1. Proposed basal ganglia-thalamocortical circuits. Parallel organization of the five basal ganglia-thalamocortical circuits. Each circuit engages specific regions of the cerebral cortex, striatum, pallidum, substantia nigra, and thalamus. *ACA*, anterior cingulate area; *APA*, arcuate premotor area; *CAUD*, caudate; (*b*), body; (*h*), head; *DLC*, dorsolateral cortex; *EC*, entorhinal cortex; *FEF*, frontal eye fields; *GPi*, internal segment of globus pallidus; *HC*, hippocampal cortex; *ITG*, inferior temporal gyrus; *LOF*, lateral obitofrontal cortex; *MC*, motor cortex; *MDpl*, medialis dorsalis pars paralamellaris; *MDmc*, medialis dorsalis pars magnocellularis; *MDpc*, medialis dorsalis pars parvocellularis; *PPC*, posterior parietal cortex; *PUT*, putamen; *SC*, somatosensory cortex; *SMA*, supplementary motor area; *SNr*, substantia nigra pars reticulata; *STG*, superior temporal gyrus; *VAmc*, ventralis anterior pars magnocellularis; *Vapc*, ventralis anterior pars parvocellularis; *VLm*, ventralis lateralis pars medialis; *VLo*, ventralis lateralis pars oralis; *VP*, ventral pallidum; *VS*, ventral striatum; *cl*−, caudolateral; *cdm*−, caudal dorsomedial; *dl*−, dorsolateral; *1*−, lateral; *ldm*−, lateral dorsomedial; *m*−, medial; *mdm*−, medial dorsomedial; *pm*, posteromedial; *rd*−, rostrodorsal; *rl*−, rostrolateral; *rm*−, rostromedial; *vm*−, vetromedial; *vl*−, ventrolateral. (Reproduced from [5])

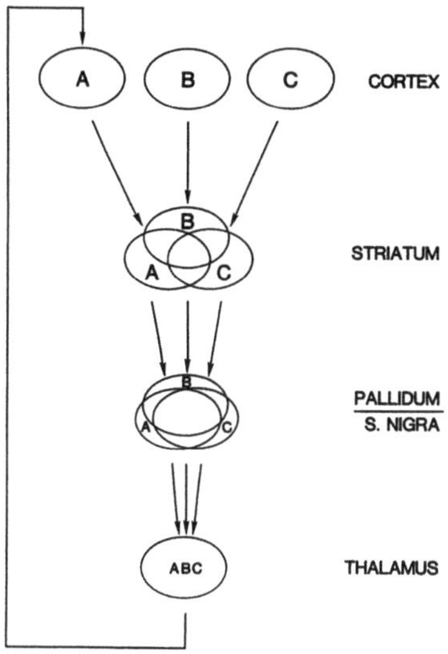

Fig. 2. Generalized basal ganglia-thalamocortical circuit. "Skeleton" diagram of the proposed basal ganglia-thalamocortical circuits. Each circuit receives output from several functionally-related cortical areas (*A, B, C*) that send partially overlapping projections to a restricted portion of the striatum. These striatal regions send further converging projections to the globus pallidus and substantia nigra, which in turn project to a specific region of the thalamus. Each thalamic region projects back to one of the cortical areas that feeds into the circuit, thereby completing the "closed loop" portion of the circuit. (Reproduced from [5])

figured so prominently in earlier views. According to the present view, however, such "funneling" occurs only within the segregated functional pathways. Since the present review is focused on the issue of sensorimotor integration as it pertains to the basal ganglia, emphasis will be placed upon the "motor" circuit, about which we have the most detailed information. For a discussion of the other proposed circuits ("oculomotor", "dorsolateral prefrontal", "lateral orbitofrontal" and "anterior cingulate" see [5]).

The "Motor" Circuit

The basal ganglia "motor" circuit takes origin from the pre- and postcentral sensorimotor fields. At the level of the striatum, this circuit primarily involves the putamen, which receives somatotopically organized projections from motor cortex (MC), premotor cortex (PMC), the supplementary motor area (SMA) and primary somatosensory cortex (SC), as well as from the superior parietal lobule (SPL, area 5) [24, 26, 34]. The "leg" areas of each of these cortical fields project to a dorsolateral sector of the putamen, the "face" areas to a ventromedial sector, and the "arm" areas to a region lying between the leg and face areas.

The somatotopic organization suggested by the topography of these corticostriate projections to the putamen has been confirmed in physiological studies of the sensorimotor response properties of putamen neurons in awake, behaving primates [4, 7, 28, 29] and by the technique of microstimulation [4]. The latter have also revealed microexcitable zones within the putamen from which movements of individual body parts can be evoked [4]. These zones appear to correspond to clusters of functionally related putamen neurons [8, 28].

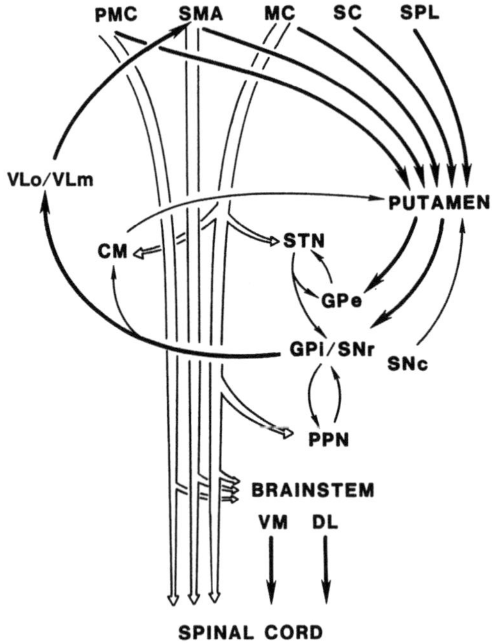

Fig. 3. The basal ganglia-thalamocortical "motor" circuit, shown in relation to several subsidiary circuits that serve to modify its output. *VM* and *DL*, ventromedial and dorsolateral brainstem pathways. Remaining abbrevations as in text and Fig. 1e

A somatotopic organization of movement-related neurons has also been found in both segments of the globus pallidus [9, 14]. The portion of GPi that receives putamen input appears to project in turn to the oral portion of the ventrolateral thalamic nucleus (VLo), which has been shown recently to project to the SMA [33, 38]. Thus the basal ganglia-thalamocortical "motor" circuit appears to be somatotopically organized throughout. Somatotopically organized corticostriate influences arising from the SMA, PMC, MC, and SC are transmitted through the "motor" circuit and ultimately projected back upon the SMA. By means of projections to both MC and the spinal cord, the SMA provides two potential routes whereby the basal ganglia can influence the segmental motor apparatus. Several lines of evidence indicate that the SMA may play an important role in the programming and control of movement.

Subsidiary Basal Ganglia Circuits

Several subsidiary circuits appear to modulate the activity within the principal striopallidal pathways. These subsidiary circuits are centered upon the STN, the intralaminar nuclei of the thalamus, the pedunculopontine nucleus (PPN), and the dopaminergic nuclei of the mesencephalic tegmentum (substantia nigra pars compacta and the ventral tegmental area). Additional inputs to the basal ganglia arise from the amygdala, the nuclei of the dorsal raphe, and the locus ceruleus with each of these projections being directed primarily to the striatal nuclei (caudate, putamen, nucleus accumbens). It is beyond the scope of this review to consider these circuits in any detail. It should be noted, however, that the topographic features characteristic of the basal ganglia-thalamocortical pathways are also found in these subsidiary circuits, and that both the centromedian nucleus (CM) [25] and the STN [21] appear to be somatotopically organized.

Neuronal Activity in the "Motor" Circuit

Neuronal activity in the basal ganglia of behaving primates has been shown to be phasically related to a wide variety of movements (see [10, 13]), including (a) spontaneously occurring movements of individual body parts, (b) self-paced alternating movements, (c) visually- and kinesthetically triggered movements, (d) postural adjustments to body tilt, (e) isometric muscular contractions, (f) slow "ramp" movements, and (g) rapid "ballistic" movements. It is conceivable that the basal ganglia might serve to control specific parameters that are common to all such movements. In primates performing a visuomotor tracking task, it has been shown that neuronal activity in the putamen and globus pallidus is related to specific parameters of limb movements, including direction, amplitude (or velocity) and force [8, 20, 29]. Moreover, the directional specificity of movement-related neuronal discharge in the putamen appears to be relatively independent of patterns of muscular activity [8]. These studies suggest a role of the basal ganglia in the control of movement direction and in the scaling of movement amplitude or velocity.

Recent studies have also provided direct evidence of participation of neurons in the "motor" circuit not only in the execution of movement but also in the preparation for it. These studies [2] have revealed a population of neurons in the primate putamen that show instruction-dependent changes in discharge related to "motor set" that are similar to those found in MC [35], PMC [37] and the SMA [36]. Given these findings and the recent evidence that the basal ganglia "motor" circuit projects to the SMA, it is likely that future studies will focus increasingly on the role of the basal ganglia in motor programming and more complex aspects of motor control.

A significant proportion of movement-related neurons in the putamen [7] and the GP [14] have been found to respond to natural somatosensory stimulation. For the putamen, 41% of arm movement-related cells responded to somatosensory stimuli. Essentially all parts of each limb are represented by ensembles of putamen neurons with response areas ranging in size from a single joint to several joints. Driving from the promixal portions of the arm was more commonly observed than from the distal. The preponderance of somatosensory responses in both putamen and GP was from deep rather than superficial structures, and the majority (82%) responded to joint rotation. Only 5% of neurons in the putamen had cutaneous receptive fields on the glabrous skin of the hand, and none responded to light touch of the hairy skin of the arm. The responses of neurons to controlled passive displacements of the elbow produced by application of load during a behavioral task have also been studied. Of putamen neurons which responded to passive manipulations of the elbow or shoulder, 74% responded to load application at latencies between 25 and 50 ms; neuronal response latencies were somewhat later in GP. Given the neuronal response latencies to perturbations observed in the sensory and motor cortices [15] and the slow conduction velocity of corticostriatal axons [27], these short-latency responses are consistent with relay of "sensory" input to the putamen from the cerebral cortex. Recently it has been shown [29] that many (44%) putamen neurons that discharge during active wrist movements are also driven by passive movements (of the same joint) that are similar in amplitude and velocity but do not evoke EMG activity. Taken together, these findings indicate that the putamen receives somatosensory information of a specific and spatially restricted nature. How such input may be used remains a topic of considerable uncertainty. Sensory input might be used to initiate, control or monitor ongoing movements, In this regard, it is noteworthy that patients with Parkinson's disease, in whom the intrastriatal processing of these signals is presumably disrupted, have been shown to be abnormally dependent upon visual input for the guidance of limb movements, suggesting that they may have an impairment in the proprioceptive mechanisms normally involved in motor control [18, 19, 30].

Based on the finding that pallidal neurons have large, disk-shaped dendritic arborizations which are oriented in a plane orthogonal to incoming striatal fibers, it has recently been suggested that the globus pallidus might serve to integrate diverse influences from the striatum, with a resultant loss of specificity and degradation of information content in favor of a more global function [32]. On the basis of the available physiologic evidence indicating maintained specificity of neuronal functional properties within the globus pallidus, this does not appear to be the case. Although a single pallidal neuron appears to integrate the output of many putamen neurons, the physiologic findings suggest that at the neuronal level such integration within the "motor" circuit is carried out along lines of individual body parts, without loss of the somatotopic and functional specificity observed in the putamen [13, 20].

The evidence for maintained somatotopy and neuronal specificity suggests that the "motor" circuit may be composed of multiple, parallel subcircuits or "channels" concerned with movements of individual body parts. Thus, in the generalized circuit depicted in Fig. 2, it may be appropriate, depending on the level of analysis, to consider A, B, and C as representing information from different cortical areas or from somatotopically corresponding subregions of three separate, but functionally related cortical areas (e.g., the "arm" representations of the primary motor, somatosensory, and supplementary motor areas). Furthermore, within each broad somatotopic area (e.g., the "arm" area) of putamen, neurons appear to be grouped into multiple functional clusters (with dimensions of 200–1000 μm) that represent a single body part (e.g., the wrist), and often a specific movement of that part (e.g., flexion) [4, 8, 28, 29]. These observations suggest that the "motor" circuit may be subdivided not only into three broad somatotopic channels representing "leg", "arm" and "face", but perhaps also into functionally defined channels of an even finer grain, based upon an individual body part or even a specific movement of an individual body part. It is furthermore conceivable that distinct subsets of neurons within the putamen may subserve a particular class of movement, e.g., self-initiated, or stimulus-triggered. The finding that cells related to motor set in the putamen are distinct from movement-related cells is also consistent with this notion. It remains to be determined whether the functional subunits (neuronal clusters) of the putamen project selectively to comparable subunits in the pallidum (and similarly for the pallidothalamic and thalamocortical projections). The answer to this question would help to clarify the "fine structure" and the nature of integration within the somatotopic channels of the "motor" circuit. Such information may prove helpful in analyzing the nature of information processing within the other basal ganglia-thalamocortical circuits.

Due to the striking impairment of movement initiation in diseases such as Parkinson's disease and in animals with striatal dopamine depletion, the basal ganglia are widely considered to play a role in the initiation of movement. On the other hand, studies of neuronal activity in the basal ganglia of the behaving monkey do not support such a role, since the onset of changes in neuronal discharge in relation to movement onset appears, on the average, to occur after that observed in the motor cortex and premotor areas [1, 8, 20, 29]. Movement-related discharge within the basal ganglia appears to be sufficiently early, however, to permit a role of these structures in controlling the build-up of activity in agonist muscles. Thus the motor circuit, while not implicated in the initiation of stimulus-triggered movements, may nevertheless play a role in scaling the amplitude and/or velocity of such movements. It is noteworthy in this regard that stimulation of the basal ganglia in monkeys trained on a visuomotor reaction-time task has been shown to have an effect on movement speed if delivered during the period of EMG activation (100 ms prior to movement onset), but has no effect if given earlier in the reaction time period [22].

While the basal ganglia do not appear to play a central role in the initiation of stimulus-triggered movements, it is possible that they may play a more direct role in the initiation of other types of movement, such as internally generated movements [17]. This seems particularly plausible in view of the fact that depletion of striatal dopamine in Parkinson's disease may be associated with akinesia, which is characterized by a striking loss of the ability to initiate voluntary movements. While basal ganglia neurons have been shown to discharge in relation to self-paced movements [9, 31], it has yet to be determined whether the onset of neuronal discharge in relation to such internally generated movements is sufficiently early (with respect to the onset of activity in other CNS motor structures) to suggest that the basal ganglia might play a role in the initiation of this type of movement.

Summary

Recent anatomical and physiological findings suggest the existence of segregated basal ganglia-thalamocortical circuits. Each circuit appears to receive multiple, partially overlapping corticostriate inputs whose influences are ultimately directed back to a single cortical area. The "motor" circuit takes origin from the pre- and postcentral sensorimotor fields, engages the putamen and topographically related portions of the globus pallidus and ventrolateral nucleus of the thalamus and terminates in the supplementary motor area. Integration of sensory and motor information within the somatotopically organized "motor" circuit appears to be carried out progressively at striatal, pallidal and thalamic levels. The possible functions of the "motor" circuit are discussed.

References

1. Aldridge JW, Anderson RJ, Murphy JT (1980) The role of the basal ganglia in controlling a movement initiated by visually presented cue. Brain Res 192:3–16
2. Alexander GE (1987) Selective neuronal discharge in monkey putamen reflects intended direction of planned limb movements. Exp Brain Res (tentatively, vol 46, pp 1–12) (in press)
3. Alexander GE, DeLong MR (1985a) Microstimulation of the primate neostriatum. II Somatotopic organization of striatal microexcitable zones and their relation to neuronal response properties. J Neurophysiol 53:1417–1430
4. Alexander GE, DeLong MR (1985b) Microstimulation of the primate neostriatum. I. Physiological properties of striatal microexcitable zones. J Neurophysiol 53:1401–1416
5. Alexander GE, DeLong MR, Strick PL (1986) Parallel organization of functionally segregated circuits linking basal ganglia and cortex. Ann Rev Neurosci 9:357–381
6. Allen GI, Tsukahara N (1984a) Cerebrocerebellar communication systems. Physiol Rev 54:957–1006
7. Crutcher MD, DeLong MR (1984a) Single cell studies of the primate putamen. I. Functiona organization. Exp Brain Res 53:233–243
8. Crutcher MD, DeLong MR (1984b) Single cell studies of the primate putamen. II. Relation to direction of movement and pattern of muscular activity. Exp Brain Res 53:244–258
9. DeLong MR (1971) Activity of pallidal neurons during movement. J Neurophysiol 34:414-427
10. DeLong MR, Georgopoulos AP (1981) Motor functions of the basal ganglia. In: Brookhar JM, Mountcastle VB, Brooks VB (eds) Handbook of physiology. The nervous system. Bethesda, MD, pp 1017–1061
11. DeLong MR, Georgopoulos AP, Crutcher MD (1983) Cortico-basal-ganglia relations an coding of motor performance. Exp Brain Res [Suppl] 7:30–40
12. DeLong MR, Alexander GE, Georgopoulos AP, Crutcher MD, Mitchell SJ, Richardson RT (1984a) Role of basal ganglia in limb movements. Human Neurobiology 2:235–244
13. DeLong MR, Georgopoulos AP, Crutcher MD, Mitchell SJ, Richardson RT, Alexander GE (1984b) Functional organization of the basal ganglia: contributions of single-cell recordin studies. In: Evered D, O'Connor M (eds) Functions of the basal ganglia. Ciba Foundatio Symposium 107, London, pp 64–82
14. DeLong MR, Crutcher MD, Georgopoulos AP (1985) Primate globus pallidus and subthalamic nucleus: functional organization. J Neurophysiol 53:530–543
15. Evarts EV (1973) Motor cortex reflexes associated with learned movement. Science 179:501-503
16. Evarts EV, Thach WT (1969) Motor mechanism of the CNS: cerebrocerebellar interrelations. Ann Rev Physiol 31:451–498
17. Evarts EV, Wise SP (1984) Basal ganglia outputs and motor control. In: Evered D, O'Connor M (eds) Functions of the basal ganglia. CIBA Foundation Symposium 107, London, p 83–96
18. Flowers K (1978a) Lack of prediction in the motor behavior of parkinsonism. Brain 101:35-52
19. Flowers K (1978b) Some frequency response characteristics of parkinsonism on pursu tracking. Brain 101:19–34

20. Georgopoulos AP, DeLong MR, Crutcher MD (1983) Relations between parameters of step-tracking movements and single cell discharge in the globus pallidus and subthalamic nucleus of the behaving monkey. J Neurosci 3:1586–1598
21. Hartmann-von-Monakow K, Akert K, Künzle H (1978) Projections of the precentral motor cortex and other cortical areas of the frontal lobe to the subthalamic nucleus in the monkey. Exp Brain Res 33:395–403
22. Horak FV, Anderson ME (1984) Influence of globus pallidus on arm movements in monkeys. II. Effects of stimulation. J Neurophysiol 52:305–322
23. Kemp JM, Powell TPS (1971) The connections of the striatum and globus pallidus: synthesis and speculation. Philos Trans R Soc Lond [Biol] 262:441–457
24. Künzle H (1975) Bilateral projections from precentral motor cortex to the putamen and other parts of the basal ganglia. An autoradiographic study in *Macaca fascicularis*. Brain Res 88:195–209
25. Künzle H (1976) Thalamic projections from the precentral motor cortex in *Macaca fascicularis*. Brain Res 105:253–267
26. Künzle H (1977) Projections from the primary somatosensory cortex to basal ganglia and thalamus in the monkey. Exp Brain Res 30:481–492
27. Liles SL (1974) Single-unit responses of caudate neurons to stimulation of frontal cortex, substantia nigra and entopeduncular nucleus in cats. J Neurophysiol 37:254–265
28. Liles SL (1979) Topographic organization of neurons related to arm movement in the putamen. In: Chase TN, Wexler NS, Barbeau A (eds) Advances in Neurology. New York, pp 155–162
29. Liles SL (1985) Activity of neurons in putamen during active and passive movements of wrist. J Neurophysiol 53:217–236
30. Martin JP (1967) The basal ganglia and posture. Great Britain
31. Neafsey EJ, Hull CD, Buchwald NA (1978) Preparation for movement in the cat. II. Unit activity in the basal ganglia and thalamus. Electroencephal. Clin Neurophysiol 44:714–723
32. Percheron G, Yelnik J, Francois C (1984) A Golgi analysis of the primate globus pallidus III. Spatial organization of the striatopallidal complex. J Comp Neurol 227:214–227
33. Schell GR, Strick PL (1984) The origin of thalamic inputs to the arcuate premotor and supplementary motor areas. J Neurosci 4:539–560
34. Selemon LD, Goldman-Rakic PS (1985) Longitudinal topography and interdigitation of cortico-striatal projections in the rhesus monkey. J Neurosci 5:776–794
35. Tanji J, Evarts EV (1976) Anticipatory activity of motor cortex neurons in relation to direction of movement. J Neurophysiol 39:1062–1068
36. Tanji J, Taniguchi K, Saga T (1980) Supplementary motor area: neuronal response to motor instructions. J Neurophysiol 43:60–68
37. Weinrich M, Wise SP (1982) The premotor cortex of the monkey. J Neurosci 2:1329–1345
38. Wiesendanger R, Wiesendanger M (1985) The thalamic connections with medial area 6 (supplementary motor cortex) in the monkey (macaca fascicularis). Exp Brain Res 59:91–104

Facets of Akinesia in Parkinson's Disease

B. L. Day [1], J. C. Rothwell [1], and C. D. Marsden [1]

Akinesia is the principal negative symptom of Parkinson's disease. Used in its widest clinical sense, the term encompasses a wide variety of different phenomena. Patients react slowly and move slowly, are rapidly fatigued by repetitive movements, and they have difficulties in performing two complex movements at the same time or in switching from one movement to another.

Bradykinesia, or slowness of movement, can be observed in the simplest of movements involving flexion or extension of a single joint [7]. Furthermore, it is apparent even when performing small movements which do not require active postural stabilisation of proximal joints [3]; for example, when flexing the top joint of the thumb through a small angle of 5° with the arm totally supported by an external framework.

Hallett and Khoshbin [8] examined the electromyographic (EMG) pattern in simple elbow flexion movements and showed that fast movements often were not performed with the usual triphasic pattern of activity in the agonist and antagonist muscles. Additional cycles of EMG activity were necessary for patients with Parkinson's disease to complete the movement. One possible reason for this is that the size of the first agonist EMG burst becomes saturated in the parkinsonian patient so that large EMG bursts cannot be generated. We have tested this idea by measuring the size of the first agonist burst in a group of patients under conditions which normally affect these variables. The first agonist burst associated with rapid wrist flexion normally increases in size when movement amplitude is increased or when a large load is applied to oppose the movement [2]. Although the first agonist burst of the patients was smaller and the movements slower when compared to those of normal subjects, it was clear that the patients were able to modulate their first agonist burst in the normal manner. Thus, both an increase in movement amplitude and an increase in load resulted in a larger agonist burst size in the patients with Parkinson's disease. In other words, it appeared that the burst size was not limited because of saturation effects. Rather, it was as if the first burst was inappropriately scaled to the parameters of movement.

This inappropriate scaling of the first agonist burst could be due to a deficit in transmission of motor commands at some site downstream of the motor cortex. The technique of brain stimulation devised by Merton and Morton [9] has allowed this hypothesis to be tested directly. The technique involves the application of a short-duration, high-voltage electrical impulse to the scalp. With the subject exerting a small background muscle contraction, such a shock produces a descending volley which is sufficient to evoke a sizeable muscle twitch. The ques-

[1] Institute of Psychiatry, Department of Neurology, De Crespigny park, Denmark Hill, London SE5 8AF, UK

Clinical Aspects of Sensory Motor Integration
Ed. by A. Struppler and A. Weindl
© Springer-Verlag Berlin Heidelberg 1987

tion is whether in Parkinson's disease there is an abnormal response to such a stimulus. Three carefully selected patients were studied both when immobile *off* drugs and mobile *on* drugs. The important result obtained from these patients was that the thresholds and latencies for motor cortex stimulation to excite biceps and thenar muscles and the resulting mechanical twitches were the same in both *on* and *off* conditions [5]. We can conclude from these data that large-diameter, rapidly conducting corticospinal motor pathways are normal in Parkinson's disease. This indicates that the inappropriate scaling of the first agonist burst, which results in slow movements, is due to abnormal commands being delivered to the motor cortex.

These observations on simple movements go some way towards explaining why patients with Parkinson's disease move slowly. However, such slowing of movements made at single joints does not correlate well with overall clinical measures of akinesia. Clearly, normal everyday movements demand more of the motor system than does a simple discrete flexion movement around a single joint. It is necessary, therefore, to study more complex motor acts.

Schwab et al. [10] graphically described the difficulties experienced by patients with Parkinson's disease who try to execute complex motor tasks. The principal deficit that they noted was an inability to perform two different movements at the same time. Thus, a patient walking across a hotel lobby to pay his bill "reached into his inside pocket with his left hand to get his wallet. At once he stopped walking, standing immobile before the strangers. Becoming aware of this, he then resumed walking but his left hand remained in his inside pocket, suggesting perhaps a planned holdup." We have studied this type of problem in a much simpler combination of tasks. Eight patients with Parkinson's disease were required to perform rapid elbow flexion movements through an angle of 15° and to squeeze an isometric force transducer (with a force of 30 N) between thumb and forefinger. The tasks were performed with the same arm in the patient's own time either separately or simultaneously and compared with the results in a group of 8 control patients. The only instruction was to make each movement as rapidly as possible.

As expected, the separate movements of both "flex" and "squeeze" were slower in the patients than in the normals (flexion movement time 349 ms patients, 229 ms normals; $P<0.001$; squeeze movement time 220 ms patients, 156 ms normals; $P<0.001$). However, when the two movements had to be performed simultaneously, there was a further prolongation of movement time in the patients (flexion 557 ms; squeeze 309 ms) [1]. In normals, simultaneous movements were performed at the same speed as the separate movements alone. Unlike the analysis of simple single-joint movements, the clinical akinesia rating of the patients was well correlated to the increase in movement time in the simultaneous task.

It might be argued that this extra slowing in simultaneous movements is simply because more muscles are needed to perform two tasks than one. If the patients had a limited amount of "energy" to activate their muscles (cf. [8]), then the more muscles used in a task, the slower it might be expected to be. This is not the case. Berardelli et al. [3] have shown that flexion movements of the terminal phalanx of the thumb against a small load are equally slow in patients whether the thumb is completely supported, and only the flexor pollicis longus need be used to produce the movement, or the whole arm is entirely free, when postural activity must occur in many muscles of the wrist, elbow and shoulder. It would appear more likely that the effect seen in simultaneous movements has a different explanation. We favour the idea that another facet of parkinsonian akinesia is the inability to execute two motor programmes at the same time.

An example of a more complex motor act is the execution of movement sequences which last for some seconds. Again, for experimental simplicity such movements may be restricted to a single joint moving in one plane. The conventional way of studying this aspect of motor control is to employ a tracking task whereby movement at a joint is transduced via a lever and potentiometer and displayed on a cathode ray oscilloscope (CRO). A target is also displayed on the CRO and the subject is asked to follow a movement pattern by tracking the target with appropriate lever movements. Usually, if the target motion is not known to the subject, then corrective movements of the joint are made based on the perceived error between the lever and target position. However, if the target pattern is known in advance, then tracking error and lag may be reduced by moving the lever in anticipation of the target. Thus, the subject moves in response to an internal model of target movement rather than in direct response to external cues.

Flowers [7] from his study of tracking behaviour in patients with Parkinson's disease suggested that the basal ganglia are involved with such predictive motor control. If this is true, then the parkinsonian patient would rely heavily on external cues and feedback for the execution of all movements. He would therefore be constrained to move slowly, checking progress at every step.

We have tested the ability of patients to utilise a predictive motor strategy by getting them to track a target. For the first 50 trials the subjects were unaware that each movement pattern was identical. In the second 50 trials they were made aware of the repetition. Normal subjects, when made aware of the repetition, responded by learning the target motion and subsequently reducing their tracking lag and tracking error. In this test, parkinsonian patients also reduced their tracking lag, although their tracking error did not improve very much [4]. We can conclude from this experiment that the parkinsonian patient is capable of learning a target motion and moving according to an internal plan in a predictive fashion. However, the predictive strategy does not improve tracking accuracy as much as for normal subjects. Therefore, patients may choose to give up the benefits of predictive action in order to achieve the accuracy of slower corrective movements.

In conclusion, the broad spectrum of clinical deficits encompassed by the general term akinesia reflects an equally broad spectrum of physiological abnormalities. In patients with Parkinson's disease there is an increase in reaction time and a decrease in movement velocity in simple tasks. These are compounded by further difficulties in performing two different tasks at the same time, and in producing accurate, predictive movements. Such deficits appear to occur independently in different individual patients [6] and therefore represent a group of related, but not identical, physiological abnormalities underlying the clinical picture of parkinsonian akinesia.

Summary

Patients with Parkinson's disease reveal different types of abnormality when faced with different motor tasks. Rapid, self-paced movements about a single joint tend to be performed slowly regardless of the postural requirements of the task. Experimental evidence points to the conclusion that slowness results from abnormal commands being delivered to the motor cortex. The more complex task of performing two independent movements simultaneously produces additional slowing of the movements. This suggests a deficit in executing two or more motor programmes at the same time. When tracking a target with single-joint arm movements, patients are capable of learning a novel movement pattern and using this information to move in anticipation of the target motion. However, such predictive movements are performed inaccurately. It is concluded that these separate abnormalities contribute to the clinical picture of akinesia in Parkinson's disease.

References

1. Benecke R, Dick JPR, Rothwell JC, Day BL, Marsden CD (1985) Performance of simultaneous motor acts in normal subjects and in patients with Parkinson's disease. Electroencephalogr Clin Neurophysiol 61:541
2. Berardelli A, Rothwell JC, Day BL, Kachi T, Marsden CD (1984a) Duration of the first agonist burst in ballistic arm movements. Brain Res 304:183–187
3. Berardelli A, Rothwell JC, Day BL, Marsden CD (1984b) Movements not involved in posture are abnormal in Parkinson's disease. Neurosci Lett 47:47–50
4. Day BL, Dick JPR, Marsden CD (1984) Patients with Parkinson's disease can employ a predictive motor strategy. J Neurol Neurosurg Psychiat 47:1299–1306
5. Dick JPR, Cowan JMA, Day BL, Berardelli A, Kachi T, Rothwell JC, Marsden CD (1984) The corticomotoneurone connexion is normal in Parkinson's disease. Nature 310:407–409
6. Evarts EV, Teravainen H, Calne DB (1981) Reaction time in Parkinson's disease. Brain 104:167–186
7. Flowers KA (1976) Visual "closed-loop" and "open-loop" characteristics of voluntary movements in patients with parkinsonism and essential tremor. Brain 99:269–310
8. Hallett M, Khoshbin S (1980) A physiological mechanism of bradykinesia. Brain 103:301–314
9. Merton PA, Morton HB (1980) Stimulation of the cerebral cortex in the intact human subject. Nature 185:227
10. Schwab RS, Chafetz ME, Walker S (1954) Control of two simultaneous voluntary motor acts in normals and parkinsonism. Arch Neurol 75:591–598

Immunohistochemical Studies on Neurotransmitters in Rat Basal Ganglia

W. H. OERTEL [1] and A. STRUPPLER [1]

Numerous neurotransmitters are present in the basal ganglia. Knowledge of their distribution is one of the prerequisites for understanding the neurochemical organization of the basal ganglia and may eventually help to develop new strategies in the treatment of basal ganglia disorders. Based on (immuno-) histochemical studies, this article summarizes the known distribution of classical neurotransmitters and neuropeptides and the coexistence of neurotransmitters and neuropeptides in neuronal cell bodies of the rat basal ganglia. (For the distribution of neurotransmitters in afferents to the basal ganglia see the recent review by McGeer et al. [15]).

The rat basal ganglia consist of two large and two small areas. The large areas are the *striatum* [nucleus caudatus/putamen (CPU), nucleus accumbens (ACB), tuberculum olfactorium (TO)] and the *pallidum* [globus pallidus (GP), nucleus entopeduncularis (EP), pallidum ventrale (VP), substantia nigra pars reticulata (SNR)]. The small areas are the nucleus subthalamicus (STN) and the substantia nigra pars compacta/area ventralis tegmenti of Tsai (SNC/AVT).

Studies were performed on untreated rats and rats that had received a stereotaxic injection of colchicine, a blocker of axonal transport, in various areas of the basal ganglia in order to enhance cell body staining. After 24 h the anesthetized animals were killed by cardiac perfusion with formaldehyde fixatives.

With single and double immunohistochemistry, dopaminergic, cholinergic and GABAergic neurons were localized by means of antisera to tyrosine hydroxylase (TH) [5], cholinacetyltransferase (ChAT) [7], and glutamic acid decarboxylase (GAD) [19], the biosynthetic enzymes for dopamine, acetylcholine, and GABA, respectively. Opioid peptide-like (OPL) immunoreactivity was visualized with a mouse monoclonal antibody to the N-terminal of β-endorphin [11]. The immunohistochemical distribution of other neuropeptides in neuronal cell bodies of the rat basal ganglia is reviewed.

Since the studies with monoamine histofluorescence [3] and with immunohistochemistry for TH [12], *dopaminergic* neurons are known to occur abundantly in SNC/AVT, but are scattered and few in number in SNR. These dopaminergic neurons diffusely innervate the dorsal striatal areas (CPU) and the entral striatal areas (ACB, TO). The neuropeptide cholecystokinin is colocalized in a subpopulation of the mesencephalic dopaminergic neurons projecting to the ventral striatum [13].

In the *striatum* (CPU, ACB, TO) only a few neurons of medium to large size are immunoreactive for ChAT, i.e., *cholinergic*, in agreement with Sofroniew et

[1] Neurologische Klinik und Poliklinik der Technischen Universität München, Möhlstr. 28, D-8000 München 80, FRG.

Clinical Aspects of Sensory Motor Integration
Ed. by A. Struppler and A. Weindl
© Springer-Verlag Berlin Heidelberg 1987

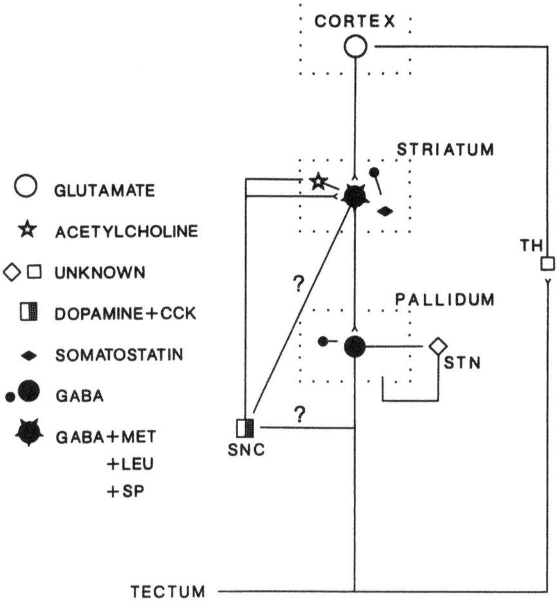

Fig. 1. Simplified diagram of the distribution and connectivity of neurochemically identified nerve cell populations in rat basal ganglia. Symbols are explained on the *left*. A subpopulation of dopaminergic neurons contain the neuropeptide cholecystokinin (*CCK*). Subpopulations of striato-pallidal GABAergic neurons contain leuenkephalin (*LEU*), metenkephalin (*MET*), or possibly substance P (*SP*). For details see text

al. [23]. In contrast, the overwhelming majority of striatal neurons is immunoreactive for GAD, i.e., of *GABAergic* nature. According to cell size and staining properties, two types of GABAergic neurons can be distinguished in all the striatal areas: a small proportion (5%) of the striatal neurons display very intense GAD immunoreactivity. These cells are of medium to large size and presumably represent local circuit neurons. The vast majority of the principal striatal nerve cells, i.e., the medium-size spiny projection neuron, exhibit intermediate GAD positivity [17, 18].

Recently, dopaminergic mesencephalo-striatal neurons have been shown to form monosynaptic contacts with these medium-size spiny (presumably GABAergic) striatofugal neurons [9].

With respect to *neuropeptides* in the striatum, somatostatin-like immunoreactivity is localized in a small population of medium-size (inter?-)neurons [6]. In striatal projection neurons several neuropeptides have been identified such as metenkephalin, leuenkephalin, substance P (for review see [22]), and dynorphin [25]. Subsequently, the *coexistence* of opioid peptides [20] such as metenkephalin or leuenkephalin [1, 4] and of substance P [1] has been described in GABAergic striatal neurons (Fig. 1).

In the *pallidum* (GP, VP, EP, SNR) the vast majority of neurons are GABAergic [17]. Most of these GAD-immunoreactive neurons are of medium to large size and presumably represent projection neurons. In addition, some small

GAD-positive, presumably local circuit neurons have been observed. At the ultrastructural level the GABAergic neurons in the pallidum are monosynaptically innervated by a high number of GABAergic nerve terminals [21]. These GABAergic axon endings correspond in morphology and peridendritic/perisomatic patterns to striatally derived nerve endings [8].

Thus, the rat basal ganglia contain two monosynaptically linked vast populations of GABAergic projection neurons: Activation of striato-pallidal GABAergic cells would lead to inhibition of GABAergic pallido-fugal cells and therefore to *disinhibition* of the target cells of pallidal projection neurons.

As neuropeptides appear to coexist with GABA in striato-fugal neurons, the massive GABAergic striato-pallidal pathway may consist of several GABAergic subprojections, containing different peptides. In contrast, coexistence of neuropeptides in pallido-fugal neurons has so far not been demonstrated. At the ventral border of GP, throughout GP and VP and around EP, scattered large ChAT-positive neurons are present [23]. (These cells are neither GAD-immunoreactive nor TH-immunoreactive.) These cholinergic nerve cells presumably represent displaced cells of the nucleus basalis magnocellularis. In the SNR, scattered and few clustered neurons remain unstained for GAD- or ChAT-immunoreactivity, however they exhibit TH-positivity, as expected from the work of Hökfelt et al. [12].

In STN, numerous GABAergic nerve terminals are observed and stem from pallidosubthalamic fibers [24]. In contrast, very few GABAergic neuronal cell bodies are seen in the STN [18]. In addition, staining for ChAT, TH or for any known neuropeptide has so far failed to reveal a neurotransmitter candidate for the subthalamo-pallidal projection neurons.

Numerous questions are unanswered. For example, the morphological relations of cholinergic striatal interneurons with mesencephalo-striatal dopaminergic fibers or with striatal GABAergic neurons and their axon collaterals are unknown. It remains to be elucidated to which extent dopaminergic neurons in SNC/AVT are innervated by pallido-fugal GABAergic nerve terminals, as suggested by electrophysiological studies [10], and by striato-fugal fibers, as shown by Wassef et al. [26] see Fig. 1, question marks. The transmitter of the subthalamo-pallidal neurons and pallido-striatal neurons is still to be revealed.

With respect to classical transmitters, cholinergic and dopaminergic neurons represent small neuronal populations, whereas GABAergic neurons constitute most of the neurons in the basal ganglia. Compounds used in the treatment of basal ganglia disorders predominantly interfere with cholinergic and dopaminergic transmission. If the data from the rat were transferable to the human basal ganglia, the multiplicity and interaction of the different and in part vast populations of GABAergic neurons would make it difficult to predict an effect of drugs acting on GABAergic transmission. Most of the compounds interfering with GABA transmission have by and large failed to show any beneficial effect in basal ganglia disorders [14]. Only recently the GABAmimetic drug progabide has been reported to ameliorate tardive dyskinesia [16]. Whether compounds interfering with the different neuropeptidergic systems [2] can improve basal ganglia dysfunction in man remains an open issue.

Summary

The immunohistochemical distribution of neurotransmitters and neuropeptides in intrinsic neurons of rat basal ganglia is reviewed. Dopaminergic neurons in substantia nigra pars compacta/area ventralis tegmenti project to the striatum and contain in part the neuropeptide cholecystokinin. In the striatal areas (nucleus caudatus/putamen, nucleus accumbens, tuberculum olfactorium) few large neurons represent cholinergic interneurons, few cells contain somatostatin, and few medium-to-large size neurons are GABAergic. The principal striatal neurons are medium-size spiny cells, which project to the pallidum and whose vast majority is GABAergic. Subclasses of these GABAergic striatopallidal neurons may contain the neuropeptide metenkephalin, leuenkephalin, or substance P. The vast majority of neurons in the pallidal areas (globus pallidus, nucleus entopeduncularis, pallidum ventrale, substantia nigra pars reticulata) are GABAergic. Scattered cholinergic neurons in the pallidal areas may correspond to displaced neurons of the nucleus basalis magnocellularis. The transmitter of the projection neurons in the nucleus subthalamicus is not identified.

Acknowledgment. This research was supported by DFG Grant Oe 95/2-1.

References

1. Penny GR, Afsharpour S, Kitai ST (1986) The glutamate decarboxylase-, leucine enkephalin-, methionine enkephalin- and substance P-immuno reactive neurons in the neostriatum of the rat and cat: evidence for partial population overlap. Neurosci 17:1011–1045
2. Agid Y, Javoy-Agid F (1985) Peptides and Parkinson's disease. Trends in Neuroscience 8:30–35
3. Andén NE, Carlsson A, Dahlström A, Fuxe K, Hillarp NÅ, Larsson K (1964) Demonstration and mapping out of nigro-neostriatal dopamine neurons. Life Sci 3:523–530
4. Aronin N, DiFiglia M, Graveland GA, Schwartz WJ, Wu J-Y (1984) Localization of immunoreactive enkephalins in GABA synthesizing neurons of the rat neostriatum. Brain Res 300:376–380
5. Berod A, Hartmann BK, Keller R, Joh TH, Pujol JF (1982) A new double labeling technique using tyrosine hydroxylase and dopamine-beta-hydroxylase immunohistochemistry. Evidence for dopaminergic cells lying in the pons of the beef brain. Brain Res 240:235–243
6. DiFiglia M, Aronin N (1982) Ultrastructural features of immunoreactive somatostatin neurons in the rat caudate nucleus. J Neurosci 9:1267–1274
7. Eckenstein F, Barde YA, Thoenen H (1981) Production of specific antibodies to choline acetyltransferase purified from pig brain. Neuroscience 6:993–1000
8. Fox CA, Andrade AN, Lu Qui IJ, Rafols JA (1974) The primate globus pallidus: a Golgi and electron microscope study. J Hirnforsch 15:75–93
9. Freund TF, Powell JF, Smith AD (1984) Tyrosine hydroxylase-immunoreactive boutons in synaptic contact with identified striatonigral neurons with particular reference to dendritic spines. Neuroscience 13:1189–1215
10. Grace AA, Bunney BS (1979) Paradoxical excitation of nigral dopaminergic cells: indirect mediation through reticulata inhibitory neurons. Eur J Pharmacol 59:211–218
11. Gramsch C, Meo T, Riethmüller G, Herz A (1983) Binding characteristics of a monoclonal β-endorphin antibody recognizing the N-terminus of opioid peptides. J Neurochem 40:1220–1226

12. Hökfelt T, Johansson O, Fuxe K, Goldstein M, Park D (1976) Immunohistochemical studies on the localization and distribution of monoamine neuron systems in the rat brain. I. Tyrosine hydroxylase in the mes- and diencephalon. Med Biol 54:427–453
13. Hökfelt T, Skirboll L, Rehfeld JF, Goldstein M, Markey K, Dann O (1980) A subpopulation of mesencephalic dopamine neurons projecting to limbic areas contains a cholecystokinin-like peptide: evidence from immunohistochemistry combined with retrograde tracing. Neuroscience 5:2093–2124
14. Marsden CD, Sheehy MP (1981) GABA and movement disorders. In: Di Chiara G, Gessa GI (eds) GABA and the basal ganglia. Raven, New York, pp 225–234
15. McGeer E, Staines WA, McGeer PL (1984) Neurotransmitters in the basal ganglia. Can J Neurol Sci 11:89–99
16. Morselli PL, Fournier V, Bossi L, Musch B (1985) Clinical activity of GABAergic agonists in neuroleptic and L-DOPA induced dyskinesia. In: Casey DE, Chase TN, Christensen AV, Gerlach J (eds) Dyskinesia: research and treatment. Springer, Berlin Heidelberg New York Tokyo, pp 128–136
17. Mugnaini E, Oertel WH (1985) An atlas of the distribution of GABAergic neurons and terminals in the rat CNS as revealed by GAD-immunohistochemistry. In: Björklund A, Hökfelt T (eds) Handbook of chemical neuroanatomy, vol 4, GABA and neuropeptides in the CNS, part I. Elsevier, Amsterdam, pp 436–608
18. Oertel WH, Mugnaini E (1984) Immunocytochemical studies of GABAergic neurons in rat basal ganglia and their relations to other neuronal systems. Neurosci Lett 47:233–238
19. Oertel WH, Schmechel DE, Tappaz ML, Kopin IJ (1981) Production of a specific antiserum to rat brain glutamic acid decarboxylase by injection of an antigen-antibody complex. Neuroscience 6:2689–2700
20. Oertel WH, Riethmüller G, Mugnaini E, Schmechel DE, Weindl A, Gramsch C, Herz A (1983) Opioid peptide-like immunoreactivity in GABAergic neurons of rat neostriatum and central amygdaloid nucleus. Life Sci [Suppl I] 33:76–79
21. Oertel WH, Nitsch C, Mugnaini E (1984) The immunocytochemical demonstration of the GABAergic neurons in rat globus pallidus and nucleus entopeduncularis and their GABAergic innervation. Adv Neurol 40:91–98
22. Penney JB Jr, Young AB (1982) Speculations on the functional anatomy of basal ganglia. Ann Rev Neurosci 6:73–94
23. Sofroniew MV, Eckenstein F, Thoenen H, Cuello AC (1982) Topography of choline acetyltransferase-containing neurons in the forebrain of the rat. Neurosci Lett 33:7–12
24. Staines WA, Oertel WH, Wu I-Y, Hökfelt T (1984) Immunohistochemistry of amino acid transmitter enzymes. Acta Neurol Scand 69:306–312
25. Vincent S, Hökfelt T, Christensson I, Terenius L (1982) Immunohistochemical evidence for a dynorphin immunoreactive striatonigral pathway. Europ J Pharmacol 85:251–252
26. Wassef M, Berod A, Sotelo C (1981) Dopaminergic dendrites in the pars reticulata of the rat substantia nigra and their striatal input. Combined immunocytochemical localization of tyrosine hydroxylase and anterograde degeneration. Neuroscience 6:2125–2139

CNS Peptides in Huntington's Chorea

P. C. Emson [1], D. Dawbarn [1], and M. E. de Quidt [1]

Introduction

Huntington's chorea is a hereditary disease inherited as an autosomal dominant gene which produces progressive degeneration of the basal ganglia and, in particular, the caudate nucleus and putamen. As these CNS regions in other mammals are particularly rich in neuropeptides such as substance P and methionine-enkephalin, it was logical to look at the content of these peptides in Huntington's disease to see if they could be used as biochemical markers for the degenerative changes characterizing this condition (see [5]).

Initial studies demonstrated that many CNS peptides are apparently relatively stable post mortem and not affected by agonal status, so that it was feasible to measure and visualize peptides such as substance P and methionine-enkephalin in the human CNS; such studies demonstrated massive losses of substance P and methionine-enkephalin content (Table 1) in Huntington's disease. These changes confirm that as in experimental animals the human CNS contains peptidergic striato-pallidal and striato-nigral pathways which are damaged in Huntington's disease. A variety of studies have demonstrated that the major cell type in the caudate-putamen is the medium-sized spiny neurone (see [7]) and that in the rat at least all these may contain the inhibitory amino acid (GABA), as demonstrated by their content of the biosynthetic enzyme glutamic acid decarboxylase (GAD) (see W. Oertel, this volume). As these cells are known to be the major cell type damaged in Huntington's disease, it was therefore not surprising that Bird and Iversen [2] and others had already reported losses of GABA and GAD in Huntington's disease. What was less expected was the demonstration that the peptides such as substance P or methionine enkephalin coexist with GAD (see W. Oertel, this volume), as it had originally been supposed that GABA and substance P provided separate outputs to the substantia nigra [10].

The discovery of additional families of peptides such as the tachykinins (substance P and substance K; cf. [11]) or further opiate-like peptides [9] provided additional biochemical markers for striato-nigral/striato-pallidal outputs [4], but what was more interesting was the discovery that the peptides neuropeptide Y and somatostatin are increased in Huntington's disease, probably reflecting the selective survival of an interneurone population containing these peptides [1, 3]. In this brief overview we summarize our biochemical findings in Huntington's disease and review the evidence that many peptides are stable post mortem.

[1] MRC Group, Department of Neuroendocrinology, Institute of Animal Physiology, Babraham, Cambridge, CB2 4AT, UK.

Clinical Aspects of Sensory Motor Integration
Ed. by A. Struppler and A. Weindl
© Springer-Verlag Berlin Heidelberg 1987

Table 1. Regional distribution of substance P, methionine-enkephalin and dynorphin B in normal human brain and in Huntington's disease

Region	Substance P (pmol/g)		Methionine-enkephalin (pmol/g)		Dynorphin B (pmol/g)	
	C	HD	C	HD	C	HD
Caudate nucleus	138 ± 14 (13)	158 ± 28 (10)	116 ± 40 (10)	156 ± 54 (10)	16.6 ± 6.2 (10)	2.1 ± 0.36[a] (9)
Putamen	112 ± 29 (6)	77 ± 20 (10)	200 ± 71 (12)	184 ± 68 (12)	25.6 ± 7.2 (9)	8.3 ± 3.0[a] (6)
Globus pallidus (lateral)	197 ± 99 (6)	18 ± 9[a] (10)	1163 ± 216 (20)	527 ± 102[a] (30)	11.1 ± 1.9 (8)	2.8 ± 1,9[a] (6)
Globus pallidus (medial)	877 ± 253 (6)	178 ± 102[a] (10)	675 ± 168 (20)	317 ± 48[a] (40)	18.4 ± 4.9 (9)	2.5 ± 0.91[a] (6)
Substantia nigra (pars compacta)	1264 ± 239 (27)	158 ± 41[a] (24)	577 ± 103 (15)	229 ± 71[a] (15)	539.9 ± 58.2 (5)	67.0 ± 10.2[a] (9)
Substantia nigra (pars reticulata)	1535 ± 117 (30)	86 ± 25[a] (25)	661 ± 145 (15)	230 ± 54[a] (16)	55.5 ± 25.2 (7)	12.6 ± 6.6[a] (6)

[a] $p < 0.05$; C, control; HD, Huntington's disease.

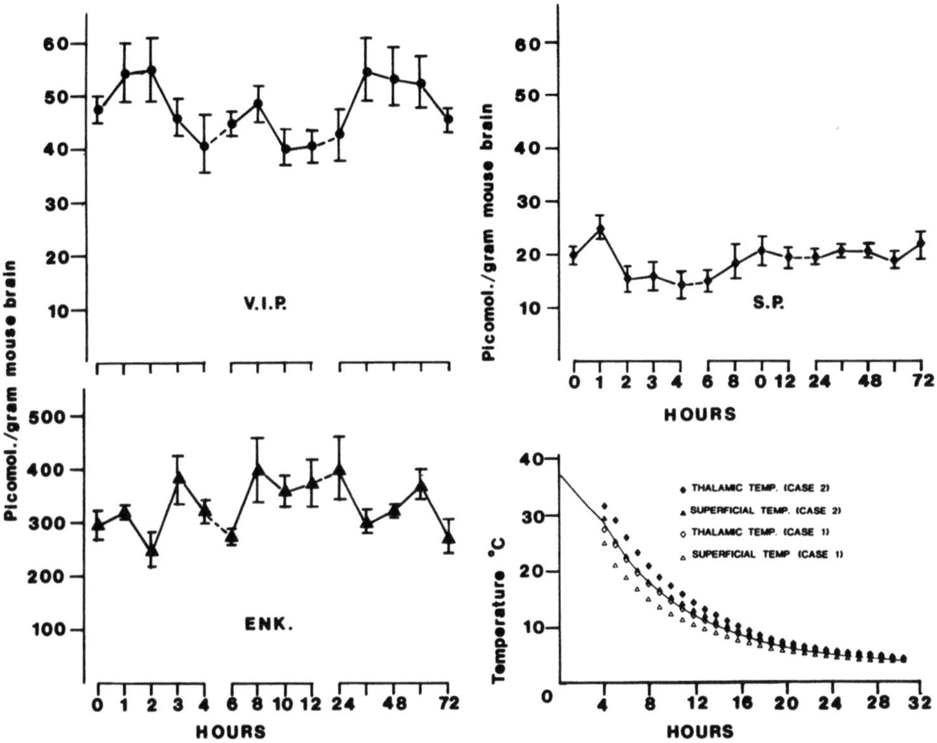

Fig. 1. Postmortem stability of vasoactive intestinal polypeptide (*VIP*), methionine-enkephalin (*ENK*) and substance P (*SP*) immunoreactivity in mouse brains. Mice were killed and placed in a cooling incubator programmed to simulate the human brain cooling curve for cadavers placed in a 4 °C mortuary refrigerator

Post-mortem Stability

In our initial investigations [6] we were concerned to establish whether the CNS peptides we could measure in experimental animals were present and stable in post-mortem human brain. In order to investigate the peptide stability we used an animal model to mimic the post-mortem cooling curve of human brain (Fig. 1) and sampled the peptide content of the mouse brain at different times post mortem. These studies revealed that several peptides were surprisingly stable post mortem and suggested that examination of human brain peptide content might be feasible. This possibility has been amply confirmed, and our detailed studies of various peptides over a period of years have indicated that in general the material detected in post-mortem human brain extracts is indistinguishable from the relevant synthetic peptide on gel or high-pressure liquid chromatography. Further immunological studies using sequence-directed antisera have confirmed as far as can be judged without peptide sequencing that the peptides detected correspond to the intact peptides and not fragments.

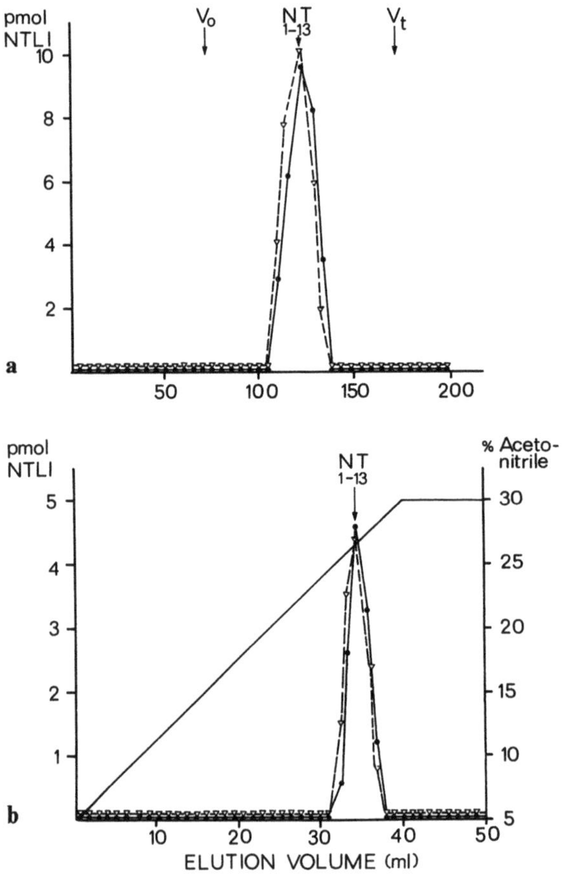

Fig. 2 a, b. Fractionation of neurotensin-like immunoreactivity (NTLI) **a** on Sephadex G25 or **b** a reverse-phase HPLC using a linear 5–30% acetonitrile gradient. Fractions were assayed using antiserum directed against the neurotensin amino-terminus (●) and carboxy-terminus (▽). Note that there is only one peak of neurotensin immunoreactivity separating at the position expected for synthetic neurotensin (NT_{1-13})

In the example shown here (Fig. 2), neurotensin immunoreactivity extracted from CNS tissue separates at the position expected for synthetic bovine neurotensin on gel chromatography (Fig. 2a) or high-pressure liquid chromatography (Fig. 2b), and the material has both amino- and carboxy-terminal immunoreactivity.

An additional argument in favour of the post-mortem stability of CNS peptides, and hence their use as biochemical markers in CNS disorders, is provided by their precise histochemical localisation (Fig. 3). Note that despite post-mortem delays the immunoreactivity detected here is precisely localised to human basal ganglia neurones and that the finer details of the neuronal processes can be readily demonstrated. The exact reason for this stability is not known but it may reflect the compartmentalisation of CNS peptides into vesicles where they are resistant to cytoplasmic peptidase action.

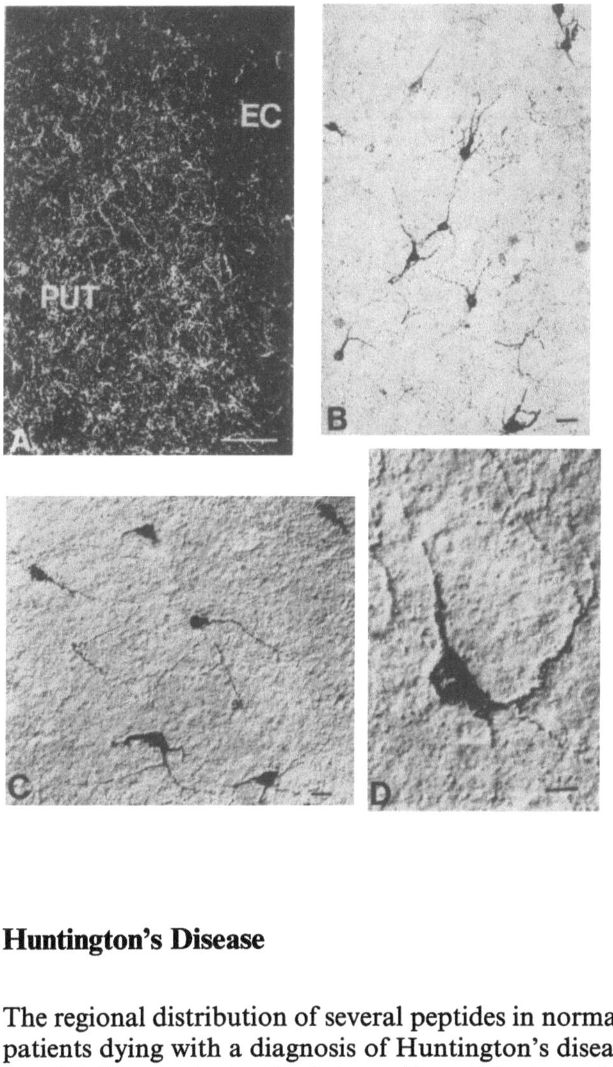

Fig. 3 A–D. Striatal neurones in Huntington's disease stained by the PAP technique for NPY. **A** Darkfield photomicrograph of the putamen (*PUT*) and external capsule (*EC*). Note the dense fibre staining in the putamen and lack of staining in the external capsule (scale bar = 50 μm). **B** Phase contrast photomicrograph of neurones in the caudate nucleus. **C, D** Differential interference photomicrographs taken of cells in the putamen (scale bar for **B, C,** and **D** = 10 μm)

Huntington's Disease

The regional distribution of several peptides in normal brain and in the brains of patients dying with a diagnosis of Huntington's disease is summarized in Tables 1 and 2. The contents of substance P, methionine enkephalin and dynorphin immunoreactivity are depleted in Huntington's disease consistent with their location in striato-pallidal and striato-nigral output neurones. These locations are also consistent with histochemical observations of choreic material [4, 12]. As in other mammalian brains, substance P is particularly concentrated in striato-nigral neurones (as is substance K; H. Arai and P. Emson 1985, unpublished work) whilst the majority of enkephalin immunoreactivity is concentrated in the lateral globus pallidus, and only small enkephalin- and dynorphin-containing projections descend to the substantia nigra (see Table 1). In contrast to the tachykinin and endorphin peptides, neuropeptide Y and somatostatin are increased in choreic brain material (Table 2). Gel chromatography of control and Huntington's disease caudate revealed no differences in the processing of somatostatin peptides, so that there was no selective increase of the large somatostatin form

Table 2. Concentration of neuropeptide Y and somatostatin in different brain areas in control and Huntington's disease

Brain region	NPY concentration (pmol/g)		SRIF concentration (pmol/g)	
	C	HD	C	HD
Brodmann area 10	14.6±1.3 (9)	16.5±1.8 (9)	44.6± 6.6 (9)	51.2± 5.0 (9)
Caudate	24.4±3.1 (9)	47.6±6.0 (9)[a]	71.2±12.3 (9)	125.3± 16.1 (9)[a]
Globus pallidus medial	0.8±0.2 (6)	1.5±0.5 (7)	27.2± 8.4 (4)	21.0± 5.9 (5)
Hippocampus	10.9±1.4 (5)	16.1±4.0 (5)	68.6±10.9 (5)	67.3± 13.7 (5)
Putamen	11.0±1.7 (12)	28.1±2.9 (12)[a]	95.6±16.6 (12)	199.9± 28.2 (12)[a]
Hypothalamus	5.2±1.2 (8)	4.8±1.7 (8)	686.0±88.0 (8)	793.0±118.0 (8)
Substantia nigra pars compacta	0	0	26.4± 4.7 (4)	31.9± 12.2 (3)

[a] $p<0.05$. C, controls; HD, Huntington's disease.
Values are means and SEM. Values in brackets indicate number of samples.

(SRIF-28) as compared with smaller product somatostatin 14 (SRIF-14) (Fig. 4). Similarly, HPLC indicated that neuropeptide Y-like (NPY) immunoreactivity was indistinguishable from that of synthetic NPY on reverse phase [3].

The earlier observations of Aronin et al. [1] and the report that NPY and SRIF may coexist in the same neurones prompted us to investigate whether SRIF-LI and NPY-LI positive neurones are preserved in Huntington's disease. NPY staining of caudate or putamen samples from patients dying with a diagnosis of Huntington's disease showed that NPY-positive neurones could be demonstrated in formalin-fixed material (see Fig. 3), and that relative numbers of these neurones were significantly increased in Huntington's disease (Fig. 5). So far we have not carried out double staining of NPY and SRIF-LI neurones, but evidence available suggests that the coexistence is probably one to one in the basal ganglia [14]. It is, of course, possible that the result illustrated in Fig. 5 reflects merely increased content per cell in Huntington's disease, making visualisation of these cells easier and raising some cells above the detection limit by the immunohistochemical techniques used.

The reason for survival of the NPY-SRIF cells in Huntington's disease and the death of the major cell type – the medium-sized spiny neurones – may be found when the genetic deficit is revealed [8]. Whatever the deficit, it will be fascinating for neuroscientists to understand how one particular cell type can be selectively damaged with apparently minimal damage to other neurones. One reason for the survival of the NPY-SRIF cells may be their high content of NADPH-diaphorase [13], which apparently renders these cells more resistant to noxious substances. Whatever the cause of survival of these cells, it is clear that the measurements of CNS peptides can be used to elucidate the circuitry and pathology of the human brain.

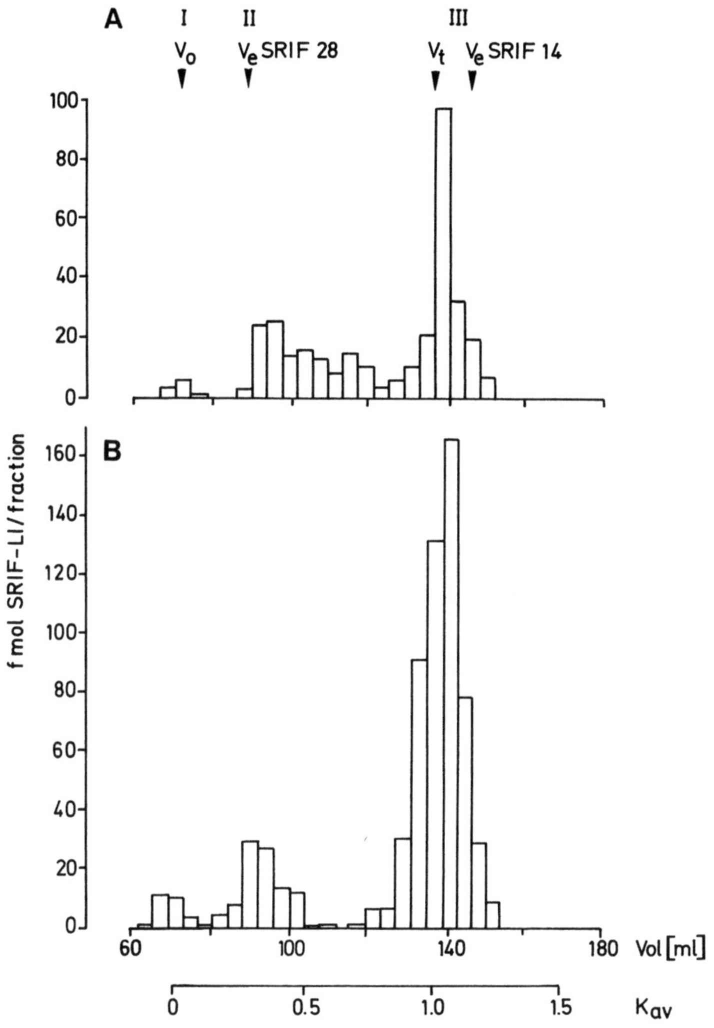

Fig. 4 A, B. Gel filtration on Sephadex G25 of **A** extracted control caudate and **B** Huntington's disease caudate followed by C-terminal SRIF radioimmunoassay

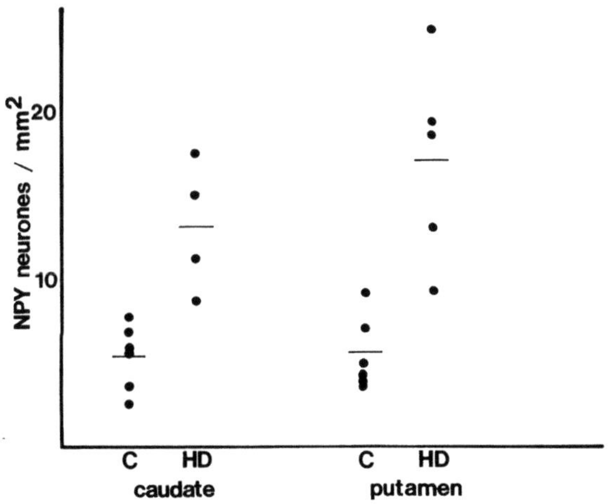

Fig. 5. Number of NPY positive neurones per mm^2 in the caudate and putamen from control brains (*C*) and Huntington's disease brains (*HD*)

Summary

The post-mortem stability of several neuropeptides allows these peptides to be used as chemical markers for pathological changes occurring in neurological illness. This possibility is illustrated by consideration of the biochemical changes found in the basal ganglia of patients dying with a diagnosis of Huntington's disease, where depletions of substance P, enkephalins and other markers reflect degeneration of descending striato-nigral pathways. In contrast, the peptides somatostatin and neuropeptide Y act as markers for a population of basal ganglia interneurones which are relatively preserved in Huntington's disease.

Acknowledgments. David Dawbarn was a Hereditary Disease Foundation Fellow, and Marion de Quidt was an MRC Scholar.

References

1. Aronin N, Cooper PE, Lorenz LJ, Bird ED, Sagar SM, Leeman SE, Martin JB (1983) Somatostatin is increased in the basal ganglia in Huntington's disease. Ann Neurol 13:519–526
2. Bird ED, Iversen LL (1974) Huntington's chorea. Postmortem measurement of glutamic acid decarboxylase, choline acetyltransferase and dopamine in basal ganglia. Brain 97:457–472
3. Dawbarn D, De Quidt ME, Emson PC (1985) Survival of basal ganglia neuropeptide Y-somatostatin neurones in Huntington's disease. Brain Res 340:251–260
4. Dawbarn D, Zamir N, Waters CM, Hunt SP, Emson PC, Brownstein MJ (1986) Decreased dynorphin-like immunoreactivity in Huntington's disease. Brain Res 372:155–158
5. Emson PC (1986) Neuropeptides and the pathology of Huntington's disease. In: Emson PC, Rossor MR, Tohyama M (eds) Peptides and neurological disease. Elsevier/North-Holland, Amsterdam. Progress in Brain Research, vol 66, pp 91–105
6. Emson PC, Arregui A, Clement-Jones V, Sandberg BEB, Rossor M (1980) Regional distribution of methionine-enkephalin and substance P-like immunoreactivity in normal human brain and in Huntington's disease. Brain Res 199:147–160
7. Graybiel AM (1984) Neurochemically specified subsystems in the basal ganglia. In: Evered D, O'Connor M (eds) CIBA Foundation Symposium 107:114–149
8. Gusella JF, Wexler NS, Conneally PM, Naylor SL, Anderson MA, Tanzi RE, Watkins PC, Othna K, Wallace MR, Sagaguchi AY, Young AB, Shoulson I, Bouilla E, Martin JB (1983) A polymorphic DNA marker genetically linked to Huntington's disease. Nature 306:234–238
9. Höllt V (1985) Multiple endogenous opioid peptides. In: Bousfield D (ed) Neurotransmitters in action. Elsevier Biomedical, Amsterdam, pp 188–193
10. Jessell TM (1983) Substance P. In: Iversen LL, Iversen SD, Snyder SH (eds) Handbook of Psychopharmacology. Plenum, New York, vol 16, pp 1–56
11. Nawa H, Hirose T, Takashima H, Inagama S, Nakanishi S (1983) Nucleotide sequences of clones cDNA for two types of bovine brain substance P precursor. Nature 306:32–36
12. Rossor M, Emson PC (1982) Neuropeptides in degenerative disease of the central nervous system. Trends in Neurosci 5:399–341
13. Thomas E, Pearse AGE (1964) The solitary active cells' histochemical demonstration of damage resistant nerve cells with a TPN-diaphorase reaction. Acta Neuropathol 3:238–249
14. Vincent SR, Johansson O, Hökfelt T, Skirboll L, Elde RP, Terenius L, Kimmel J, Goldstein M (1983) NADPH-diaphorase: a selective histochemical marker for striatal neurones containing both somatostatin- and avian pancreatic polypeptide (APP)-like immunoreactivities. J Comp Neurol 217:252–263

Neuropeptides in Central Movement Disorders of Man

A. WEINDL [1], J. UNGER [2], M. SCHWARTZBERG [1], J. TRIEPEL [1], W. LANGE [1], and A. STRUPPLER [1]

Introduction

Of the more than 35 neuropeptides presently known, a major part is found in brain areas involved in the control of movement. The neural circuits and the neuroactive substances of basal ganglia of the rat have been surveyed in this volume [37]. In the cerebral cortex of rat and other mammals small neurons contain somatostatin, neuropeptide Y (NPY), vasoactive intestinal polypeptide (VIP), cholecystokinin (CCK), enkephalin [9, 19, 21, 26, 30, 31]. In some cortical VIP neurons a coexistence with choline acetyltransferase, the biosynthetic enzyme of acetylcholine, has been shown [19]. Cortical neurons projecting to the striatum contain glutamate [34]. In the striatum the majority of neurons is GABA-ergic; most perikarya are medium-size spiny striato-fugal projection neurons, and a minority are medium- to large-size cells representing local circuit neurons. Few striatal neurons are cholinergic [45]. Somatostatin is contained in medium-size aspiny interneurons [17]. Met-, leuenkephalin, substance P [40] and dynorphin [49] are contained in striatal neurons projecting to the pallidum. A coexistence of met-, leuenkephalin and substance P has been found in GABA-ergic striatal neurons [2, 4]. In the pallidum the majority of neurons is GABA-ergic, most of them being projection neurons. GABA-ergic projection neurons of the striatum terminate in the substantia nigra at dopaminergic neurons [29]; some of these dopamine neurons contain CCK [30]. These neurons reciprocally project to GABA-ergic and cholinergic neurons in the striatum. GABA-ergic pallidal neurons project further to thalamus, tectum, subthalamic nucleus and possibly to the substantia nigra pars compacta. Unknown are the transmitters of the neurons projecting from the thalamus back to the cortex, and from the subthalamic nucleus back to the thalamus.

Here we report on some peptides involved in the neural circuitry of basal ganglia as revealed by immunohistochemistry, on the quantitative distribution of some peptides in early postmortem tissue of control persons, and on changes observed in Parkinson's disease. In order to obtain some insight into altered mechanisms of peptidergic neurons in the living patient we have measured peptide concentrations in the CSF of patients with various motor disturbances.

[1] Neurologische Klinik und Poliklinik der Technischen Universität München, Möhlstr. 28, D-8000 München 80, FRG.
[2] Present address: Department of Neurology, University of Rochester, Rochester, NY 14642, USA.

Clinical Aspects of Sensory Motor Integration
Ed. by A. Struppler and A. Weindl
© Springer-Verlag Berlin Heidelberg 1987

Material and Methods

Brain tissue was obtained either of control persons by early postmortem autopsy (2½–12 h) from the Department of Forensic Medicine University of Munich or of deceased neurological patients from the Department of Pathology, Technical University of Munich.

After dissection, tissue samples were either fixed for immunohistochemical studies or were rapidly frozen for quantitative determinations of peptides by radioimmunoassay. Chemically fixed blocks were processed for paraffin sectioning and for the immunoperoxidase method of Sternberger [47] using antisera against somatostatin and NPY. For the demonstration of the colocalization of somatostatin and NPY, neighboring 1–2 μm paraffin sections were incubated with the respective antisera.

The specificity of the antisera was tested immunohistochemically as well as by enzyme-linked immunosorbent assay and radioimmunoassay after preadsorption with structurally related peptides.

For quantitative measurements of peptides in human tissue, samples were brought into 0.1 N acetic acid, then homogenized and extracted. The protein content was determined with the Lowry method [33].

Somatostatin and VIP were determined using specific and sensitive radioimmunoassays described elsewhere [48, 51].

Cerebrospinal fluid was obtained by routine lumbar puncture at 9–11 a.m.; while the first and second portions were used for routine laboratory tests the third portion was immediately frozen and kept at −80 °C until further processing.

Observations

Immunohistochemistry

Somatostatin is widely distributed in neurons of the human forebrain, including cerebral cortex, hippocampus, amygdala and basal ganglia (Fig. 1). In the caudate nucleus and putamen, somatostatin is contained within medium-size aspiny neurons (Fig. 2a). Similarly, NPY is present in cerebral cortex, hippocampus, amygdala and basal ganglia. In medium-size aspiny neurons of the striatum NPY and somatostatin are colocalized within the same neuron, as can be demonstrated on neighboring 1–2 μm sections through the same perikaryon (Fig. 2a and b). VIP neurons are present in the cerebral cortex, but only in small numbers in basal ganglia in contrast to somatostatin and NPY.

Radioimmunoassay

Quantitative measurements of somatostatin in early postmortem *tissue* of control persons (Table 1) revealed high values in hypothalamus (18.32 ± 3.08 ng/mg pro-

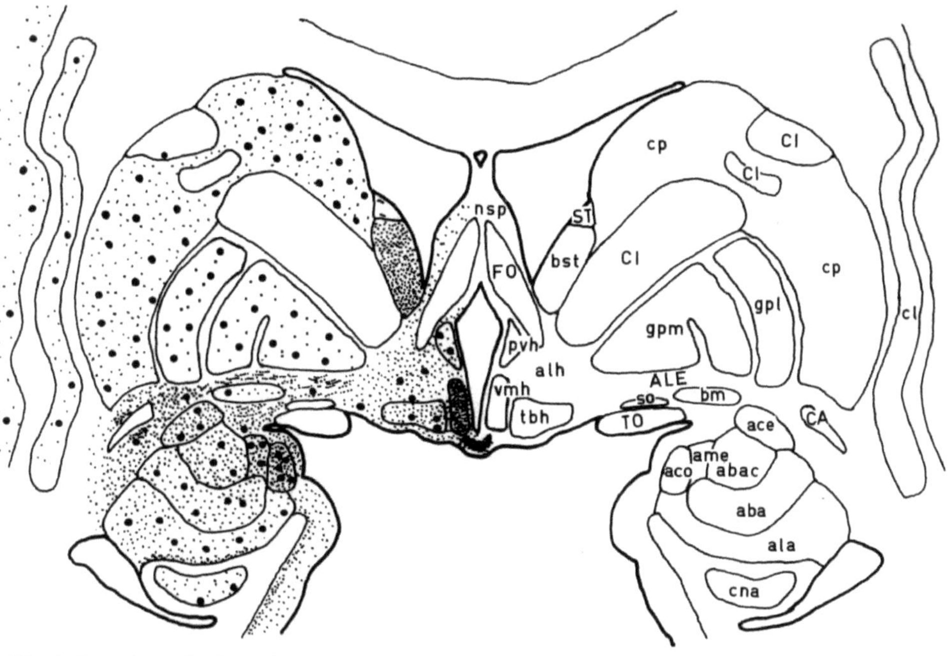

Fig. 1. Drawing of a frontal section through the brain of a control person at the level of the hypothalamus. Somatostatin immunoperoxidase reaction. The density of somatostatin perikarya (*large dots*), fibers and terminals (*small dots*) is illustrated on the left side. Somatostatin perikarya and fibers are present in hypothalamus, amygdala, caudatoputamen (*cp*), lateral (*gpl*) and medial (*gpm*) pallidum, claustrum (*cl*) and cortex. High densities of somatostatin fibers are present in infundibulum, ventromedial hypothalamic nucleus (*vmh*), bed nucleus of the stria terminalis (*bst*), central (*ace*), medial (*ame*) and cortical amygdaloid nucleus (*aco*). *aba* nu., amygdalae basalis; *abac* nu., amygdalae basalis accessorius; *ala* nu., amygdalae lateralis; *ALE*, ansa lenticularis; *alh*, area lateralis hypothalami; *bm* nu., basomedialis; *CA*, commissura anterior; *CI*, capsula interna; *cna*, cornu ammonis; *FO*, fornix; *nsp* nu., septi; *pvh* nu., paraventricularis hypothalami; *so* nu., supraopticus; *ST*, stria terminalis; *tbh* nu., tuberis hypothalami; *TO*, tractus opticus

tein; $n=3$). Values in caudate nucleus were higher (2.67 ± 1.02 ng/mg protein; $n=3$) than in putamen (1.50 ± 0.60 ng/mg protein; $n=3$), thalamus (1.46 ± 0.67 ng/mg protein; $n=3$), frontal (1.72 ± 0.82 ng/mg protein; $n=4$), central 1.61 ± 0.46 ng/mg protein; $n=4$), temporal (1.87 ± 0.32 ng/mg protein; $n=3$) and occipital cortex (1.41 ± 0.28 ng/mg protein; $n=4$). In a patient with Parkinson's disease and dementia (Table 1) values were significantly lower in frontal (1.12 ng/mg protein) and temporal cortex (0.58 ng/mg protein) while other regions did not show major changes.

VIP measurements in control persons (Table 1) showed low values in hypothalamus (0.60 ± 0.11 ng/mg protein; $n=3$), thalamus (0.44 ± 0.14 ng/mg protein; $n=3$), caudate (0.95 ± 0.37 ng/mg protein; $n=2$), putamen (0.81 ± 0.07 ng/mg protein; $n=2$) and higher values in frontal (1.81 ± 0.17 ng/mg protein; $n=2$), central (2.94 ± 0.40 ng/mg protein; $n=3$), occipital (3.83 ± 0.84 ng/mg protein; $n=3$) and temporal cortex (3.59 ± 0.16 ng/mg protein; $n=2$). In the brain of the Parkinson patient (Table 1) no significant changes were observed. Brain specimens of

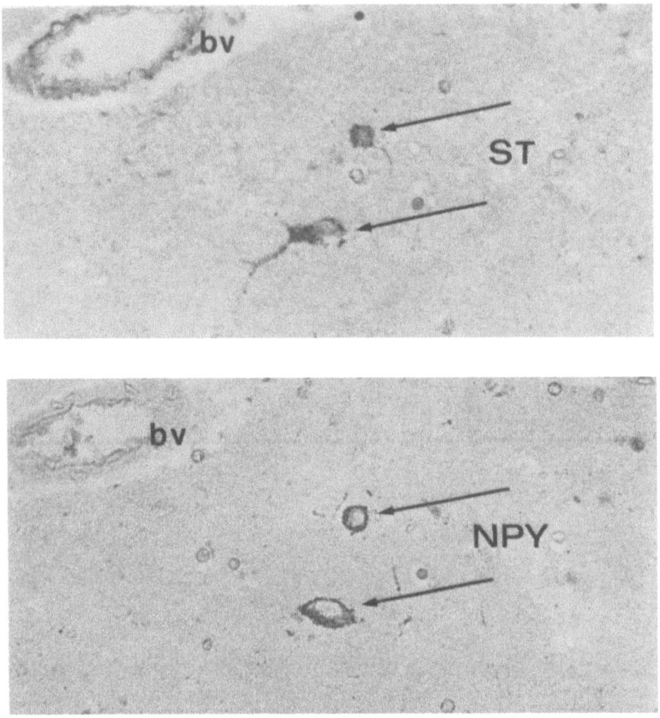

Fig. 2. Colocalization of somatostatin (*ST*) and *NPY* in two perikarya of the human striatum demonstrated by immunoperoxidase staining with **a** somatostatin and **b** *NPY* antisera on neighbouring 1–2 μm serial sections. In both neurons somatostatin and NPY coexist. *bv*, blood vessel. 250 ×

Table 1. Concentrations of somatostatin and VIP in brain regions of controls and one parkinsonian subject.

Peptide concentrations are expressed as ng/mg protein; the number of control samples is indicated in parentheses

Brain region	Somatostatin		Vasoactive intestinal polypeptide	
	Controls (mean ± SEM)	Parkinsonian subject	Controls (mean ± SEM)	Parkinsonian subject
Frontal cortex	1.72 ± 0.82 (4)	1.12	1.81 ± 0.17 (2)	1.42
Central cortex	1.61 ± 0.46 (4)	1.53	2.94 ± 0.40 (3)	2.76
Temporal cortex	1.87 ± 0.32 (3)	0.58	3.59 ± 0.16 (2)	3.64
Occipital cortex	1.41 ± 0.28 (4)	0.23	3.83 ± 0.84 (3)	2.26
Caudate nucleus	2.67 ± 1.02 (3)	2,31	0,95 ± 0.37 (2)	1.11
Putamen	1.50 ± 0.60 (3)	1.64	0.81 ± 0.07 (2)	1.09
Thalamus	1.46 ± 0.67 (3)	0.76	0.44 ± 0.14 (3)	0.61
Hypothalamus	18.32 ± 3.08 (3)	23.16	0.60 ± 0.11 (3)	0.82

Protein content was detected according to the Lowry method (1951).

Table 2. Somatostatin (ST) concentrations in the CSF of patients with movement disorders and Alzheimer's disease. In control persons no significant differences between age groups below and above 50 years are observed. Significant decreases are found only in Parkinson's, Wilson's, and Alzheimer's disease.

Concentrations of somatostatin in the cerebrospinal fluid of neurological patients

	n	Age (years) mean ± SEM	ST (pg/ml) mean ± SEM
Controls (age < 50)	32	34 ± 2	150.2 ± 9.3
Controls (age > 50)	30	63 ± 2	143.3 ± 7.5
Huntington's chorea	29	42 ± 2	78.1 ± 10.4
Parkinson's disease	18	64 ± 3	44.8 ± 7.1
Wilson's disease	2	28; 32;	40; 28;
Alzheimer's dementia	12	58 ± 4	51.5 ± 11.4
Amyotrophic lateral-sclerosis	8	57 ± 3	114.3 ± 18.1

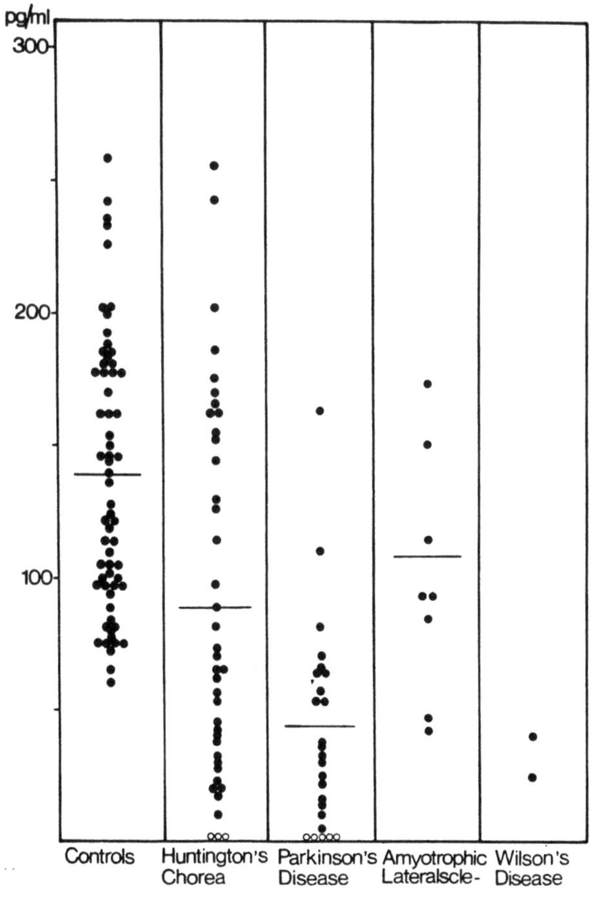

Fig. 3. Somatostatin concentrations in human CSF. Decreased levels are found in Parkinson's and Wilson's disease in contrast to amyotrophic lateral sclerosis

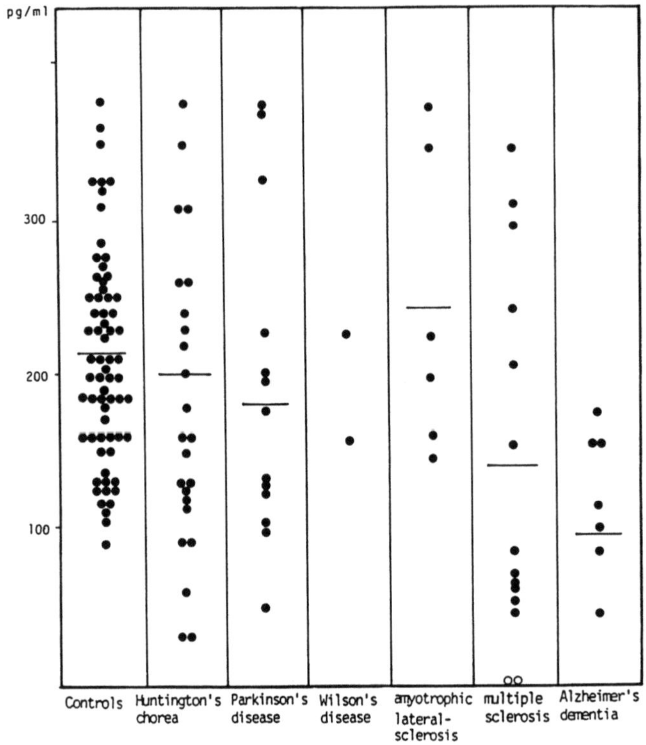

Fig. 4. VIP concentrations in human CSF. Alterations are found neither in Parkinson's, Wilson's, Huntington's disease nor in amyotrophic lateral sclerosis, in contrast to Alzheimer's disease

patients with Huntington's disease and amyotrophic lateral sclerosis are under investigation.

Quantitative radioimmunological measurements of somatostatin (Table 2; Fig. 3) and VIP (Table 3; Fig. 4) were carried out in the *cerebrospinal fluid* of control persons and of patients with various movement disorders as well as with Alzheimer's disease. Somatostatin values in control persons are 150.2±9.3 pg/ml ($n=32$) in the age group below 50 years and 143.3±7.5 pg/ml ($n=30$) in the age group above 50 years. There was a significant decrease in Parkinson's disease (44.8±7.1 pg/ml; $n=18$), Wilson's disease (40; 28 pg/ml) as well as in Alzheimer's disease (51.5±11.4 pg/ml; $n=12$). In Huntington's disease the values showed a greater degree of spreading, suggesting that there may be a subgroup of patients in which somatostatin values are not decreased. In amyotrophic lateral sclerosis there was only a slight, but not significant decrease of values (114.3±18.1 pg/ml; $n=8$).

VIP values in the cerebrospinal fluid of control persons were 208.3±11.9 pg/ml ($n=35$) in the age group below 50 years and 201.6±9.3 pg/ml ($n=32$) in that above 50 years.

In contrast to somatostatin, VIP values were not significantly different from control persons in patients with Parkinson's disease (183.8±29.6 pg/ml; $n=12$), Wilson's disease (230; 146 pg/ml), Huntington's disease (201.4±25.7 pg/ml; n=25). In patients with Alzheimer's disease the values were significantly decreased (101.7±19.9 pg/ml; $n=9$), although values were overlapping with controls.

Table 3. VIP concentrations in the CSF of patients with movement disorders, Alzheimer's disease and multiple sclerosis. In control persons below and above 50 years of age no significant differences are observed. Values in patients with movement disorders affecting mainly basal ganglia (Huntington's, Parkinson's, or Wilson's disease) do not differ from control persons. Decreased values are found in Alzheimer's disease.

Concentrations of VIP in the cerebrospinal fluid of neurological patients

	n	Age (years) mean ± SEM	VIP (pg/ml) mean ± SEM
Controls (age < 50)	35	34 ± 2	208.3 ± 11.9
Controls (age > 50)	32	63 ± 2	201.6 ± 9.3
Huntington's chorea	25	42 ± 2	201.4 ± 25.7
Parkinson's disease	12	61 ± 3	183.8 ± 29.6
Wilson's disease	2	28; 32;	230; 146;
Alzheimer's dementia	9	59 ± 4	101.7 ± 19.9
Multiple sclerosis	14	47 ± 4	145.5 ± 31.0

Discussion

Peptides have been recognized as neuroactive substances which are used in neurotransmission in addition to classical transmitters. While classical transmitters such as acetylcholine, glutamate, GABA and other amino acid transmitters are known for a very rapid onset and very short duration of action, the peptides appear to modify transmitter actions by a longer duration of activity. The demonstration of a coexistence of a classical transmitter with a peptide in the same neuron has been interpreted in the sense of a synergistic coactivation [28].

Our observation of somatostatin in cortical neurons as well as in medium-size neurons of the striatum of the human are in agreement with literature reports of Geola et al. [26] and Martin [35]. In accordance with findings in the rat [50] we have found a colocalization of somatostatin and NPY in medium-size striatal neurons of the human brain. Coexistence of somatostatin and NPY was further demonstrated in rat and human cortical and rat hypothalamic neurons [10]. These neurons possess the enzyme nicotinamide adenine dinucleotide phosphate diaphorase (NADPH-d), which supposedly has a protective function against the damaging process leading to the loss of striatal neurons in Huntington's disease [24], possibly caused by quinolinic acid [6]. NADPH-d positive neurons are relatively spared in Huntington's disease [6, 24]. In contrast to the depletion of substance P and enkephalin, somatostatin and NPY neurons appear to be relatively preserved in Huntington's disease [15, 20, 24]. Quantitative measurements of somatostatin and NPY in Huntington's disease did not reveal a decrease of both peptides in striatum [3, 11, 20, 36] and Nu. accumbens [8].

Quantitative measurements of peptide concentrations in postmortem tissue of patients with basal ganglia disorders have been carried out by several laboratories

[1, 20, 22, 35, 41]. Our observations of a decreased somatostatin concentration in the frontal cortex of a Parkinson patient with mental impairment is in accordance with literature reports of a decreased somatostatin content in Parkinson's disease plus dementia [1, 22, 41]. This appears to be related to a loss of cortical somatostatin neurons and fibers.

Somatostatin measurements in the CSF of control persons [27, 39] and of patients with neurodegenerative diseases [5, 7, 12, 13, 18, 25, 32, 43, 46, 53] have been carried out by several groups. Our observation of decreased levels of somatostatin in Parkinson's disease plus dementia and in Alzheimer's disease are in agreement with reports from other groups [16, 38] and can be correlated with measurements in brain tissue [1, 14, 22, 42]. Decreased levels of somatostatin in Wilson's disease (hepatolenticular degeneration) have not been reported so far and may reflect involvement of striatal somatostatin neurons in the process of hepatolenticular degeneration due to abnormal copper storage.

Literature reports of somatostatin in the CSF of Huntington patients are controversial. While Cramer et al. [12] reported decreased levels, Beal et al. [7] found normal or elevated levels. The variation of values observed in our patients suggests that there is a subgroup with normal and a subgroup with decreased somatostatin levels.

VIP has been shown to be present in the CSF of control persons [23, 44]. No significant changes of VIP in the CSF of patients with basal ganglia diseases including Huntington's disease were found; VIP measurements in brain tissue [21] did not reveal changes. Decreased levels of VIP in the CSF of Alzheimer patients [52] appear to correlate with the cortical localization of VIP and the possibly disturbed turnover of VIP. The observation of unaltered somatostatin and VIP concentrations in the CSF of ALS patients may be explained by the fact that somatostatin and VIP neurons are not a major component of the corticospinal tract, which in addition to the lower motoneuron system is basically affected in this disease.

The observations presented indicate that quantitative changes in regional tissue concentrations of some peptides are associated with some degenerative CNS diseases, including movement disorders. CSF measurements of peptide concentrations reveal patterns in some neurodegenerative disorders.

Quantitative measurements of peptide alterations may provide in the living patient a diagnostic window for gaining some insight into altered mechanisms of peptidergic neurons.

Summary

Several peptides are present and partially coexist with classical transmitters in neurons of cortical and subcortical neural circuits involved in movement control. The role of two peptides, somatostatin and vasoactive intestinal polypeptide (VIP), has been examined in degenerative motor disturbances. RIA measurements in early postmortem brain tissue revealed decreased somatostatin concentrations in frontal and temporal cortex, and no changes of VIP concentrations in

Parkinson's disease. RIA measurements in the CSF revealed decreased somatostatin concentrations in Parkinson's, Wilson's as well as in Alzheimer's disease, and decreased concentrations of VIP in Alzheimer's disease only. Changes of peptide concentrations in the CSF in some neurodegenerative diseases may be correlated with their immunohistochemical distribution in relation to the preferential impairment of specific functional systems in neurodegenerative diseases, such as central movement disorders. CSF measurements of peptides in neurologic system disorders may provide, even in the living patient, a diagnostic window into the neural circuitry of peptidergic neurons.

Acknowledgment. This work was supported by DFG grant We 608-7. The valuable cooperation of Dr. T. Gilg, Department of Forensic Medicine, University of Munich, and of Dr. M. Schiessl, Department of Pathology, Technical University of Munich, as well as the excellent technical assistance of Ms. I. Wild and S. Mühlsiemer are acknowledged.

References

1. Agid Y, Javoy-Agid F (1985) Peptides and Parkinson's disease. Trends Neurosci 8:30–35
2. Afsharpour S, Penny GR, Kitai ST (1984) Glutamic acid decarboxylase, leucine enkephalin, and substance P immunoreactive neurons in the neostriatum of the rat and cat. Soc Neurosci Abstr 10:702
3. Aronin N, Cooper PE, Lorenz LJ, Bird ED, Sagar S, Leeman SE, Martin JB (1983) Somatostatin is increased in the basal ganglia in Huntington's disease. Ann Neurol 13:519–526
4. Aronin N, DiFiglia M, Graveland GA, Schwartz WJ, Wu J-Y (1984) Localization of immunoreactive enkephalins in GABA-synthesizing neurons of the rat neostriatum. Brain Res 300:376–380
5. Beal MF, Crowdon JH, Mazurek MF, Martin JB (1986) CSF somatostatin-like immunoreactivity in dementia. Neurology 36:294–297
6. Beal MF, Kowall NW, Ellison DW, Mazurek MF, Swartz KJ, Martin JB (1986) Replication of the neurochemical characteristics of Huntington's disease by quinolinic acid. Nature 321:168–171
7. Beal MF, Mazurek MF, Crowdon JH, Black P, Martin JB (1985) CSF somatostatin-like immunoreactivity in neurologic disease. J Neurol [Suppl] 232:176
8. Beal MF, Bird ED, Langlais PJ, Martin JB (1984) Somatostatin is increased in the nucleus accumbens in Huntington's disease. Neurology 34:663–666
9. Chan-Palay V, Allen YS, Lang W, Haesler U, Polak JM (1985) I. Cytology and distribution in normal human cerebral cortex of neurons immunoreactive with antisera against neuropeptide Y. J Comp Neurol 238:382–389
10. Chronwall BM, Chase TN, O'Donohue TL (1984) Coexistence of neuropeptide Y and somatostatin in rat and human cortical and rat hypothalamic neurons. Neurosci Lett 32:213–217
11. Cooper P, Aronin N, Bird E, Leeman SE, Martin JB (1981) Increased somatostatin in the basal ganglia of Huntington's disease. Neurology 31:64
12. Cramer H, Kohler VJ, Oepen G, Schomburg G, Schröter E (1981) Huntington's chorea – measurements of somatostatin, substance P, and cyclic nucleotides in the cerebrospinal fluid. J Neurol 225:183–187
13. Cramer H, Schaudt D, Rissler K, Strubel D, Warter J-M, Kuntzmann F (1985) Somatostatin-like immunoreactivity and substance P-like immunoreactivity in the CSF of patients with senile dementia of Alzheimer type, multi-infarct syndrome and communicating hydrocephalus. J Neurol 232:346–351
14. Davies P, Katzman R, Terry RD (1980) Reduced somatostatin-like immunoreactivity in cerebral cortex from cases of Alzheimer's disease and Alzheimer's senile dementia. Nature 288:279–280

15. Dawbarn D, de Quidt ME, Emson PC (1985) Survival of basal ganglia neuropeptide Y/ somatostatin neurones in Huntington's disease. Brain Res 340:252–260
16. Delfs JR (1985) Somatostatin and Alzheimer's disease: possible pathophysiological associations. In: Senile dementia of the Alzheimer type. Alan R. Liss Inc., pp 243–261
17. DiFiglia M, Aronin N (1982) Ultrastructural features of immunoreactive somatostatin neurons in the rat caudate nucleus. J Neurosci 2:1267–1274
18. Dupont E, Christensen SE, Hansen AP, de Fine Olivarius B, Orskov H (1982) Low cerebrospinal fluid somatostatin in Parkinson's disease. Neurology 32:312–314
19. Eckenstein F, Baughman RW (1984) Two types of cholinergic innervation in cortex, one colocalized with vasoactive intestinal polypeptide. Nature 309:153–155
20. Emson PC, Dawbarn D, de Quidt ME CNS peptides in Huntington's chorea (this volume)
21. Emson PC, Fahrenkrug J, Spokes EGS (1979) Vasoactive intestinal polypeptide (VIP): distribution in normal human brain and Huntington's disease. Brain Res 173:174–178
22. Epelbaum J, Ruberg M, Moyse E, Javoy-Agid F, Dubois B, Agid Y (1983) Somatostatin and dementia in Parkinson's disease. Brain Res 278:376–379
23. Fahrenkrug J, Schaffalitzky de Muckadell OB, Fahrenkrug A (1977) Vasoactive intestinal polypeptide (VIP) in human cerebropsinal fluid. Brain Rcs 124:581–584
24. Ferrante RJ, Kowall NW, Beal MF, Richardson EP Jr, Bird ED, Martin JB (1985) Selective sparing of a class of striatal neurons in Huntington's disease. Science 230:561–563
25. Francis PT, Bowen DM, Neary D, Palo J, Wikstrom J, Olney (1984) Somatostatin-like immunoreactivity in lumbar cerebrospinal fluid from neurohistologically examined demented patients. Neurobiol Aging 5:183–186
26. Geola FL, Yamada T, Warwick RJ, Tourtelotte WW, Hershman JM (1981) Regional distribution of somatostatin-like immunoreactivity in the human brain. Brain Res 229:35–42
27. Hansen AB, de Fine Olivarius SE, Pederson E, Sorensen K, Dupont E, Ingenslee I, Christensen SE, Orskov H (1984) Cerebrospinal fluid SLI: on its origin and pathology in neurological disease. In: Rosenthal J (ed) Proceed 2nd Symp Somatostatin, Athens, University Press Tübingen
28. Hökfelt T, Johansson O, Goldstein M (1984) Chemical anatomy of the brain. Science 225:1326–1334
29. Hökfelt T, Johansson O, Fuxe K, Goldstein M, Park D (1976) Immunohistochemical studies on the localization and distribution of monoamine neuron systems in the rat brain. I. Tyrosine hydroxylase in the mes- and diencephalon. Med Biol 54:427–453
30. Hökfelt T, Skirboll L, Rehfeld JF, Goldstein M, Markey K, Daun O (1980) A subpopulation of mesencephalic dopamine neurons projecting to limbic areas contains a cholecystokinin-like peptide: evidence from immunohistochemistry combined with retrograde tracing. Neurosci 5:2093–2194
31. Khachaturian H, Lewis ME, Schäfer MKH, Watson SJ (1985) Anatomy of the opioid systems. Trends Neurosci 8:111–119
32. Kohler J, Schröter E, Cramer H (1982) Somatostatin-like immunoreactivity in the cerebrospinal fluid of neurological patients. Arch Psychiatr Nervenkr 231:503–508
33. Lowry HO, Rosebrough NJ, Farr AL, Randall RJ (1951) Protein measurement with the Folin reagent. J Biol Chem 53:265–275
34. Madl JE, Larson AA, Beitz AJ (1986) Monoclonal antibody specific for carbodiimide-fixed glutamate: immunocytochemical localization in the rat CNS. J Histochem Cytochem 34:317–326
35. Martin JB (1984) Huntington's disease: new approaches to an old problem. Neurology 34:1049–1072
36. Nemeroff CB, Youngblood WW, Manberg PJ, Prange AJ (1983) Regional brain concentrations of neuropeptides in Huntington's chorea and schizophrenia. Science 221:972–975
37. Oertel WH, Struppler A Immunohistochemical studies on neurotransmitters in rat basal ganglia (this volume)
38. Oram JJ, Edwardson J, Millard PH (1982) Investigation of cerebrospinal fluid neuropeptides in idiopathic senile dementia. Gerontology 27:216–223
39. Patel YC, Rao K, Reichlin S (1977) Somatostatin in human cerebrospinal fluid. N Engl J Med 296:529–533
40. Penney JB Jr, Young AB (1982) Speculation on the functional anatomy of basal ganglia. Ann Rev Neurosci 6:73–94

41. Rinne UK, Rinne JO, Rinne JK (1984) Brain neurotransmitters and neuropeptides in Parkinson's disease. Acta Physiol Pharmacol Latinoam 34:287–299
42. Rossor MN, Emson PC, Mountjoy CQ, Roth M, Iversen LL (1980) Reduced amounts of immunoreactive somatostatin in temporal cortex in senile dementia of Alzheimer type. Neurosci Lett 20:373–377
43. Serby M, Richardson SB, Twente S, Siekierski J, Corwin J, Rotrosen J (1984) CSF somatostatin in Alzheimer's disease. Neurobiol Aging 5:187–189
44. Sharpless NS, Thal LJ, Perlow MJ et al. (1984) Vasoactive intestinal peptide in cerebrospinal fluid. Peptides (Fayetteville) 5:429–433
45. Sofroniew MV, Eckenstein F, Thoenen H, Cuello AC (1982) Topography of choline acetyltransferase-containing neurons in the forebrain of the rat. Neurosci Lett 33:7–12
46. Soininen HS, Jolkonen JT, Reinikainen KJ, Halonen TO, Riekkinen PJ (1984) Reduced cholinesterase activity and somatostatin-like immunoreactivity in the cerebrospinal fluid of patients with dementia of the Alzheimer type. J Neurol Sci 63:167–172
47. Sternberger LA (1979) Immunocytochemistry, 2nd edn. Wiley, New York Chichester
48. Unger J, Weindl A, Pitzl H, Schrell U, Lange W (1984) Antiserumproduktion gegen Somatostatin (ST): Titerkontrolle mit ELISA und Radioimmunoassay zur Bestimmung der quantitativen Verteilung von ST im Rattengehirn. Verh Anat Ges 78:515–517
49. Vincent SR, Hökfelt T, Christensson I, Terenius L (1982) Immunohistochemical evidence for a dynorphin immunoreactive striatonigral pathway. Eur J Pharmacol 85:251–252
50. Vincent SR, Johansson O, Hökfelt T, Skirboll L, Elde RP, Terenius L, Kimmel J, Goldstein M (1983) NADPH-diaphorase: a selective histochemical marker for striatal neurons containing both somatostatin and avian pancreatic polypeptide (APP)-like immunoreactivities. J Comp Neurol 217:252–263
51. Weindl A, Unger J, Gnahn H, Lange H, Lange W, Struppler A (1985) Neuropeptide bei degenerativen ZNS Erkrankungen – Radioimmunologischer Nachweis von Somatostatin und vasoaktivem intestinalem Polypeptid im Liquor cerebrospinalis. In: Gänshirt H, Berlit P, Haack G (eds) Kardiovaskuläre Erkrankungen und Nervensystem. Neurotoxikologie. Probleme des Hirntodes. Springer, Berlin Heidelberg New York Sydney Tokyo (Verh Dtsch Ges Neurol, vol 3, pp 935–940)
52. Wikkelsø C, Fahrenkrug J, Blomstrand C, Johansson BB (1985) Dementia of different etiologies: vasoactive intestinal polypeptide in CSF. Neurology 35:592–595
53. Wood PL, Etienne P, Lal S, Gauthier S, Cajal S, Nair NPV (1982) Reduced lumbar CSF somatostatin levels in Alzheimer's disease. Life Sci 31:2073–2079

Analysis of Extrapyramidal Motor Symptoms from Stereoencephalotomy

H. Narabayashi [1]

Introduction

For many years after the first description of the disease by James Parkinson in 1817, tremor and rigidity have been known as two main motor symptoms of paralysis agitans. In recent years akinesia has also been recognized as being equally or perhaps even more important than these two symptoms in the disease's symptomatology, although the term "akinesia" is still not clearly defined. Whether or not tremor and rigidity are separately understood in regard to their generating mechanisms has not been sufficiently investigated.

Careful analysis of the cases of Parkinson's disease having undergone stereotaxic thalamotomy has raised the question that tremor and rigidity may be differently located within the thalamus. It has been the common experience for many years that the anterior lesion of the ventrolateral nucleus of the thalamus (VL) abolishes mainly rigidity but is often followed by recurrence of tremor, and also that the posterior VL lesion abolishes tremor completely and sustainedly. Stereotaxic lesion in the internal segment of the pallidum was reported to produce sustained improvement of rigidity, but not of tremor [7, 18]. Pallidal projection flows into the ventroanterior nucleus (VA) and VL of the thalamus, suggesting that the projection is concerned mainly with the generation of rigidity rather than of tremor. All these observations seem to suggest that different substrates and generating mechanisms may exist for both tremor and rigidity in the depth of the human brain [10].

Smith [24] studied neuropathologically 45 successfully operated brains and observed that rigidity tended to be relieved by lesions involving the pallidothalamic projection, and that tremor was better relieved by more posterior lesions involving cerebellar projection to the thalamus.

Brief Technical Description

With the introduction of the microelectrode technique during the stereotaxic procedure, it became possible to demarcate and closely identify each of the targeted structures.

The microelectrode installed at the tip of the stereotaxically inserted needle is of 5–10 μm in size. It might be better termed a semi-microelectrode, but it is small

[1] Department of Neurology, Juntendo University School of Medicine, 2-1-1, Hongo, Bunkyo-ku, Tokyo 113, Japan.

Clinical Aspects of Sensory Motor Integration
Ed. by A. Struppler and A. Weindl
© Springer-Verlag Berlin Heidelberg 1987

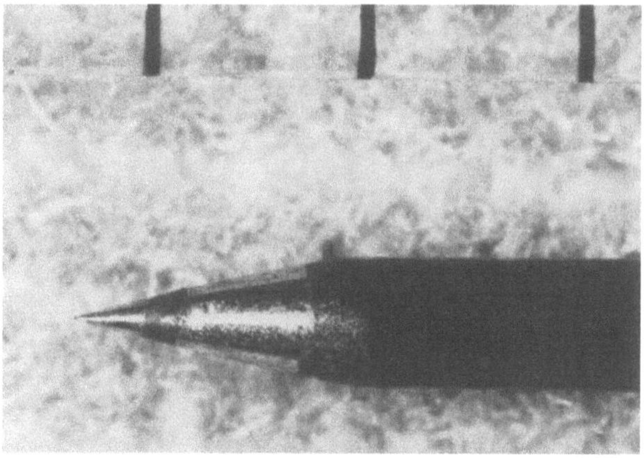

Fig. 1. Tip of microelectrode needle, bipolar and concentric (scale 1 mm)

enough to record isolated unitary activities extracellularly (Fig. 1). The sheath electrode is installed about 0.5 mm behind the tip electrode, and bipolar recording or localized bipolar stimulation of depth structures is possible through these concentric electrodes. Other instruments for recording are the same as those used in the neurophysiology laboratory.

The neuronal activities obtained from the structures which the microelectrode needle passes provide information on (a) whether it is a fiber (white matter) or a cellular (gray matter) area, (b) the density of the neuron population and the size of neurons or fibers (large, medium, or small), and (c) discharge patterns of unitary activities.

Dewulf [4] published a detailed study about the size and population of neurons of the thalamus, which are different in each subnucleus and even in each smaller subdivision within the subnuclei. Background noises through the microelectrode in the course of needle insertion indicate overall electrical activities arising from cells and fibers around the electrode and are therefore different in and specific to each structure [5]. To avoid terminological confusion, in the following description the term VO complex (VOA, ventralis oralis anterior; VOP, ventralis oralis posterior) will be used instead of VL, of which it is almost a synonym in American terminology [23].

Change of noise level from one structure to another, e.g., from the caudate nucleus to the internal capsule, from the internal capsule to the dorsal thalamic nuclei, then to the VO complex, and finally to the nucleus ventralis intermedicus (VIM) (Fig. 2), gives us exact information about the location of the tip electrode. It is a delicate and reliable procedure which has been routinely used in the author's surgical theater since 1971 on about 750 cases of various involuntary movements.

In addition to these noise level changes, another advantage of the technique is that it can register the isolated unitary neuronal activities of various patterns [13]. In patients with tremor, rhythmic burst discharges synchronous with peripheral tremor are recorded in VIM, but not in VO complex or ventroposterior nucleus (VP), as reported by many authors [1–3, 7, 14, 21]. Evoked activities for pe-

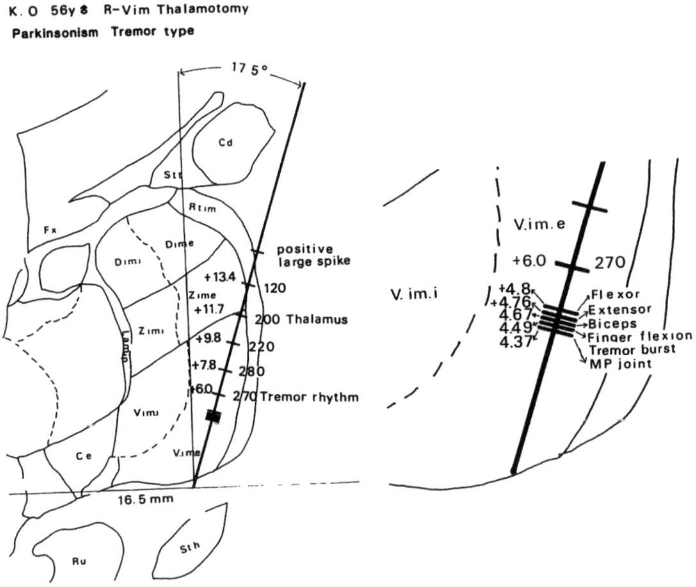

Fig. 2. Tracking of microelectrode needle to VIM. Noise level is shown on the *right* of the needle track. Distance in mm to the radiologically determined target is shown on the *left* of the needle track. In the *right figure,* areas of neurons that responded to peripheral kinesthetic stimuli of several muscles and joints are mapped

ripheral proprioceptive or kinesthetic stimuli such as passive muscle stretch, pressure on muscle belly, or passive or active movements of joints, are detectable in VIM, similarly as in VP neurons for tactile stimuli [6, 20]. The above-described tremor-synchronous burst discharges in VIM, which can be detected in about 85–90% of the cases with tremor, seem to be the result of afferent projections of peripheral tremorous movement, and are delicately organized in a somatotopic fashion.

All these features make it possible to identify and configurate the VIM, thus making it possible to locate exactly other structures adjacent to the VIM, such as VO complex, VP, or center medianum (CM). The ventral border of the thalamus to the subthalamic fiber area is easily figured out.

Tremor and Rigidity – Parkinsonism

After determining VIM neurons as described, high-frequency stimulation, 1 ms pulse, 40–60 Hz, is given to VIM, which almost always suppresses, but sometimes desynchronizes or facilitates tremor without changing rigidity much [14]. VO complex stimulation mainly enforces muscular hypertonus, slightly influencing tremor secondarily, depending on the stimulation parameter.

Electrocoagulation lesion of the size of 3–5 mm in diameter is then produced in the neuron area by controlled hyperthermia at 70 °C for 30–60 s. Electrocoagulation of VIM around the tremor-synchronous neurons abolishes tremor completely and sustainedly without changing rigidity much. On the other hand, rigidity can be abolished well by VO complex coagulation, but not satisfactorily enough by VIM coagulation. Therefore, the VO complex and the pallidothalamic projection system are understood as being concerned with the rigidity-generating mechanism, and VIM with that of tremor. In surgery on parkinsonism, these two subnuclei are always lesioned.

Neuroanatomically, the VO complex receives pallidal projection, but the VIM does not [9]. On the other hand, VIM receives projection from cerebellar deep nuclei, as well as proprioceptive inflow from the periphery, presumably via the spinothalamic tract.

Based on these observations, tremor- and rigidity-related neurons in parkinsonism are interpreted as being slightly differently located at the level of thalamic subnuclei, i.e., tremor in VIM and rigidity in VO complex. VIM is understood as the key structure where sensorimotor integration for tremor modulation occurs, but not the only center for producing tremor. Tremor seems to be a phenomenon based on the circuitry involving both central and peripheral mechanisms [12, 16].

Analysis of Various Involuntary Movements

The author also tried to classify various kinds of so-called extrapyramidal involuntary movements other than parkinsonism into two groups, i.e., symptoms depending on the VO complex and on the VIM, from the results of surgery targeting on either VO complex or VIM.

VIM Group

1. Various types of pathological tremor other than parkinsonism, such as brachial or flapping tremor, tremor in syndrome of Benedikt, or midbrain tremor due to head injury, are also alleviated by VIM lesion. The existence of rhythmic burst discharges in VIM that are synchronous with tremor and the effects of electrical stimulation and of surgical lesion are almost the same as in parkinsonian tremor. The papers already published may be referred to for further details [11, 12, 16]. In these conditions, involvement of cerebellar efferent pathways to the ventral thalamus is suspected, which is assumed to cause some functional changes in VIM neurons, resulting in tremorous movement. Hirai et al. [8] suggested that a slightly larger lesion was often necessary for relief of cerebellar tremor than of parkinsonian tremor.

A well-known characteristic feature of cerebellar tremor – namely, that it appears only when muscles are in action – seems important. Essential and familial tremor are interpreted in a similar way as appearing only in movement or in posture, although the pathology is still not known. When these tremors failed to respond to pharmacological therapy and when the degree of tremor is hazardous

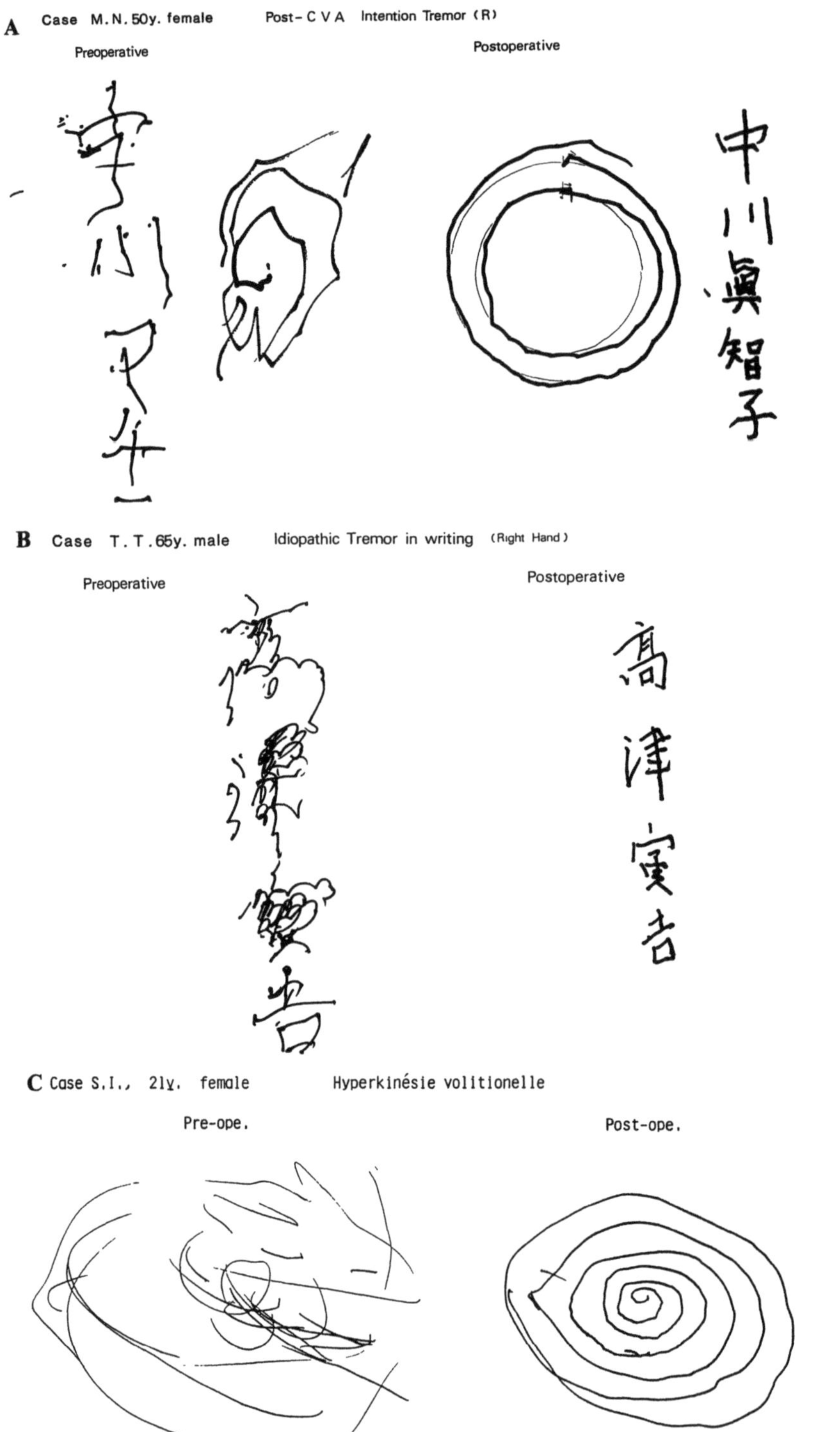

Fig. 3 A–C. Tremors (**A** and **B**) and hyperkinésies volitionnelles (**C**) are markedly improved after VIM surgery

enough to the patient's activity in average day life (ADL), surgery was tried. The improvement in handwriting in these tremor cases is shown in Fig. 3 A and B. Almost complete relief of tremor is usually obtained, and VIM is always the target structure [12, 15, 16].

2. *Hyperkinésie volitionnelle* is a peculiar, gross, rapid, abnormal movement or irregular shaking, mostly of the arm, suddenly appearing in voluntary movement or posturing. Pathological reports of the condition are few, but its clinical features are similar to the superior red nucleus syndrome [22]. As in cerebellar tremor cases, the effective lesion site is the VIM, but the lesion size needs to be a little larger, perhaps because of greater involvement of proximal shoulder muscles. Severe involuntary shaking usually disappears or is markedly reduced after surgery, but slight abnormal posture of wrist and fingers often remains. Figure 3 C shows the handwriting of a patient with hyperkinésie volitionnelle before and after surgical procedure on the VIM.

VO Complex Group

1. *Rigidity in parkinsonism:* It may not be necessary to state repeatedly that rigidity in parkinsonism is totally and sustainedly alleviated by lesion within the inter-

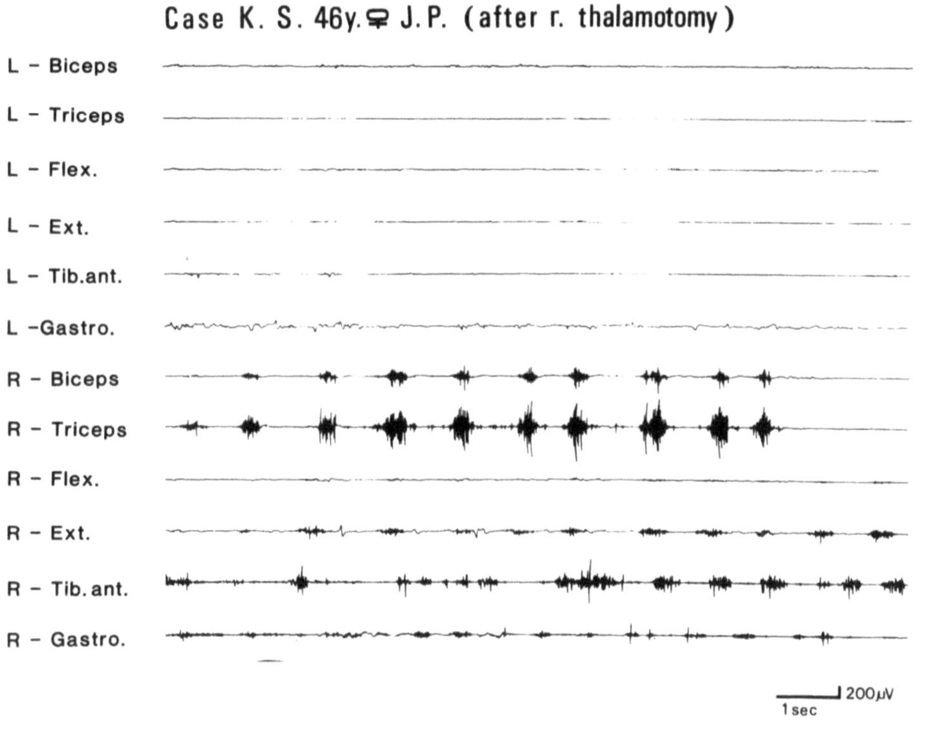

Fig. 4. L-dopa-induced dyskinesia on the operated and unoperated side. *Biceps*, biceps brachii muscle; *Triceps*, triceps brachii muscle; *Flex.*, forearm flexor muscle; *Ext.*, forearm extensor muscle; *Tib. ant.*, *tibialis anterior muscle; L*, left; *R*, right

nal pallidum or VO complex, which was established many years ago. Pharmacologically, rigidity is one of the symptoms resulting from striatal dopamine deficiency.

2. Levodopa-induced *dykinesia* is the reverse side effect in the long-term levodopa therapy for parkinsonism, appearing in about 30–40% of cases. It is an abnormal movement presenting choreic, athetoid, or sometimes ballistic movements of extremities in younger patients with juvenile parkinsonism (JP) [25], and more abnormal rostral (neck and jaw) movement in elderly patients. It is interpreted as being the result of overdosage of L-dopa or supersensitivity of striatal neurons to dopamine, and its appearance together with an up-down oscillation of the levodopa effect often inhibits the smooth continuation of levodopa therapy. These dyskinesias can be abolished by surgery on VO complex but not on VIM, as recently reported [13]. Figure 4 represents dyskinesia on the operated and unoperated side extremities in a case of JP. It is therefore suggested that levodopa-induced dyskinesia and rigidity in parkinsonism take the same anatomical route in their generation and that each of them are reversible to the other by pharmacological manipulation of the dopamine (DA) level within the striatum.

3. *Idiopathic dyskinesia* in the neck and face area, i.e., oromandibular or buccolingual dyskinesia, is a condition unknown in pathology but is also analyzed in a similar way. The condition is usually aggravated by levodopa and suppressed or reduced to various degrees by a DA antagonist, such as haloperidol or tiapride. Unilateral or bilateral surgery on VO complex markedly reduced abnormal movements in two cases with this symptom [19].

Two Groups of Involuntary Movements

As explained above, in several conditions the author tried to categorize abnormal involuntary movements of an extrapyramidal nature into two main groups depending on their structural basis, as shown in Table 1 A and B.

A classification of the extrapyramidal involuntary movements was proposed about 50–60 years ago by A. Jacob, C. and O. Vogt, and O. Foerster as hypertonic-hypokinetic and hypotonic-hyperkinetic groups. This dynamic idea for explaining the pathophysiology of these symptoms has now proved true with respect to rigidity and its reverse phenomenon, i.e., levodopa-induced dyskinesia, which can be manipulated pharmacologically. In addition, oromandibular dyskinesia and choreo-ballistic movements are interpreted on a similar basis, since they are known to be reduced by DA antagonists such as haloperidol, chlorpromazine, and tiapride. All these movements are improved by surgery on VO complex and are therefore interpreted as based on the pallido-VO complex projection system.

However, various types of tremor arising from a pathology of the cerebellum, or of its projection to the thalamus, and from hyperkinésie volitionnelle are of a different nature and are markedly improved by surgery on VIM. Thus, it is proposed that in addition to the pallido-VO complex group of symptoms, other group of symptoms based on the pathophysiology of VIM should be considered.

Table 1A, B. Two groups of involuntary movements, classified from the effect of stereotaxic surgery on either VO complex or VIM

Table 1A

By surgery on VO complex (*pallidum int.→ VO complex*):

A. *Highly improved*
   1. Rigidity in parkinsonism
   2. Dopa-induced dyskinesia (DID)
   3. Oromandibular dyskinesia (OMD)
   4. Hemiballism (*Luysian-pallido-VL thalamus*)

B. *Relatively improved*
   1. Dystonic athetoid cerebral palsy

Table 1B

By surgery on VIM:

A. *Highly improved*
   1. Parkinsonian tremor
   2. Cogwheeling
   3. Intention or action tremor
   4. Flapping tremor (brachial tremor)
   5. Postural tremor
   6. Familial or essential tremor
   7. Writer's cramp

B. *Relatively improved*
   1. Hyperkinésies volitionnelles
   2. Extrapyramidal myoclonus (Hassler)

However, choreic movement in Huntington's disease and idiopathic generalized dystonia are still not clearly understood in terms of their generation and conduction pathways, from experience with human stereotaxic surgery.

Summary

Various extrapyramidal involuntary movements are reviewed from the results of stereotaxic thalamotomy using microelectrode recording, which makes it possible to identify the neuronal discharge pattern of each subnucleus in the base of the thalamus, such as VIM or VO complex (synonymous to VL), during surgical procedure. They are divided into two groups of symptoms, which are well relieved by surgery either on VIM or on VO complex of the thalamus. Surgery on the VIM nucleus abolished almost completely various types of tremor, including parkinsonism and hyperkinésie volitionnelle. On the other hand, surgery on VO complex, which receives pallidal afferent projections, markedly reduced or eliminated parkinsonian rigidity, levodopa-induced dyskinesia, idiopathic dyskinesia, and hemiballism.

References

1. Albe-Fessard D, Arfel G, Guiot G (1963) Activités électriques caractéristiques de quelques structures cerebrales chez l'homme. Ann Chir 17:1185–1214
2. Bates JAV (1969) The significance of tremor phasic units in the human thalamus. In: Gillingham FJ, Donaldson IMD (eds) Third symposium on Parkinson's disease. Livingstone, London, pp 118–124
3. Bertrand G, Jasper H (1969) Microelectrode recording of unit activity in the human thalamus. Confin Neurol 26:205–208
4. Dewulf A (1971) Anatomy of the normal human thalamus. Elsevier, Amsterdam London New York
5. Fukamachi A, Ohye Ch, Saito Y, Narabayashi H (1977) Estimation of the neural noise within the human thalamus. Acta Neurochir [Suppl] (Wien) 24:121–136
6. Goto A, Kosaka K, Kubota K, Nakamura R, Narabayashi H (1968) Thalamic potentials from muscle afferents in the human. Arch Neurol 19:302–309
7. Hardy J (1966) Electrophysiological localization and identification. J Neurosurg 24:410–414
8. Hirai T, Miyazaki M, Nakajima H, Shibazaki T, Ohye Ch (1983) The correlation between tremor characteristics and the predicted volume of effective lesion in stereotaxic nucleus ventralis intermedius thalamotomy. Brain 106:1001–1018
9. Mehler WR, Nauta WJH (1974) Connections of the basal ganglia and of the cerebellum. Confin Neurol 36:205–222
10. Narabayashi H (1968) Functional differentiation in and around the ventrolateral nucleus of the thalamus based on experience in human stereoencephalotomy. Johns Hopkins Med J 122:295–300
11. Narabayashi H (1981) Surgical approach to tremor. In: Marsden CD, Fahn S (eds) Movement disorders. Butterworth, London, pp 292–299
12. Narabayashi H (1982) Tremor mechanism. In: Schaltenbrand G, Walker AR (eds) Stereotaxy of the human brain, 2nd edn. Thieme, Stuttgart New York, pp 510–514
13. Narabayashi H (1984) Levodopa-induced dyskinesia and thalamotomy. J Neurol Neurosurg Psychiatry 47:831–839
14. Narabayashi H, Ohye C (1978) Parkinsonian tremor and nucleus intermedius of the human thalamus. In: Desmedt JE (ed) Progress in clinical neurophysiology, vol 5. Karger, Basel, pp 165–172
15. Narabayashi H, Ohye C (1980) Importance of microstereoencephalotomy for tremor alleviation. Appl Neurophysiol 43:222–227
16. Narabayashi H, Ohye C (1983) Pathological tremor: its generating mechanism and treatment. In: Yahr MD (ed) Current concepts of Parkinson's disease and related disorders. Excerpta Medica, Amsterdam Tokyo, pp 47–55
17. Narabayashi H, Okuma T (1953) Procaine oil blocking of the globus pallidus for the treatment of rigidity and tremor of parkinsonism. Proc Jpn Acad 29, No 3
18. Narabayashi H, Okuma T, Shikiba S (1956) Procaine oil blocking of the globus pallidus. Arch Neurol Psychiatry 75:36–48
19. Narabayashi H, Yokochi F, Nakajima Y (1985) Idiopathic oromandibular dyskinesia treated by Vo complex microstereotactic thalamotomy. Appl neurophysiol 48:309–314
20. Ohye Ch, Narabayashi H (1979) Physiological study of presumed ventralis intermedius neurons in the human thalamus. J Neurosurg 50:290–297
21. Ohye Ch, Saito Y, Fukamachi A, Narabayashi H (1974) An analysis of the spontaneous rhythmic and non-rhythmic burst discharge in the human thalamus. J Neurol Sci 22:245–259
22. Rondot P, Said G, Ferrey G (1972) Les hyperkinésies volitionnelles. Étude électrologique – classification. Rev Neurol 126:415–426
23. Schaltenbrand G, Bailey P (1959) Introduction to stereotaxis with an atlas of the human brain. Thieme, Stuttgart New York
24. Smith MC (1967) Stereotactic operations for Parkinson's disease. Anatomical observations. Mod Trend Neurol 4:21–52
25. Yokochi M, Narabayashi H, Iizuka R, Nagatsu T (1984) Juvenile parkinsonism. Some clinical, pharmacological, and neuropathological aspects. In: Hassler RG, Christ JF (eds) Advances in neurology, vol 40. Raven, New York, pp 407–413

Stimulation for the Treatment of Motor Disorders

P. L. GILDENBERG [1]

Chronic Spinal Cord Stimulation

Stimulation of various parts of the nervous system has been employed for treatment of several motor disorders for some time. Although directly implanted dorsal cord stimulators have been used for the management of pain since 1967 [86], the first published report of the use of spinal cord stimulation in motor disorders did not appear until 1973 when Cook et al. [7] reported improvement in spasticity in patients with multiple sclerosis for whom they were using spinal cord stimulation for pain. It proved to be effective also in subsequent patients without pain [6, 8]. Shortly afterward, Dooley [27, 29] reported beneficial effects in patients with olivopontocerebellar atrophy and Friedreich's ataxia. The author had the opportunity to report the use of high-frequency dorsal cord stimulation in spasmodic torticollis, in a series which had begun in 1971 [36, 37].

Although there are a number of conditions involving spasticity which have been treated with spinal cord stimulation, multiple sclerosis remains the most common indication [8, 28, 43, 55, 69, 81–83, 88]. It is difficult to assess the results in multiple sclerosis, since the disease has a variety of components and may fluctuate over time. Depending on the criteria used, approximately two-thirds of patients have an encouraging enough response to temporary stimulation to warrant the implantation of a permanent stimulator [85]. Improvement may last for a number of years, suggesting that exacerbation of the multiple sclerosis may be uncommon in patients undergoing spinal cord stimulation [27, 85].

If one considers overall function of the patient, the amount of assistance required may decrease significantly on assessment by both staff and patients. In one series of 36 cases with long-term follow-up [85], only one patient was autonomous before implantation, but 11 were afterwards. Of the 6 patients who were bedridden prior to implantation, only one remained so. Improvement in spasticity occurred in two-thirds of the implanted patients, and 27 of 30 patients with impaired bladder function had improvement following implantation. In a larger series that included 166 implanted patients [8], 99 or 60% were considered to have a satisfactory result, as were 50% in another series [27].

Other degenerative diseases have responded favorably to spinal cord stimulation. Spastic paraplegia of Strümpell-Lorrain is improved in 70% of cases [85], spinocerebellar and cerebellar ataxia in 59%, and olivopontocerebellar dystrophy in 50% [26], in various series.

Results in spasmodic torticollis have been variable. One problem with evaluating spasmodic torticollis patients is that a large percentage of them appear

[1] 6560 Fannin – Suite 1530, Houston, TX 77030, USA.

Clinical Aspects of Sensory Motor Integration
Ed. by A. Struppler and A. Weindl
© Springer-Verlag Berlin Heidelberg 1987

to have significant superimposed emotional problems [37, 38]. Once a therapeutic effect is established, however, it usually persists [37, 52, 95]. Although some authors [80] have not been encouraged, others [37, 52, 95] have noted improvement in one-third [37, 52] to two-thirds [95] of patients evaluated. The observation has been made at three separate centers [36–38, 52, 93–95] that many patients respond best to high-frequency stimulation between 1100 and 1500 Hz.

Patients with spasticity from partial lesions of the spinal cord may benefit from spinal cord stimulation [85], and one series reports improvement with spasticity of cerebral origin [86].

Although most series report little improvement in function in patients with cerebral palsy, amyotrophic lateral sclerosis, or dystonia musculorum deformans, Waltz [95] reported improvement in all those conditions, especially with his 4-electrode system.

Interestingly, spinal cord stimulation, when used for a variety of motor disorders (cerebral palsy, dystonia, multiple sclerosis, and torticollis), sometimes resulted in psychological improvement in self-image, anxiety (except in torticollis and multiple sclerosis), and depression, whether or not there was benefit to motor function [56, 57].

With such impressive results, it is difficult to see how some investigators demonstrated no significant improvement in function, unless they were perhaps too free to attribute any subjective improvement to the placebo effect [73, 99] or to an improvement in motivation [40].

Since it is often difficult to assess improvement in neurologic function in a group of patients whose clinical manifestations may vary greatly in regard to both quantity and quality of symptoms, a number of objective methods of assessment have been recommended, ranging from review of motion pictures or video tapes by uninvolved observers [49] to comprehensive electrophysiologic recordings. Changes on spinal cord stimulation in pattern of muscle discharge, as recorded by polyelectromyography, have been documented by several different groups [3, 23, 25] and appear to correlate with change in spasticity. Improvement of strength for voluntary activity, with electromyographic monitoring, may demonstrate a decrease in spasticity, return of voluntary control, or improvement in synergy on spinal cord stimulation [51, 71, 83]. Spasticity can be measured by use of a mechanical device which measures resistance to stretch simultaneously with sinusoidal stretching of the arm in the rotational axis [83, 84], the superimposition of voluntary activity, or compensation following abrupt active unloading [87]. Such techniques are quite sensitive to changes in spasticity, and have provided objective evidence for physiological improvement after spinal cord stimulation.

Schemes for assessing impairment of various individual motor activities or general disability scores, such as the Kurtzke Disability Scale, have been employed, in hopes that they would allow closer correlation of objective and subjective signs of improvement and provide a numerical index for the effect of spinal cord stimulation [88], but they have not provided data which would be directly correlated with overall clinical improvement.

Why should there be such disagreement about the effects of spinal cord stimulation in movement disorders?

In reviewing reports on the effect of spinal cord stimulation on movement disorders, it is distressing to see how often the authors fail to indicate the parameters of stimulation or even the precise spinal cord level stimulated. The frequency may be anywhere from 33 Hz [79] to 1400 Hz [70] for the treatment of spasticity, although most reports did not specify. In some studies different combinations of electrodes are used at various frequencies for individual patients [80, 93–95]. It is logical to assume that different conditions might respond better to different parameters of stimulation, but it does not appear that a consistent pattern of stimulation parameters can be correlated with specific motor problems.

A number of factors may be modified by spinal cord stimulation. Spasticity may be improved. There may be improvement in synergy in parts not paralyzed. There may be an improvement in muscle strength because of release from inhibiting reflexes. There may be improvement in or return of voluntary control, or the threshold for various conventional types of sensory stimuli may be decreased [83].

To further confound the analysis of the effects of spinal cord stimulation, it has been demonstrated that epidural electrodes which appear on AP fluoroscopy to be dorsal to the spinal cord may actually lie anterior to it, although such placement can generally be identified by physiologic means [80].

A study was performed to determine which structures within the spinal cord are stimulated by dorsal cord stimulation at various thresholds [24]. Quantitative electrically induced sensation and muscle twitches were investigated with increasing spinal cord stimulation currents to find the sequence in which specific structures of the spinal cord are affected by stimulation, in order to help define the underlying neurophysiologic mechanisms involved with modification of motor control. As the stimulus strength was gradually increased, first the dorsal columns were stimulated, then the dorsal root entry zones, and finally the corticospinal and dorsal spinocerebellar tracts [24]. Anterior placement of the electrode presented a different pattern of stimulation of intramedullary structures.

The electrical impulse spreads throughout the spinal cord tissues in an uneven distribution, depending on the relative impedance of the various tissues in the area. The current may spread particularly along pathways, nerve fibers, or vascular structures or may find the surrounding cerebrospinal fluid a particularly efficient volume conductor. As the current spreads through the tissues, neurones of varying thresholds are encountered, further complicating the pattern in which effective stimulation is distributed [53].

Although the use of spinal cord stimulation in the treatment of pain was originally based on theoretical considerations (which may or may not actually apply), we have no good explanation for why stimulation of the spinal cord should have an effect in various motor disorders. In thinking about such mechanisms, several considerations should be mentioned.

Stimulation does not always stimulate. The application of an unphysiologic, repetitive burst of electrical energy does not simulate normal activity, and may or may not facilitate physiologic activity. The driven synchronous discharging of a pool of neurones may interrupt the orderly organized interplay between neural structures and block or inhibit their usual function. High-energy repetitive bursts of discharging neurones may release an excess of neurotransmitter to inhibit ongoing activity and influence the function of an area long after the stimulation has ceased. Stimulation at a higher frequency than the neurones can follow may cause prolonged depolarization and inhibition.

Stimulation may affect the mechanisms compensating for the pathology (which may in themselves be detrimental to normal function) or may increase central excitation. On the other hand, stimulation may either facilitate an area or enhance an inhibitory function which may exert itself locally or distantly.

How can stimulation of the spinal cord specifically affect spasticity [62]?

1. The most important result of spinal cord stimulation is a decrease in muscle tone and an improvement in voluntary movement.
2. There is an improvement in the coordination of movements and improvement in the synergy of movements [6].
3. The H-reflex may be inhibited on spinal cord stimulation [50, 75].
4. The H-reflex inhibition by stimulation of cutaneous afferents, which is often absent in patients with multiple sclerosis, may reappear on spinal cord stimulation [42].
5. Spinal cord stimulation may influence both cervical somatosensory evoked potentials and brain stem auditory-evoked potentials [78].

It has been documented in decerebrate cats that dorsal cord stimulation at 50 Hz applied for 30 s causes a reduction of monosynaptic reflexes elicited by electrical dorsal root stimulation for up to 10 min. The firing frequency of lumbar motoneurones decreased and the firing of Renshaw cells increased [84].

Spinal cord stimulation consequently seems to be expressed as an effect on spinal interneurones which integrate motor and sensory information prior to their control of motor neurones through the mechanisms of reciprocal innervation, and it probably restores the balance between excitation and inhibition by a facilitation of the inhibitory pathways, possibly related to excitation of the Ia inhibitory interneurone by the corticospinal tract [47], or long-loop corticospinal pathways [46]. In addition, brain stem control of spinal motoneurones is also mediated by vestibulospinal and reticulospinal pathways, and, in experimental preparations, stimulation can cause of both vestibulospinal and reticulospinal pathways to produce a coactivation of alpha and gamma motoneurones [5]. A shorter propriospinal pathway originating from cells located in the C-3 and C-4 segments, which receive monosynaptic excitation from corticospinal and rubospinal pathways and from cervical primary afferents [58], may be especially affected by spinal cord stimulation. Such long propriospinal pathways may be responsible for a coactivation of lumbar alpha and gamma motoneurones observed when the contralateral forelimb is flexed or by the stretch of a single contralateral dorsal neck muscle [61], effects which last a long time after the cessation of the stimuli, perhaps through spinobulbar connections.

Increased sensitivity to stretch of the muscle spindles has been demonstrated in spasticity, presumably because of a reduction of gamma motoneurone inhibition [74], so that exaggeration of the stretch reflexes reflects central hyperreactivity to normal spindle input or failure of inhibition to such input, which may be suppressed by spinal cord stimulation. Long-loop reflexes [22], demonstrated by the effect of contraction of lower limbs on facilitation of reflexes in the upper limbs and vice versa, may have an inhibitory action which may be activated by spinal cord stimulation.

Consequently, the existence of long tracts which may be affected by chronic dorsal column stimulation suggests that stimulation may excite ascending and descending pathways for a net facilitation of the reciprocal inhibitory mechanisms of the spinal segments and monosynaptic reflexes in spasticity [89]. Beneficial effects have been demonstrated whether the spinal cord is stimulated in thoracic or in cervical levels, although stimulation at higher levels appears to be more effective in motor disorders.

Additional evidence of the physiologic effect of spinal cord stimulation is derived from the measurement of cervical somatosensory evoked potentials (SSEP) and brain stem evoked potentials (BSEP) in patients with multiple sclerosis. The flattening of the SSEP curve frequently seen in multiple sclerosis may return to a more normal configuration with spinal cord stimulation [79], which may correlate with clinical improvement. Similar results are seen in correction of the decrease in amplitude and prolongation of latency of the BSEP seen in approximately one-third of multiple sclerosis patients [79]. This is a particularly interesting finding, since the demonstration of altered synaptic activity at brain stem levels enhances the possibility that the effect of spinal cord stimulation on motor activity may take place at higher levels than the spinal cord [48], even though the final expression concerns segmental reflex activity. On the other hand, visual evoked potentials are unchanged by spinal cord stimulation [78].

Stimulation for torticollis, in contrast, may depolarize the afferent nerves concerned with proprioception of the neck to disrupt the abnormal tonic neck reflex pattern [36, 37].

It has been demonstrated that electrical nerve stimulation of radial or median nerve also can suppress contralateral clonus for as long as 3 h after cessation of stimulation in subjects with spasticity. Since the stimulation of the nerve in the wrist can also suppress ankle clonus [92], a segmental influence cannot be entertained, but the mechanism must involve centrally mediated inhibition.

Similarly, stimulation of the cauda equina with epidural electrodes may not only improve paraplegic spasticity, but might also improve regulation of bowel programs, induce piloerection and sweating below the level of the lesion, or influence morning erections [78]. Since such spasticity may be diminished by decreasing cutaneous or extraneous input, one might speculate that the effect is either from disrupting the organization of the peripheral input or by the facilitation of inhibitory pathways within the isolated segment of the spinal cord.

In summary, one might conclude that although the physiologic mechanism for the effect of spinal cord stimulation has not been documented, there is sufficient evidence of clinical improvement to justify its use in a number of conditions, especially multiple sclerosis. Other degenerative diseases, such as Strümpell-Lorrain disease, may also benefit significantly, but there has been little effect in amyotrophic lateral sclerosis. Spasticity may be improved in partial lesions of the spinal cord, which appears to be more amenable to spinal cord stimulation than that of cerebral origin. The pain, intolerable pulling sensation, and altered head position of spasmodic torticollis may be improved in some patients. One investigator also reports encouraging results for dystonia musculorum deformans and stroke [94].

Chronic Cerebellar Stimulation

Chronic cerebellar stimulation (CCS) in humans began in 1973 when Cooper [9] first implanted an array of stimulating electrodes over the surface of the anterior lobe of the cerebellum in patients with epilepsy [9, 11, 14]. He had been impressed by a report by Morruzzi [60] in 1950 that stimulation of the paleocerebellum had great inhibitory influences on postural tone and cortical activity in the cat. Dow et al. [30] reported similar findings in the unanesthetized rat.

Although only a few epileptics were thus treated, the hardware and technique were also used in patients with cerebral palsy, in hopes that the inhibitory effects of cerebellar stimulation would diminish spasticity [10]. Since that time, it is estimated that over 1000 chronic cerebellar stimulators have been implanted in patients with cerebral palsy, but there is still little consensus about the effectiveness [64]. Some reports indicate that there is little effect [34, 44, 45, 96, 97], and others that there is a benefit [35, 59, 72, 98], some even with great enthusiasm [12, 16, 18, 19, 21, 63, 66]. Reports involving objective testing or double-blind evaluation tend to show equivocal or negative results [63–67].

In general, tests of neurophysiologic function, improvement in activities of daily living, signs of spasticity, and tests of individual motor function have documented changes [63, 64]. A curious observation was noted in one double-blind study where cerebral-palsied children were stimulated with an external transmitter which was changed monthly, sometimes being set to be functional and sometimes used as a placebo. Some children became less spastic and developed increased mental alertness and improved clarity of speech. Curiously, these improvements occurred during placebo periods as often as with stimulation [34].

In those reports that indicate improvement, the greatest effect of chronic cerebral stimulation in cerebral palsy is on spasticity. It can be subjectively [19] and objectively [90] demonstrated that there is lowering of spastic muscle tone in many of the patients. It is also observed that some patients with athetosis show improvement. Less drooling is a common observation, and many patients are felt to have better feeding, dressing and ambulation, as well as clearer speech [19]. In some patients stimulation can reduce rigidity and coactivation of muscles, sometimes quite promptly [66].

Cooper followed 124 patients from 6 months to several years [10] and reported that spasticity improved in 73% and athetosis in 61%; somewhat less than half showed improvement in ambulation and self-care.

Other reports disagree as to whether chronic cerebellar stimulation helps speech in cerebral palsy. In one study [97] involving direct evaluation of speech, no difference was found. In another [68] involving structured listening tasks evaluated in a blind fashion by speech pathologists, there was no significant improvement in speech intelligibility or articulatory accuracy. However, strain/strangle vocal quality improved considerably. Another study [72] noted increase in duration of vowel phonation by a significant amount, and 4 of 10 patients showed improved articulation, particularly of consonants. Many of the subjective reports or reports of families, however, suggest that improvement in speech does occur [10, 19, 20]. My own anecdotal experience concerns a 22-year-old Mexican engineer who could not speak well enough to use a telephone and communicated inefficiently with his coworkers. Two months after implantation of a cerebellar stimulator, he phoned from Mexico to demonstrate how he was able to use a telephone for the first time.

There seems to be little doubt that chronic cerebellar stimulation exerts a physiologic influence by decreasing passive reflexes and improving abnormal patterns of co-contraction [31]. Cortical somatosensory evoked responses are significantly reduced in amplitude in many patients, which may correlate with clinical benefits [16]. There are changes in H-reflexes [90], stretch reflexes [13], primitive reflex patterns, and coordination of muscle groups of respiration [39, 64].

Then why is there so much confusion and disagreement about the use of chronic cerebellar stimulation? Patients who are treated vary considerably, and it is difficult to assign baseline criteria from which subsequent measurements can be made. Many factors influence the performance of cerebral-palsied children. Their families are often biased in their assessment and driven with dedication to the patient. They are often so impressed that "something is being done" that they may have unrealistic expectations and convey them to the patients, which may in turn influence the patients' performances. The enthusiasm, personality, and attitude of the neurosurgeon also play a role in that the neurosurgeon who believes enthusiastically will certainly influence the patients' efforts to exceed prior performances.

To add to the confusion, animal experiments indicate that slight variations in placement, current density, and frequency may provide significant differences in the effect of chronic cerebellar stimulation [64]. Yet there is considerable variability from one study to another in clinical stimulation parameters, particularly the intensity of stimulation.

There appears to be a correlation between the clinical benefit and the effects of chronic cerebellar stimulation on the thalamic component of somatosensory evoked potentials (SSEP) after stimulation on the median nerve [13, 66, 91]. This led to the suggestion that "biocalibration" [17] be used to determine the most effective stimulus strength. The SSEP is recorded while the amplitude of cerebellar stimulation is adjusted to the threshold for SSEP changes. This is usually below the current density at which the patient perceives stimulation from sensation of the tentorium, so-called tentorial paresthesia, which is used by some investigators to regulate current [34, 65]. It is, however, generally near the charge density that is recommended for safe and effective stimulation between 0.4 and 0.8 $mC/C^2/$ phase, well below the threshold for producing tissue damage. Stronger stimulation may cause variable or adverse physiological effects [20, 21].

There are several theories about the effect of chronic cerebellar stimulation. It has been suggested that the reduction in amplitude of cerebral somatosensory evoked potentials following stimulation of the cerebellum is related to Purkinje-cell-induced disfacilitation of dentate neurones [32]. Projection via the brachium conjunctivum is suggested by the observation of a bilateral reduction in amplitude of the somatosensory evoked potentials with unilateral stimulation of the cerebellum. The reduction in spasticity may not be related to the cerebral cortex, since it can be demonstrated in hemidecorticate monkeys [41]. It is also necessary to speculate that more than thalamic suppression is involved, because stereotactic lesions of the thalamus are not effective for the treatment of spasticity.

The theory of Purkinje cell stimulation is inadequate. Cooper [13] indicates that the therapeutic effect may not be due to activation of Purkinje cells but may be due to inhibition, which would also explain the reduction of amplitude of reflexes, evoked potentials, and paroxysmal discharges in the electroencephalogram.

Current distributed through the cerebellum more effectively suppresses peripheral reflex activity than stimulation confined near the surface [54], so it is possible that the effect is due to stimulation of the deeper structures.

Schvarcz et al. [76, 77] noted consistent electrophysiologic changes on stimulation of the dentate nucleus of the cerebellum. The H-reflex recovery curve in patients with spasticity due to cerebral palsy returned toward normal, not only during stimulation, but for 3 h thereafter. Unilateral stimulation affected reflexes bilaterally, resulted in improved motor function and relief of spasticity, and enabled complex voluntary movements which could not be performed prior to stimulation [76]. Speech was improved as well as posture, balance and gait.

Experimental studies show that stimulation of multiple sites in the brain stem can also have a significant influence on motor tone, as demonstrated by intense inhibition of the tonic vibration reflex in cats [33]. Positive results suggesting potential clinical applicability were found from stimulation of the medial part of the internal capsule, the contralateral precruciate cortex, and the ventral anterior nucleus of the thalamus, as well as the anterior lobe of the cerebellum.

Stimulation of other CNS structures has also been used clinically for the treatment of motor disorders. Cooper [15, 17] observed that the thalamic component of somatosensory evoked potentials was suppressed with cerebellar stimulation. He inhibited activity in that area by stimulation of the posterior part of the ventrolateral (VL) nucleus adjacent to the internal capsule. Placement of the electrode was monitored by evoked potentials. There were 49 patients with cerebral palsy, dystonia musculorum deformans, multiple sclerosis or cerebrovascular lesions who underwent stimulation, and 27 demonstrated some clinical benefit [17], of whom 20 demonstrated marked reduction of involuntary movement, spasticity, torticollis, or tremor. The worst results were seen in patients with dystonia, and the best in patients with cerebral palsy or vascular hemiparesis.

Andy [1] performed stimulation of the parafascicular-centrum median area for intractable pain in 4 patients who also had dyskinesia, and noted improvement in the dyskinesia as well as the pain. One of the patients had cerebral palsy and demonstrated marked improvement in speech and ataxia, whereas the diagnosis of the other 3 patients was somewhat obscure. An interesting observation was that the therapeutic effect was stimulus-frequency dependent. The maximum pain reduction was obtained at a faster rate than the reduction in the motor disability.

In addition, Andy [2] reports that such stimulation (mostly along the centrum median-parafascicular complex and intralaminar system) may allow improvement in other motor disorders, including Parkinson tremor and rigidity, torticollis, or choreiform movements occurring after a stroke.

Another recent report indicates suppression of severe intention tremor in multiple sclerosis by stimulation of the contralateral midbrain and basal ganglia [4].

Thus, we can see that stimulation of a number of areas of the nervous system can influence motor manifestations of neurological diseases. There is certainly a physiologic effect in some patients, but we do not know which patients or by what mechanism. The therapeutic opportunity is welcome, even though the mechanism remains obscure. Stimulating devices have improved to the point where engineers can provide almost any type of stimulation required. It is up to the clinicians working together with the neurophysiologists to determine what parameters of stimulation are needed.

Summary

It has now been over 10 years since implanted stimulators have been used, but there is still considerable uncertainty and controversy about them. Spinal cord stimulation has been documented to be beneficial in the management of spasticity, particularly that of multiple sclerosis or Strümpell-Lorrain disease. Other authors have reported improvement in spinocerebellar and cerebellar ataxia, olivopontocerebellar dystrophy, and spasmodic torticollis. There is controversy about the degree of improvement and its usefulness in other conditions, which may stem from the failure of many authors to specify the precise parameters or location of stimulation, as well as failure to control patient selection or observations used for evaluation. The physiologic mechanism has not been documented, but a number of effects are speculated.

Chronic cerebellar stimulation has engendered even more controversy. It was initially used for the management of epilepsy, which is notoriously difficult to evaluate, but its use in that condition has been largely abandoned. There are some authors who still advocate its use in cerebral palsy. Although it is estimated that over 1000 chronic cerebellar stimulators have been implanted for the management of cerebral palsy, there is still little consensus about the effectiveness. There has been no documented demonstration of improvement in individual physiologic parameters, but the impression of the clinicians as well as the patients and their families is that spasticity can be significantly diminished in many patients, and athetosis in some others. A decrease in passive reflexes and improvement in abnormal patterns of co-contraction has been demonstrated as well as changes in H-reflexes and evoked potentials. Again, some of the controversy appears to center around the parameters of stimulation, and the physiologic mechanism is poorly understood.

Stimulation of other locations within the central nervous system has been used clinically for the treatment of motor disorders, including stimulation of the thalamus in the ventrolateral nucleus and the parafascicular-centrum medianum area, although only in isolated reports.

References

1. Andy OJ (1980) Parafascicular-center median nuclei stimulation for intractable pain and dyskinesia (painful-dyskinesia). Appl Neurophysiol 43:133–144
2. Andy OJ (1983) Thalamic stimulation for control of movement disorders. Appl Neurophysiol 46:107–111
3. Barolat-Romana G, Myklebust JB, Hemmy DC, Myklebust B, Wenninger W (1985) Immediate effects of spinal cord stimulation in spinal spasticity. J Neurosurg 62:558–562
4. Brice J, McLellan L (1980) Suppression of intention tremor by contingent deep-brain stimulation. Lancet I:1221–1222
5. Carli G, Diete-Spiff K, Pompeiano O (1976) Responses of the muscle spindles and of the extrafusal fibres in an extensor muscle to stimulation of the lateral vestibular nucleus in the cat. Arch Ital Biol 103:209–242
6. Cook AW (1974) Stimulation of the spinal cord in motor neurone disease. Lancet II:230–231
7. Cook AW, Weinstein SP (1973) Chronic dorsal column stimulation in multiple sclerosis. NY State J Med 73:2868–2872
8. Cook AW, Taylor JK, Nidzgorski F (1981) Results of spinal cord stimulation in multiple sclerosis. Appl Neurophysiol 44:55–61
9. Cooper IS (1973) Effect of chronic stimulation of anterior cerebellum on neurological disease. Lancet I:206
10. Cooper IS (ed) (1978) Cerebellar stimulation in man. Raven, New York
11. Cooper IS (1984) Historical review of cerebellar stimulation. In: Davis R, Bloedel JR (eds) Cerebellar stimulation for spasticity and seizures. CRC Press, Boca Raton, pp 3–10
12. Cooper IS, Upton AR (1978) Effects of cerebellar stimulation on epilepsy, the EEG and cerebral palsy in man. Electroencephalogr Clin Neurophysiol [Suppl] 34:349–354
13. Cooper IS, Upton AR (1978) Use of chronic cerebellar stimulation for disorders of disinhibition. Lancet I:595–600
14. Cooper IS, Crighel E, Amin I (1973) Clinical and physiological effects of stimulation of the paleocerebellum in humans. J Am Geriatr Soc 21:40–43
15. Cooper IS, Upton AR, Amin I (1980) Reversibility of chronic neurologic deficits. Some effects of electrical stimulation of the thalamus and internal capsule in man. Appl Neurophysiol 43:244–258
16. Cooper IS, Upton AR, Rappaport ZH, Amin I (1980) Correlation of clinical and physiological effects of cerebellar stimulation. Acta Neurochir [Suppl] (Wien) 30:339–344
17. Cooper IS, Upton AR, Amin I (1982) Chronic cerebellar stimulation (CCS) and deep brain stimulation (DBS) in involuntary movement disorders. Appl Neurophysiol 45:209–217
18. Davis R, Cullen RF, Flitter MA, Duenas D, Engle H, Ennis B (1977) Control of spasticity and involuntary movements. Neurosurgery 1:205–207
19. Davis R, Barolat-Romana G, Engle H (1980) Chronic cerebellar stimulation for cerebral palsy – 5 year study. Acta Neurochir [Suppl] (Wien) 30:317–332
20. Davis R, Engle H, Kudzma J, Gray E, Ryan T, Dusnak A (1982) Update of chronic cerebellar stimulation for spasticity and epilepsy. Appl Neurophysiol 45:44–50
21. Davis R, Bloedel JR (1984) Cerebellar stimulation for spasticity and seizures. CRC Press, Boca Raton
22. Delwaide PJ, Toulouse P, Crenna P (1981) Hypothetical role of long-loop reflex pathways. Appl Neurophysiol 44:171–180
23. Dimitrijevic MR, Dimitrijevic MM, Sherwood AM, Faganel J (1980) Neurophysiological evaluation of chronic spinal cord stimulation in patients with upper motor neuron disorders. Int Rehab Med 2:82–85
24. Dimitrijevic MR, Faganel J, Sharkey PC, Sherwood AM (1980) Study of sensation and muscle twitch responses to spinal cord stimulation. Int Rehabil Med 2:76–81
25. Dimitrijevic MR, Faganel J, Young RR (1981) Underlying mechanisms of the effects of spinal cord stimulation in motor disorders. Appl Neurophysiol 44:133–140
26. Dooley DM, Nisonson I (1981) Treatment of patients with degenerative diseases of the central nervous system by electrical stimulation of the spinal cord. Appl Neurophysiol 44:71–76

27. Dooley DM, Sharkey J (1977) Electrostimulation of the nervous system for patients with demyelinating and degenerative diseases of the nervous system and vascular diseases of the extremities. Appl Neurophysiol 40:208–217
28. Dooley DM, Sharkey J (1981) Electrical stimulation of the spinal cord in patients with demyelinating and degenerative diseases of the central nervous system. Appl Neurophysiol 44:218–224
29. Dooley DM, Kasprak M, Stibitz M (1976) Electrical stimulation of the spinal cords in patients with demyelinating and degenerative diseases of the central nervous system. J Fla Med Assoc 63:905–909
30. Dow RS, Fernandez-Guardiola A, Manni E (1962) The influence of the cerebellum on experimental epilepsy. Electroencephalogr Clin Neurophysiol 14:393–407
31. Ebner T, Vitek J, Schwartz A, Bloedel J (1980) Effects of cerebellar stimulation on abnormal proprioceptive reflexes in spastic primates. Exp Neurol 70:721–725
32. Eccles JL, Ito M, Szentagothai J (1967) The cerebellum as a neuronal machine. Springer, Berlin Heidelberg New York
33. Fujita S (1980) Experimental studies on therapeutic effect of electrical stimulation of the brain on muscular hypertonus. Appl Neurophysiol 43:298–304
34. Gahn NH, Russman BS, Cerciello RL, Fiorentine M (1981) Chronic cerebellar stimulation for cerebral palsy. A double blind study. Neurology (Minneap.) 31:87–90
35. Galanda M, Nadvornik P, Fodor S (1980) Stereotactic approach to therapeutic stimulation of cerebellum for spasticity. Acta Neurochir [Suppl] (Wien) 30:345–349
36. Gildenberg PL (1978) Treatment of spasmodic torticollis by dorsal column stimulation. Appl Neurophysiol 41:113–121
37. Gildenberg PL (1981) Comprehensive management of spasmodic torticollis. Appl Neurophysiol 44:233–243
38. Gildenberg PL, Tasker RR (1982) Spasmodic torticollis. Contemp Neurosurg 4:1–8
39. Gray E, Davis R, Cohn M (1982) Respiration and joint compliance changes in spastic patients following chronic cerebellar stimulation. In: Davis R, Bloedel JR (eds) Cerebellar stimulation for spasticity and seizures. CRC Press, Boca Raton, pp 157–168
40. Hawkes CH, Wyke MA, Desmond A, Bultitide MI, Kanegaonkar GS (1980) Stimulation of the dorsal column in multiple sclerosis. Br Med J 1:889–891
41. Hemmy DC, Larson SJ, Sances A Jr, Millar EA (1977) The effect of cerebellar stimulation on focal seizure activity and spasticity in monkeys. J Neurosurg 46:648–653
42. Illis LS, Oygar AE, Sedgwick EM, Awadalla MAS (1976) Dorsal column stimulation in the rehabilitation of patients with multiple sclerosis. Lancet I:1383–1386
43. Illis LS, Sedgwick EM, Tallis RC (1980) Spinal cord stimulation in multiple sclerosis: clinical results. J Neurol Neurosurg Psychiatry 43:1–14
44. Ivan LP, Ventureyra EC, Wiley J, Doyle D, Pressman E, Knights R, Guzman C, Uttley D (1981) Chronic cerebellar stimulation in cerebral palsy. Surg Neurol 1:81–84
45. Ivan LP, Ventureyra EC (1982) Chronic cerebellar stimulation in cerebral palsy. Childs Brain 9:121–125
46. Jankowska E, Lundberg A, Roberts WJ, Stuart D (1974) A long propriospinal system with direct effect on motoneurons and on interneurones in the cat lumbosacral cord. Exp Brain Res 21:169–194
47. Jankowska E, Pardo Y, Tanaka R (1976) Disynaptic inhibition of spinal motoneurones from the motor cortex in the monkey. J Physiol (Lond) 258:467–487
48. Jobbling DT, Tallis RC, Sedgwick EM, Illis LS (1986) Electronic aspects of spinal cord stimulation in multiple sclerosis. Med Biol Eng Comput (to be published)
49. Klingler D, Kepplinger B (1982) Quantification of the effect of epidural spinal electrostimulation (ESES) in central motor disorders. Appl Neurophysiol 45:221–224
50. Krainick JU, Thoden U, Strassburg HM, Wenzel D. The effect of electrical spinal cord stimulation on spastic movement disorders. In: Wullenberg, Brock, Hamer, Klinger, Spoerri (eds) Advances in neurosurgery, vol 4. Springer, Berlin Heidelberg New York, pp 257–260
51. Laitinen LV, Fugl-Meuer AR (1982) Assessment of functional effect of epidural electrostimulation and selective posterior rhizotomy in spasticity. Appl Neurophysiol 45:331–334
52. Lamb S (1983) Clinical application of transcutaneous electrical nerve stimulation and high cervical cord stimulation for movement disorders. J Neurosurg Nurs 15:1–4

53. Larson SJ, Sances A Jr, Riegel DH, Meyer GA, Dallman DE, Swiontek T (1974) Neurophysiological effects of dorsal column stimulation in man and monkey. J Neurosurg 41:217–223
54. Larson SJ, Sances A Jr, Hemmy DC, Millar EA, Walsh PR (1977) Physiological and histological effects of cerebellar stimulation. Appl Neurophysiol 40:160–174
55. Lazorthes Y, Siegfried J, Broggi G (1981) Electrical spinal cord stimulation for spastic motor disorders in demyelinating diseases – a comparative study. In: Hosobuchi Y, Corbin T (eds) Indications for spinal cord stimulation. Excerpta Med, Amsterdam, pp 48–59
56. Levita E, Rilan M, Waltz JM (1981) Psychological effects of spinal cord stimulation: preliminary findings. Appl Neurophysiol 44:93–96
57. Levita E, Sorkin BA, Waltz JM (1986) Spinal cord stimulation revisited. Psychological effects. Appl Neurophysiol 49:69–75
58. Lundberg A (1966) Integration in reflex pathway. In: Granit (ed) Muscular afferents and motor control. Nobel Symposium I. Almqvist & Wiksell, Stockholm, pp 275–305
59. Miyasaka K, Hoffman HF, Forese AB (1978) The influence of chronic cerebellar stimulation on respiratory muscle coordination in a patient with cerebral palsy. Neurosurgery 2:262–265
60. Moruzzi G (1950) Problems in cerebellar physiology. Chas. C. Thomas, Springfield
61. Murthy KSK, Gildenberg PL, Marchand JE (1978) Descending long-spinal excitation of lumbar alpha and gamma motoneurones evoked by stretch of dorsal neck muscles. Brain Res 140:165–170
62. Murthy KSK (1981) Propriospinal mechanisms involved in the effects of spinal cord stimulation. Appl Neurophysiol 44:22–29
63. Penn RD (1982) Chronic cerebellar stimulation for cerebral palsy: a review. Neurosurgery 1:116–121
64. Penn RD (1985) Chronic cerebellar stimulation in humans. In: Wilkins RH, Rangachary SS (eds) Neurosurgery. McGraw-Hill, New York, pp 2537–2543
65. Penn RD, Etzel ML (1977) Chronic cerebellar stimulation and developmental reflexes. J Neurosurg 46:506–511
66. Penn RD, Gottlieb GL, Agarwal GC (1978) Cerebellar stimulation in man. Quantitative changes in spasticity. J Neurosurg 48:779–786
67. Penn RD, Myklebust BM, Gottlieb GL, Agarwal GC, Etzel ME (1980) Chronic cerebellar stimulation for cerebral palsy. Prospective and double blind studies. J Neurosurg 53:160–165
68. Ratusnik DL, Wolfe VI, Penn RD, Schewitz S (1978) Effects on speech of chronic cerebellar stimulation in cerebellar palsy. J Neurosurg 48:876–882
69. Read DJ, Matthews WB, Higson RH (1980) The effect of spinal cord stimulation on function in patients with multiple sclerosis. Brain 103:803–833
70. Reynolds AF, Oakley JC (1982) High frequency cervical epidural stimulation for spasticity. Appl Neurophysiol 45:93–97
71. Richardson RR, Cerullo LJ, McLone DG, Gutierrez FA, Lewis V (1979) Percutaneous epidural neurostimulation in modulation of paraplegic spasticity. Acta Neurochir (Wien) 49:235–243
72. Robertson LT, Meek M, Smith WL (1980) Speech changes in cerebral-palsied patients after cerebellar stimulation. Dev Med Child Neurol 22:608–617
73. Rosen JA, Barsoum AH (1979) Failure of chronic dorsal column stimulation in multiple sclerosis. Ann Neurol 6:66–67
74. Rushworth G (1980) Some pathophysiological aspects of spasticity and the search for rational and successful therapy. Int Rehabil Med 2:23–26
75. Scerrati M, Onofri M, Pola P (1982) Effects of spinal cord stimulation on spasticity: H-reflex study. Appl Neurophysiol 45:62–67
76. Schvarcz JR, Sica RE, Morita E (1980) Chronic self-stimulation of the dentate nucleus for the relief of spasticity. In: Gillingham J, Gybels J, Hitchcock E, Rossi GC, Szikla G (eds) Advances in stereotactic and functional neurosurgery 4. Acta Neurochir [Suppl 30] 351–359
77. Schvarcz JR, Sica RS, Morita E, Bronstein A, Sanz O (1982) Electrophysiological changes induced by chronic stimulation of the dentate nuclei for cerebral palsy. Appl Neurophysiol 45:55–61

78. Sedgwick EM, Illis LS, Tallis RC, Thornton ARD, Abraham P, El-Negamy E, Docherty TB, Soar JS, Spencer SC, Taylor FM (1980a) Evoked potentials and contingent negative variation during treatment of multiple sclerosis with spinal cord stimulation. J Neurol Neurosurg Psychiatry 43:15–24
79. Sedgwick EM, Thornton ARD, El-Negamy E, Jobbling DT, Tallis RC, Illis LS (1980) Electrophysiological responses and stimulus parameters associated with spinal cord stimulation. Int Rehabil Med 2:71–75
80. Sharkey PC, Dimitrijevic MM, Campos RJ, Dimitrijevic MR (1981) Remarks on spinal cord stimulation and the placebo effect. Appl Neurophysiol 44:119–125
81. Siegfried J (1976) Traitement neurochirurgical de quelques symptomes de la sclerose en plaques. Med Hyg 34:901–903
82. Siegfried J (1977) Traitement neurochirurgical de l'hypertonie musculaire. Med Hyg 35:2062–2066
83. Siegfried J (1980) Treatment of spasticity by dorsal cord stimulation. Int Rehabil Med 2:31–34
84. Siegfried J, Krainick JU, Haas H, Adorjani C, Meyer M, Thoden U (1978) Electrical spinal cord stimulation for spastic movement disorders. Appl Neurophysiol 41:134–141
85. Siegfried J, Lazorthes Y, Broggi G (1981) Electrical spinal cord stimulation for spastic movement disorders. Appl Neurophysiol 44:77–92
86. Shealy CN, Mortimer JT, Reswick JB (1967) Electrical inhibition of pain by stimulation of the dorsal columns. Preliminary clinical report. Anesth Analg 46:489–491
87. Struppler A, Struppler E (1981) Quantification and documentation of changes produced by spinal cord stimulation. Appl Neurophysiol 44:114–118
88. Tallis RC, Illis LS, Sedgwick EM (1982) The quantitative assessment of the influence of spinal cord stimulation on motor function in patients with multiple sclerosis. Int Rehabil Med 5:10–16
89. Thoden U, Krainick JU, Strassburg HM, Wenzel D (1976) Modulation of monosynaptic reflexes by dorsal column stimulation (DCS) in spasticity. Eur J Physiol 365:41
90. Upton ARM (1978) Neurophysiological aspects of spasticity and cerebellar stimulation. In: Cooper IS (ed) Cerebellar stimulation in man. Raven, New York, pp 101–110
91. Upton ARM (1984) Biocalibration of cerebellar stimulation. In: Davis R, Bloedel JR (eds) Cerebellar stimulation for spasticity and seizures. CRC Press, Boca Raton, pp 133–140
92. Walker JB (1982) Modulation of spasticity: prolonged suppression of a spinal reflex by electrical stimulation. Science 216:203–204
93. Waltz JM, Andreesen WH (1981) Multiple-level spinal cord stimulation. Technique Appl Neurophysiol 44:30–36
94. Waltz JM (1982) Computerized percutaneous multi-level spinal cord stimulation in motor disorders. Appl Neurophysiol 45:73–92
95. Waltz JM, Reynolds LO, Riklan M (1981) Multi-lead spinal cord stimulation for control of motor disorders. Appl Neurophysiol 44:244–257
96. Whittaker CK (1980) Cerebellar stimulation for cerebral palsy. J Neurosurg 52:648–653
97. Wolfe VI, Ratusnik DL, Penn RD (1981) Long-term effects on speech of chronic cerebellar stimulation in cerebral palsy. J Speech Hear Disord 46:286–290
98. Wong PK, Hoffman HJ, Froese AB, Platt LJ, Blacha S, Gear J (1979) Cerebellar stimulation in the management of cerebral palsy: clinical and physiological studies. Neurosurgery 5:217–224
99. Young RF, Goodman SJ (1979) Dorsal spinal cord stimulation in the treatment of multiple sclerosis. Neurosurgery 5:225–230

VII. Thalamocortical Contributions to Sensory Motor Integration

The Physiological Basis of VIM Thalamotomy for Involuntary Movement Disorders

R. R. Tasker [1], F. A. Lenz [1], J. O. Dostrovksy [1], K. Yamashiro [1], J. Chodakiewitz [1], and D. G. Albe-Fessard [1]

Introduction

Physiological corroboration of the lesion site for thalamotomy [12, 20] offers an opportunity to study sensorimotor integration at the thalamic level. The target [8–10, 17, 21, 25, 27, 40, 41, 46, 53, 55, 57] where a 4–6 mm diameter lesion abolishes various types of tremor in a high proportion of cases [27, 38, 57] and ameliorates other dyskinesias, lies just rostral to ventrocaudal nuclear (VC) representation of lips or manual digits at or a little above the intercommissural (ACPC) line in nucleus ventralis intermedialis (VIM) or adjacent oralis posterior (VOP) of thalamus: It remains controversial whether lesions here arrest the progress of Parkinson's disease [38, 51]. The results of macro- and microstimulation and of microelectrode recording in the target area will be interpreted in the light of sensorimotor integration.

Methods

A two-stage stereotactic procedure is employed using the Leksell frame with computer graphics display of physiological data collected under local anaesthesia along three trajectories 2 mm apart in the same sagittal plane chosen to bracket the expected target [55]. Microstimulation and microrecording are done with an electrically sharpened platinum-iridium or tungsten monopolar electrode insulated with sintered glass with an impedance of 0.2 M at 1000 Hz advanced with hydraulic microdrive. Microstimulation at 100 Hz, 0.2 ms, up to 160 μA, is performed every 0.5 to 1 mm starting 10 mm above the target site. Recording is performed more frequently once responses occur below 30 μA. Recording data, a voice channel, and multiple contralateral EMGs are recorded on tape for off-line analysis. Receptive fields (RF), effective modality, type of response, relation to voluntary movement and tremor are all identified on-line and further studied off-line [30]. For correlation with voluntary movement, histogram analysis of discriminated cells is done by aligning all records during repetition of a voluntary movement to EMG onset. Baseline thalamic cell firing is calculated for the first 8 bins, and a significant response is defined as one where mean +2 SD is exceeded. Cumulative sum and change of slope of cumulative sum are computed to identify increased neuronal firing related to and in advance of the EMG of vol-

[1] Division of Neurosurgery, Room 215, 14th floor, Eaton Wing, Toronto General Hospital, 200 Elizabeth Str., Toronto, Ontario M5G 2C4, Canada.

Clinical Aspects of Sensory Motor Integration
Ed. by A. Struppler and A. Weindl
© Springer-Verlag Berlin Heidelberg 1987

untary movement [31, 32]. For tremor, cross-correlation analysis between thalamic neurons and EMG was carried out by examining neuronal spike auto power at tremor frequency and coherence function between thalamic firing and EMG. Latency between thalamic firing and EMG was measured by averaging EMG activity time locked to the occurrence of thalamic action potentials [29, 33, 34].

Results

Macrostimulation

Macrostimulation is the simplest and most widely used method of physiological corroboration [50, 55, 59], though its precision is limited by the relatively large electrode (1.1 mm coaxial, 0.5 mm pole separation in our studies) and current spread (up to 2 mm radius with our electrode using 60 Hz, 3 ms, negative square wave trains up to 1 mA). Our own pooled results from 10000 stimulation sites [55, 56] reveal two types of somatosensory, motor, vestibular, and various dyskinetic effects in the target region. In VC and medial lemniscus just caudal to target, contralateral paraesthetic effects are referred to the face and hand. Hallucination of contralateral muscular activity in the absence of actual movement is elicited in the target in VIM. Of 84 such responses, 10 consisted of the desire to move, 40 of pulling or compression, 18 of rhythmic pulsation, jumping, flicking, throbbing, jerking, tremor, or other to-and-fro motions, 5 of continuous movement, 10 of weakness, fatigue, heaviness, or loss of control. Unsustained contraction of various portions of the contralateral musculature at the onset of a stimulus train, exhausted by repeated stimulation (unlike tetanization elicited in internal capsule rostral to the middle of the ACPC line), occurred in the rostral portion of the target in VIM and VOP. Hallucinations of rotation, "dizziness", and faintness occurred in the target in VIM (as well as in subthalamus, lateral lemniscus, and medial geniculate), and driving or suppression of the patient's dyskinesia also occurred in VIM (parkinsonian tremor nearly always suppressed, other dyskinesias either unaffected or driven at 100 Hz, most readily suppressed at 300 Hz).

Microstimulation

Results of microstimulation and microrecording will be illustrated by the findings in a single patient (Figs. 1 and 2; Table 1). In the expected location of VC, 2–12 μA induced paraesthetic responses in minute areas of the contralateral body, which sites changed less with electrode movement in the same sagittal plane than with similar movements across different coronal planes (sites 1–15, Table 1).

In the target area, where the lesion that was eventually made controlled the patient's upper-limb action tremor, 11 sites (sites 16–26, Table 1) were found where stimulation induced sensory effects with motor connotations in the absence of movement. In the rostral margin of the lesion area were 4 sites (sites 18, 22,

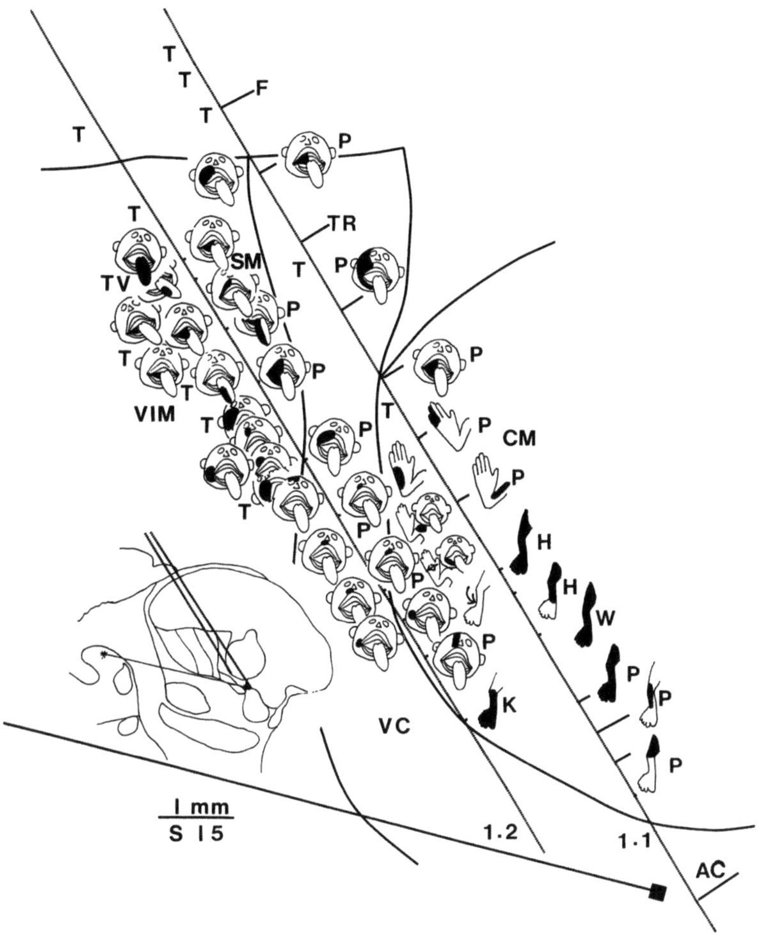

Fig. 1. Results of microstimulation and microelectrode recording in two trajectories in 15-mm (physiologically 12-mm) sagittal plane in patient with cerebellar action tremor caused by multiple sclerosis, corrected by computer to patient's own ACPC distance. Scale at *lower left* 1 mm. *Oblique line* ending in *square* = ACPC line; *square*, site of PC. Electrode trajectories shown as *dotted lines* with stimulation data on *right*, recording data on *left*. Somatotopy indicated by shading. *VIM*, ventral intermediate; *VC*, ventrocaudal nucleus of thalamus; *CM*, centre median. Stimulation symbols: *F*, faint; *TR*, tremor reduction; *P*, paraesthesiae; *H*, hot; *W*, warm; *K*, cold; *AC*, contralateral auditory; *SM*, sensorimotor; *MO*, motor effects. *Verticals* to electrode trajectories indicate stimulation threshold, the longest 160 μA, the shortest 2 μA. Recording symbols: *T*, tremor; *V*, voluntary cell. Trajectories numbered *1.2*, *1.1* below

27, 28, Table 1) where microstimulation induced contralateral motor responses. At 2 sites (sites 18, 22, Table 1) motor and sensory effects co-existed, somatotopically related at site 18. At site 22, 120 μA induced pressure in the neck, while 160 μA caused flexion of all the fingers. Albe-Fessard et al. [1, 7] have described inducing first somatotopically related motor effects, then paraesthetic effects by suprathreshold stimulation at sites where "motor" sensory effects occurred. At

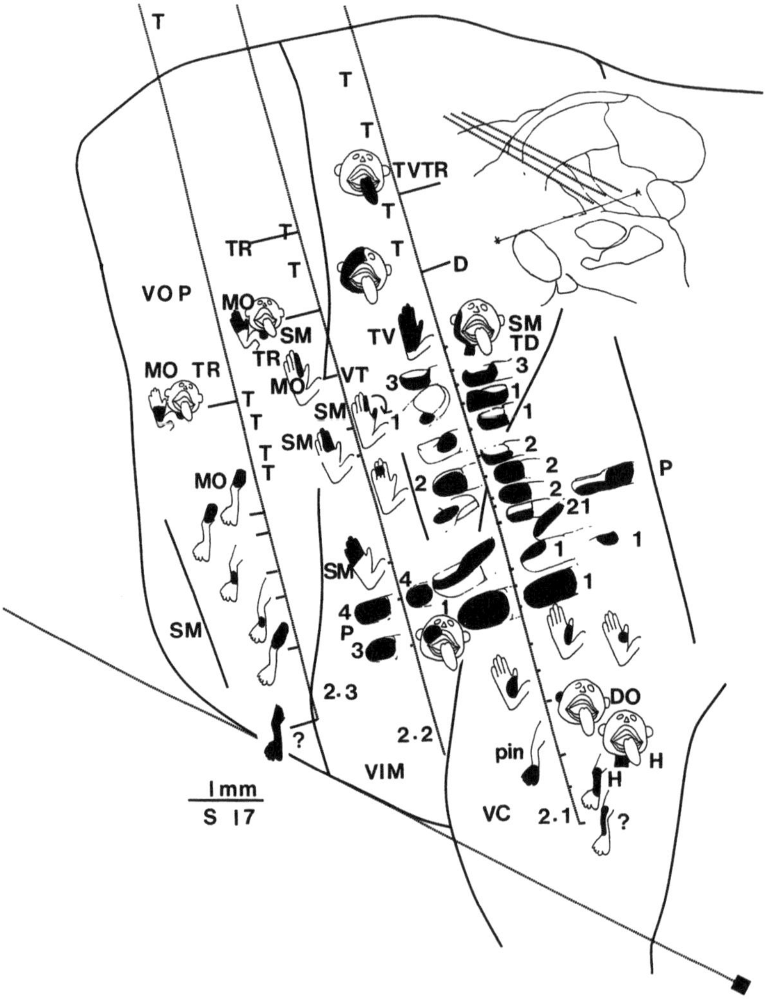

Fig. 2. Same patient as in Fig. 1, 17-mm (physiologically 14-mm) sagittal plane. *VOP*, Ventral oral nucleus of thalamus. Stimulation symbols: *TD*, tremor driving; *D*, dizziness; 1, 2, 3, 4, contralateral manual digits; *?*, indescribable feeling; *DO*, pain; *PIN*, site where response elicited by pinprick. For *trajectory 2.1* stimulation data at *right*, recording data at *left;* for 2.2 and 2.3 the reverse. Trajectories numbered *2.3, 2.2, 2.1* below

2 sites, one in VIM, feelings of lightheadedness (site 29) and faintness (site 30) were elicited, responses previously associated [55] with the rostral vestibular pathway. In another patient true vertigo was elicited in a similar site. Microstimulation altered tremor at 5 sites (sites 22, 26, 27, 31–33, Table 1). Ohye and Narabayashi et al. [17, 40, 41, 44] have recorded similar observations as well as tremor modification with suprathreshold stimulation of sites where kinaesthetic sensory units were located but where threshold responses were paraesthesiae rather than "motor" sensory effects.

Table 1. Correlation between microstimulation and microrecording by type of response

Site numbered from above	Trajectory	Microstimulation (details plotted to right unless specified)			Microrecording (details plotted to left unless specified)	
		Symbols	Threshold μA	Response	Symbols	Effective stimulus RF, SA, RA
1	1.2	P	2	"Numb" tongue, floor of mouth		Squeeze base tongue
2	1.2	P	2	"Tingle" upper lower lip		Squeeze stroke upper lip, RA
3	1.2	P	2	"Tingle" upper lip		Press upper lip
4	1.2	P	4	"Tingle" upper lip		Touch upper lip
5	2.1	P	6	"Tingle" radial edge terminal phalanx middle finger		Touch radial edge terminal phalanx middle finger, RA
6	2.1	P	3	"Tingle" radial edge terminal phalanx middle finger		Touch radial edge terminal phalanx middle finger, RA
7	2.1	P	3	"Tingle" ulnar half distal phalanx thumb		Press ulnar edge thumb nail base
8	2.1	P	3	"Tingle" radial edge terminal phalanx index		Stroke radial edge terminal phalanx index
9	2.1	P	3	"Tingle" pad terminal phalanx index		Light pressure pad terminal phalanx index, SA
10	2.1	P	3	"Tingle" ulnar side base terminal phalanx thumb		Deep pressure ulnar side base terminal phalanx thumb
11	2.1	P	4	"Tingle" pad terminal phalanx thumb		Light pressure pad terminal phalanx thumb, RA
12	2.1	P	6	"Tingle" MP joint thumb		Light touch dorsum MP joint thumb
13	2.1	P	3	"Tingle" MP joint thumb		Light pressure thenar muscle distal 2/3 MC thumb, RA
14	2.2	P	12	(left) "Tingle" terminal phalanx ring finger		(right) Pressure terminal phalanx ring finger
15	2.2	P	8	(left) "Tingle" terminal phalanx middle finger		(right) Stroking upper lip, adjacent cheek
16	1.2	SM	5	"Pressure" in mouth	TV	Non-sensory neuron fires synchronously with jaw tremor, increases firing with voluntary tongue protrusion
17	1.2	Not marked	4	"Pushing" right upper lip		Squeeze tongue, SA, synchronous with jaw tremor
18	2.3	SMMO	40	(left) "Spasm" at waist, causes hip flexion	T	(right) Non-sensory, non-voluntary neuron fires about tremor rate

Table 1 (continued)

Site numbered from above	Trajectory	Microstimulation (details plotted to right unless specified)			Microrecording (details plotted to left unless specified)	
		Symbols	Threshold μA	Response	Symbols	Effective stimulus RF, SA, RA
19	2.3	SM	40	(left) "Spasm" lower limb		Not recorded
20	2.3	SM	30	(left) "Spasm" antero-lateral ankle		Not recorded
21	2.3	SM	60	(left) "Spasm" posterior thigh		Not recorded
22	2.2	MOSMTR	120 160	(left) "Pressure" in neck Caused flexion thumb fingers reduced tremor		(right) Site surrounded by neurons firing synchronously with elbow tremor
23	2.2	SM	20	(left) "Spasm" thumb, index		(right) Passive extension manual digist 1–3 and wrist
24	2.2	SM	8	(left) "Spasm" index, middle finger		Not recorded
25	2.2	SM	6	(left) "Spasm" all fingers		Not recorded
26	2.1	SMTD	10	Temporal, neck muscles "cramped"; tremor worse	TV	Non-sensory neuron increases firing with clenching, opening fist, fires about tremor rate
27	2.3	MOTR	100	(left) Causes contraction neck muscles, dorsiflexion wrist, reduction of tremor	T	(right) Non-sensory, non-voluntary neuron synchronous with elbow tremor
28	2.2	MO	60	(left) Causes extension index	VT	(right) Non-sensory neuron fires synchronously with elbow tremor, increases firing with fist clenching
29	1.1	F	120	"Lightheaded"		Not recorded
30	2.1	D	100	"Faint"		Not recorded
31	1.1	TR	100	Reduced tremor		Not recorded
32	2.1	TR	140	Reduced tremor	TV	Non-sensory neuron synchronous with elbow tremor, increased firing with tongue protrusion
33	2.2	TR	160	(left) Reduced tremor	T	(right) Non-sensory non-voluntary cell synchronous with elbow tremor

RF = receptive field,; SA, RA = slowly rapidly, adapting.

Microelectrode Recording

Our data in this patient and a series of others [30] are in keeping with and extend published data with microelectrode recording in the human thalamus [1, 4, 7–11, 25, 35, 40, 41, 55]. In the somatosensory system in VC (sites 1–15, Table 1) superficial or deep cutaneous stimulation produced slowly adapting (SA) and rapidly adapting (RA) responses in small discrete RFs shifting less within a given sagittal plane than between sagittal planes, somatotopically medially to laterally organized. Rostral to these in the target area are found neurons responsive to muscle squeezing and tendon stretching (sites 17, 23). Ohye and Narabayashi et al. [17, 40, 41, 44] have considered these spindle afferents because they could be fired by peripheral nerve stimulation that failed to produce a conscious sensory or a motor effect [18]. Maendly et al. [37] have localized group 1 muscle afferent input here in the monkey. Rostral to these sites in VIM (in other patients more often in VOP) are found cells with altered (usually increased) firing in relation to and in advance of a particular contralateral voluntary movement [31], as also shown by Hongell et al. [22]. Examples are sites 16, 26, 28, 32, (Table 1). Voluntary cells previously described in man [6, 8–10, 15, 25] and animals [23, 24, 36, 51] lay 3 mm or more rostral to the most rostral lemniscal cell and showed a medial to lateral somatotopy. Though not seen in this patient, a number of these voluntary cells have been recorded in man [32] similar to those previously reported in animals [24, 36, 52], somatotopically medially to laterally organized with the additional feature of an RF to deep stimuli opposite in direction to that of the related voluntary movement. Voluntary cells fired 800 ms prior to the peak of EMG activity plateaued during the rest of the movement.

Most interesting are the so-called "tremor cells", first described by Albe-Fessard et al. [4]. Despite many clinical observations [1–11, 15, 17, 19, 22, 25, 29, 33–35, 40–42, 44, 45, 49], it remains controversial [29, 55] whether they fire rigorously synchronously with peripheral tremor. Crowell et al. [15] were unable to demonstrate synchrony in 8 neurons recorded by themselves but were successful in 2 recorded by Bertrand and Jasper [15]. Tremor cells have been anecdotally recorded in patients with non-parkinsonian tremor and even in those with non-tremorous dyskinesia and pain [55]. Tremor cells have been observed in many brain sites in monkeys with experimental tremor including sensorimotor cortex, whose tremor is arrested by corticospinal section [14, 26, 55]. Many sites marked T, at which neurons fired at or near peripheral tremor frequency, are noted in Figs. 1 and 2 (see especially sites 16–18, 22, 27, 28, 32, 33). A more extensive study [33, 34] revealed two types of tremor cells: (a) cells with no clear-cut firing peak in tremor EMG range, and (b) cells with high degrees of autopower at tremor frequency. The latter were of four types:

1. Sensory cells located caudally in thalamus
2. Voluntary cells located rostrally
3. Combined sensorimotor cells intermixed with voluntary cells
4. Cells with neither sensory nor motor features, of which those located more caudally showed better defined peaks of activity at tremor frequency.

This type of cell comprised most of those seen in Figs. 1 and 2. Of these the sensory and combined cells showed the greatest degrees of spike autopower and coherence with tremor EMG. Since the activity of the sensory cells is evoked, this left the combined cells as the only significant candidates for the possible role of tremor pacemaking, which also requires firing in advance of EMG bursts at a latency comparable with transmission time from thalamus to motor cortex (3 ms) through corticospinal tract to muscle (17–30 ms), requirements met by 5 out of 8 combined cells studied, but by only 1 out of 13 voluntary, by 0 out of 3 sensory, and by 3 out of 22 "no response" cells. It was also shown that tremor EMG in all contralateral upper limb muscles studied was synchronous. When pooled data from all patients studied were plotted on a single diagram so that the coronal planes of the most rostral lemniscal cells coincided, no response cells filled a gap between lemniscal cells located caudally and voluntary cells located rostrally. This raised the possibility that these might be related to the vestibular input.

Correlation Between Microstimulation and Microelectrode Recording

Table 1 correlates the results of microstimulation and microrecording in this patient with the same electrode at the same site, data amplified by experience in other patients. Microstimulation at sites where superficial or deep cutaneous units were recorded in VC induces paraesthetic responses in the RFs of these cells (sites 1–15, Table 1). Stimulation at sites in VIM where units respond to muscle belly squeezing or tendon stretching and where tremor cells often occur induces sensory effects with motor connotations in appropriate sites such as "pushing" (site 17, Table 1), "spasm" (site 23, Table 1). At some sites where "motor" sensory effects occurred, tremor cells with neither sensory nor motor effects were recorded (sites 18, 22, Table 1). Failure to record units at some sites with "motor" sensory effects may be due to technical shortcomings or the greater spread of microstimulation than range over which extracellular recording is possible. It is still unclear as to what the precise differences are between units where stimulation induces paraesthetic effects and those where stimulation induces "motor" sensory effects, the place of spindle afferents in this dichotomy, or the afferent path of "motor" sensory effects. Stimulation of sites of voluntary cells, also often associated with tremor cells, located in rostral VIM or VOP, also induced "motor" sensory effects somatotopically related to voluntary movements such as pressure (sites 16, 26, Table 1) or induced somatotopically related movement (site 28, Table 1). This is not surprising in view of their co-existence with the somatotopically related microstimulation-induced motor and sensory effects discussed above. We recognized no recording correlates of assumed vestibular sites. Association of "motor" sensory microstimulation effects and of tremor, voluntary and muscle afferent cells has been mentioned. Stimulation at tremor cell sites also may reduce tremor (sites 22, 27, 32, 33, Table 1) but not always (sites 16–18, 28, Table 1).

Overall Significance

The above observations can be interpreted in the following manner: the lesion that affects tremor lies in VIM or caudal VOP just rostral to lemniscal lip or manual digit neurons in an area where muscle afferent (? spindle afferent) units are located, whose afferent path is uncertain, by analogy with animal studies projecting to area 3A [16]. This lesion lies caudal, in turn, to cells firing in relation to a particular voluntary movement where microstimulation induces movement, by analogy with human anatomical and animal work in dentatothalamic pathway projecting to area 4 [16, 39]. Interspersed with these voluntary cells are voluntary cells with a sensory input of uncertain source also projecting to area 4 in analogy with animal studies. Of unknown significance to tremor control are the vestibular pathways through VIM. In the target area and in a relatively large surrounding zone, tremor cells of five different sorts are located where stimulation may reduce tremor, of which the combined cells alone meet the criteria of tremor pacemakers. The existence, precise identity, and role in the pathophysiology of other dyskinesias of tremor cells is uncertain, as is their homology in non-tremorous conditions. Which one or combination of cell systems found in the target area must be destroyed to control tremor or any other dyskinesia is unknown. These data as well as analogous data in animals failed to identify any thalamic cell with output to both motor and sensory cortex [16], leaving as yet unexplained the mechanism of thalamic sensorimotor correlation suggested by the evidence of Woolsey [60] and Woolsey et al. [61] for dual independent sensory and motor representation in both sensory and motor cortex. Even if the role of combined tremor cells as pacemakers for tremor is accepted, it is uncertain whether they function as pure pacemakers or as part of a long-loop reflex. If they are pacemakers, disease must cause the same cells in the target area to take on and transmit through motor cortex an autonomous pathological firing pattern, different for each dyskinesia, an unlikely possibility. Support for this contention comes from the old and dubiously significant observations of Pollack and Davis [48] that parkinsonian tremor was altered but not stopped by deafferentation and the failure to silence thalamic tremor cells in animals with experimental tremor by curarization [13] and dorsal rhizotomy [43]. Since many different types of tremor cells are known, it could be that the ones persisting after curarization were not the ones responsible for tremor. The observations of Lamarre et al. [28] suggest the monkey thalamus may possess an inherent rhythmicity at parkinsonian tremor frequency. The long-loop theory proposed by Phillips [47] and supported by the studies of Tatton and Lee [58] seems more plausible, suggesting that a basically unstable reflex arc from muscle through thalamus, motor cortex, and corticospinal tract could sense the tendency to dyskinesia produced by CNS pathology elsewhere and act as a servo loop reinforcing that disturbance whose interruption should reduce dyskinesia of various types. The cerebellar deafferentation theory mentioned by Narabayashi [40], suggesting that synchronous firing of thalamic cells results from cerebellar deafferentation, and supported by the observations of Tasker and Sogabe [54] that even in rats neocerebellar lesions converted the rapid 12-Hz tremor caused by harmaline in normal animals to a 4-Hz tremor eliminated by lesions of ventrolateral thalamus, an effect shown by Lamarre et al. [28] to be due to induction of a 3–7 Hz firing of ventrolateral thalamic cells, could be interpreted as support for either theory.

Summary

The correlation of the results of macro- and microstimulation and of microelectrode recording in the human thalamus, used to localize stereotactic surgical targets, provides some insight into sensorimotor correlation. Tremor is controlled by destroying muscle-joint receptor neurons, which are also evoked tremor cells and whose stimulation induces sensations of movement, and also by destroying voluntary cells, some with sensory input, which are located more rostrally and whose stimulation causes actual movement. These can act as autonomous tremor cells. Cutaneous receptor neurons, whose stimulation induces paraesthesiae, lie caudally. Since the receptor neurons appear to project only to sensory cortex and voluntary neurons to motor cortex, sensorimotor correlation does not appear to occur in these thalamocortical circuits. It might occur if autonomous tremor cells functioned through a "long-loop reflex".

Acknowledgments. This research was supported by the PSI Foundation, the Parkinson Foundation of Canada, and the Bertha Rosenstadt Visiting Professorship of the University of Toronto.

References

1. Albe-Fessard D (1973 a) Electrophysiological methods for the identification of thalamic nuclei. Z Neurol 205:15–28
2. Albe-Fessard D (1973 b) Physio-pathologie du Parkinson. In: Le point sur la maladie de Parkinson. Merck Sharp et Dohme, (Ed), p 110
3. Albe-Fessard D, Arfel G, Guiot G (1963) Activités électriques caractéristiques de quelques structures cérébrales chez l'homme. Ann Chir 17:1185–1214
4. Albe-Fessard D, Guiot G, Hardy J (1963 b) Electrophysiological localization and identification of subcortical structures in man by recording spontaneous and evoked activities. Encephalogr Clin Neurophysiol 15:1052–1053
5. Albe-Fessard D, Arfel G, Guiot G, Derome P, Hertzog E, Vourch G, Brown H, Aleonard P, de la Herran J, Trigo JC (1966) Electrophysiological studies of some deep cerebral structures in man. J Neurol Sci 3:37–51
6. Albe-Fessard D, Arfel G, Guiot G, Derome P, Guilbaud G (1967) Thalamic unit activity in man: Encephalogr Clin Neurophysiol [Suppl] 25:132–142
7. Albe-Fessard D, Arfel G, Derome P, Dondey M (1970) Electrophysiology of the human thalamus with special reference to trigeminal pain. In: Hassler R, Walker AE (eds) Trigeminal neuralgia, pathogenesis and pathophysiology. Saunders, Philadelphia, pp 139–148
8. Bertrand G, Jasper H (1965) Microelectrode recording of unit activity in the human thalamus. Confin Neurol 26:205–208
9. Bertrand G, Jasper H, Wong A (1967) Microelectrode study of the human thalamus: functional organization in the ventrobasal complex. Confin Neurol 29:81–86
10. Bertrand G, Jasper HH, Wong A, Mathews G (1969) Microelectrode recording during stereotactic surgery. Clin Neurosurg 16:328–355
11. Bertrand C, Martinez SN, Hardy J, Molina-Negro P, Velasco F (1973) Stereotactic surgery for parkinsonism: microelectrode recording, stimulation, and oriented sections with a leucotome. In: Krayenbühl H, Maspes PE, Sweet WH (eds) Progress in neurological surgery, vol 5. Karger, Basel, pp 79–112
12. Cooper IS, Bravo G (1958) Chemopallidectomy and chemothalamotomy. J Neurosurg 15:244–250

13. Cordeau JP (1966) Further studies in patterns of central unit activity in relation with tremor. J Neurosurg 24:213–218
14. Cordeau JP, Gybels J, Jasper H, Poirier LJ (1960) Microelectrode studies of unit discharges in the sensorimotor cortex: investigations in monkeys with experimental tremor. Neurology 10:591–600
15. Crowell RM, Perret E, Siegfried J, Villoz JP (1968) "Movement units" and "tremor phasic units" in the human thalamus. Brain Res 11:481–488
16. Friedman DP, Jones EG (1981) Thalamic input to areas 3A and 2 in monkeys. J Neurophysiol 45:59–85
17. Fukamachi A, Ohye C, Narabayashi H (1973) Delineation of the thalamic nuclei with a microelectrode in stereotaxic surgery for parkinsonism and cerebral palsy. J Neurosurg 39:214–225
18. Goto A, Kosaka K, Kubota K, Nakamura R, Narabayashi H (1968) Thalamic potentials from muscle afferents in the human. Arch Neurol 19:302–309
19. Guiot G, Hardy J, Albe-Fessard D (1962) Délimitation précise des structures sous-corticales et identification des noyaux thalamiques chez l'homme par l'électrophysiologie stéréotaxique. Neurochirurgica 5:1–18
20. Hassler R, Riechert T (1954) Indikationen und Lokalisations-Methode der gezielten Hirnoperationen. Nervenarzt 25:411–417
21. Hirai T, Shibazaki T, Nakajima H, Imai S, Ohye C (1979) Minimal effective lesion in the stereotactic treatment of tremor. Appl Neurophysiol 42:307–308
22. Hongell A, Wallin G, Hagbarth KE (1973) Unit activity connected with movement initiation and arousal situations recorded from the ventrolateral nucleus of the human thalamus. Acta Neurol Scand 49:681–698
23. Horne MK, Porter R (1980) The discharges during movement of cells in the ventrolateral thalamus of the conscious monkey. J Physiol 304:349–372
24. Horne MK, Tracey DJ (1979) The afferents and projections of the ventroposterolateral thalamus in the monkey. Exp Brain Res 36:129–141
25. Jasper H, Bertrand G (1966) Thalamic units involved in somatic sensation and voluntary and involuntary movements in man. In: Purpura DP, Yahr MD (eds) The thalamus. Columbia University Press, New York, pp 365–390
26. Joffroy AJ, Lamarre Y (1974) Single cell activity in the ventral lateral thalamus in the unanesthetized monkey. Exp Neurol 42:1–16
27. Laitinen LV (1985) Brain targets in surgery for Parkinson's disease. J Neurosurg 62:349–351
28. Lamarre Y, Dumont M, Joffroy AJ (1973) Le tremblement de type parkinsonien et l'action de l'harmaline: faits expérimentaux et mécanismes hypothétiques. Archives Italiennes de Biologie 1–2:493–503
29. Lenz FA, Tasker RR, Kwan HC, Schnider S, Kwong R, Murphy JT (1985) Single unit analyses of the ventral tier of lateral thalamic nuclei in patients with parkinsonian tremor. Soc Neurosci (Abstr) 11:1164
30. Lenz FA, Tasker RR, Dostrovsky JO, Kwan HC, Yamashiro K, Murphy JT (1986a) Single unit analysis of the human ventral thalamic nuclear group: somatosensory responses. J Neurophysiol, submitted for publication
31. Lenz FA, Tasker RR, Kwan HC, Dostrowsky JO, Murphy JT (1986b) Single unit analysis of the human thalamic ventral nuclear group: activity correlated with active movement. J Neurophysiol, submitted for publication
32. Lenz FA, Tasker RR, Kwan HC, Dostrovsky JO, Murphy JT (1986c) Single unit analysis of the human thalamic ventral nuclear group: sensory motor integration. J Neurophysiol, submitted for publication
33. Lenz FA, Tasker RR, Kwan HC, Dostrovsky JO, Murphy JT (1986d) Single unit analysis of the human thalamic ventral nuclear group: correlation of thalamic "tremor cells" with the 3–6 Hz component of parkinsonian tremor. J Neurosci, submitted for publication
34. Lenz FA, Tasker RR, Kwan HC, Dostrovsky JO, Murphy JT (1986e) Single unit analysis of the human thalamic ventral nuclear group: the 3–6 Hz component of tremor synchronous activity in functional identified neurons. J Neurosci, in preparation
35. Li C-L, Van Buren JM (1972) Micro-electrode recordings in the brain of man with particular reference to epilepsy and dyskinesia. In: Somjen GG (ed) Neurophysiology studied in man. Excerpta Medica, Amsterdam, pp 49–63

36. MacPherson JM, Rasmussen DD, Murphy JT (1980) Activities of neurons in "motor" thalamus during control of limb movement in the primate. J Neurophysiol 44:11–28
37. Maendly R, Rüegg DG, Wiesendanger M, Wiesendanger R, Lagowska J, Hess B (1981) Thalamic relay for group 1 muscle afferents of forelimb nerves in the monkey. J Neurophysiol 46:901–917
38. Matsumoto K, Shichijo F, Fukami T (1984) Long-term follow-up review of cases of Parkinson's disease after unilateral or bilateral thalamotomy. J Neurosurg 60:1033–1044
39. Mehler WR (1971) Idea of a new anatomy of the thalamus. J Psychiatry Res 8:203–217
40. Narabayashi H (1982) Tremor mechanisms. In: Schaltenbrand G, Walker AE (eds) Stereotaxy of the human brain. Thieme, Stuttgart New York, pp 510–514
41. Ohye C (1982) Depth microelectrode studies. In: Schaltenbrand G, Walker AE (eds) Stereotaxy of the human brain. Thieme, Stuttgart New York, pp 372–389
42. Ohye C, Narabayashi H (1979) Physiological study of presumed ventralis intermedius neurons in the human thalamus. J Neurosurg 50:290–297
43. Ohye C, Bouchard R, Larochelle L, Bédard P, Borcher B, Raphy B, Poirier LJ (1970) Effect of dorsal rhizotomy on postural tremor in the monkey. Exp Brain Res 10:140–150
44. Ohye C, Fukamachi A, Narabayashi H (1972) Spontaneous and evoked activity of sensory neurons and their organization in the human thalamus. Z Neurol 203:219–234
45. Ohye C, Nakamura R, Fukamachi A, Narabayashi H (1975) Recording and stimulation of the ventralis intermedius nucleus of the human thalamus. Confin Neurol 37:258
46. Ohye C, Hirai T, Miyazaki M, Shibazaki T, Nakajima H (1982) Vim thalamotomy in the treatment of various kinds of tremor. Appl Neurophysiol 45:275–280
47. Phillips CG (1969) Motor apparatus of the baboon's hand. The Ferrier Lecture 1968. Proc R Soc Lond (Biol) 173:141–174
48. Pollack LJ, Davis L (1930) Muscle tone in Parkinsonian states. Arch Neurol Psychiatry 23:303–319
49. Raeva SN (1972) Unit activity of some deep nuclear structures of the human brain during voluntary movement. In: Somjen GG (ed) Neurophysiology studied in man. Excerpta Medica, Amsterdam, pp 64–78
50. Schaltenbrand G, Wahren W (1982) Electroanatomical observations. In: Schaltenbrand G, Walker AE (eds) Stereotaxy of the human brain. Thieme, Stuttgart New York, pp 390–409
51. Scott RM, Brody JA, Cooper IS (1970) The effect of thalamotomy on the progress of unilateral Parkinson's disease. J Neurosurg 32:286–388
52. Strick PL (1976) Activity of ventrolateral thalamic neurons during arm movement. J Neurophysiol 39:1032–1044
53. Tasker RR (1976) Surgery for Parkinson's disease. In: Morley TP (ed) Current controversies in neurosurgery. Saunders, Philadelphia, pp 411–418
54. Tasker RR, Sogabe K (1973) A rat model for parkinsonian rest tremor. Trans Am Neurol Assoc 98:240–241
55. Tasker RR, Organ LW, Hawrylyshyn P (1982a) The thalamus and midbrain of man. A physiological atlas using electrical stimulation. Thomas, Springfield, chap XI
56. Tasker RR, Organ LW, Hawrylyshyn P (1982b) Investigation of the surgical target for alleviation of involuntary movement disorders. Appl Neurophysiol 45:261–274
57. Tasker RR, Siqueira J, Hawrylyshyn P, Organ LW (1983) What happened to VIM thalamotomy for Parkinson's disease. Appl Neurophysiol 46:68–83
58. Tatton WG, Lee RG (1975) Evidence for abnormal long-loop reflexes in rigid parkinsonian patients. Brain Res 100:671–676
59. Walker AE (1982) Normal and pathological physiology of the thalamus. In: Schaltenbrand G, Walker AE (eds) Stereotaxy of the human brain. Thieme, Stuttgart New York, pp 181–217
60. Woolsey CN (1958) Organization of somatic sensory and motor areas of the cerebral cortex. In: Harley HF, Woolsey CN (eds) Biological and biochemical bases of behavior. University of Wisconsin Press, Madison, pp 63–81
61. Woolsey CN, Travis AM, Barnard JW, Ostenso RS (1953) Motor representation in the postcentral gyrus after chronic ablation of precentral and supplementary motor areas. Fed Proc 12:160

SEP and Muscle Responses Related to Thalamic (VL) and Subthalamic Structures in Man

P. BIRK [1], H. RIESCHER [1], A. STRUPPLER [1], and M. KEIDEL [1]

Introduction

In addition to pharmacotherapy, parkinsonian tremor and spasmodic torticollis can effectively be treated by small stereotaxic lesions in the ventrolateral thalamus [6]. As a side effect, postoperatively a contralateral hypotonia in postural tone is observed [10, 11]. For instance, parkinsonian rigidity and tremor are abolished, but no loss of force, epicritic sensitivity or ataxia are observed, and the phasic stretch reflexes remain unchanged.

In accordance with these postoperative findings, we assume that a system involved in the control of muscle tone is affected by the lesion in a rather selective way. We suggest that muscle afferents directly projecting from the spinal level to the target area play a role in the control of muscle tone [4]. Further, the target area seems to be somatotopically organized. We argue that the interruption of these muscle afferents in combination with the correct somatotopic site of the lesion is important for therapeutic efficacy in the above-mentioned disorders.

Methods and Results

Localization Procedure

The coordinates of the target area are calculated from positive ventriculograms of the third ventricle. A series of frontal and sagittal x-ray films of the third ventricle is used. The target point mostly varies about 7.5–10 mm lateral to the wall of the third ventricle, and 4–5 mm posterior to anterior commissure-posterior commissure (CA-CP) midplane (Fig. 1 A, B). The target point differs in the laterality and anterior-posterior direction in relation to the clinical manifestation in the periphery (arm or leg).

The coordinates of the target point according to the ventriculograms are transferred to the stereotaxic atlas of Schaltenbrand and Wahren [9] proportional to the CA-CP length. Normally, the probe trajectory passes through nucleus ventralis oralis posterior (VOP), nucleus ventralis intermedius (VIM) and radiatio praelemniscalis (RP) (Fig. 2).

[1] Neurologische Klinik und Poliklinik der Technischen Universität München, Möhlstr. 28, D-8000 München 80, FRG.

Clinical Aspects of Sensory Motor Integration
Ed. by A. Struppler and A. Weindl
© Springer-Verlag Berlin Heidelberg 1987

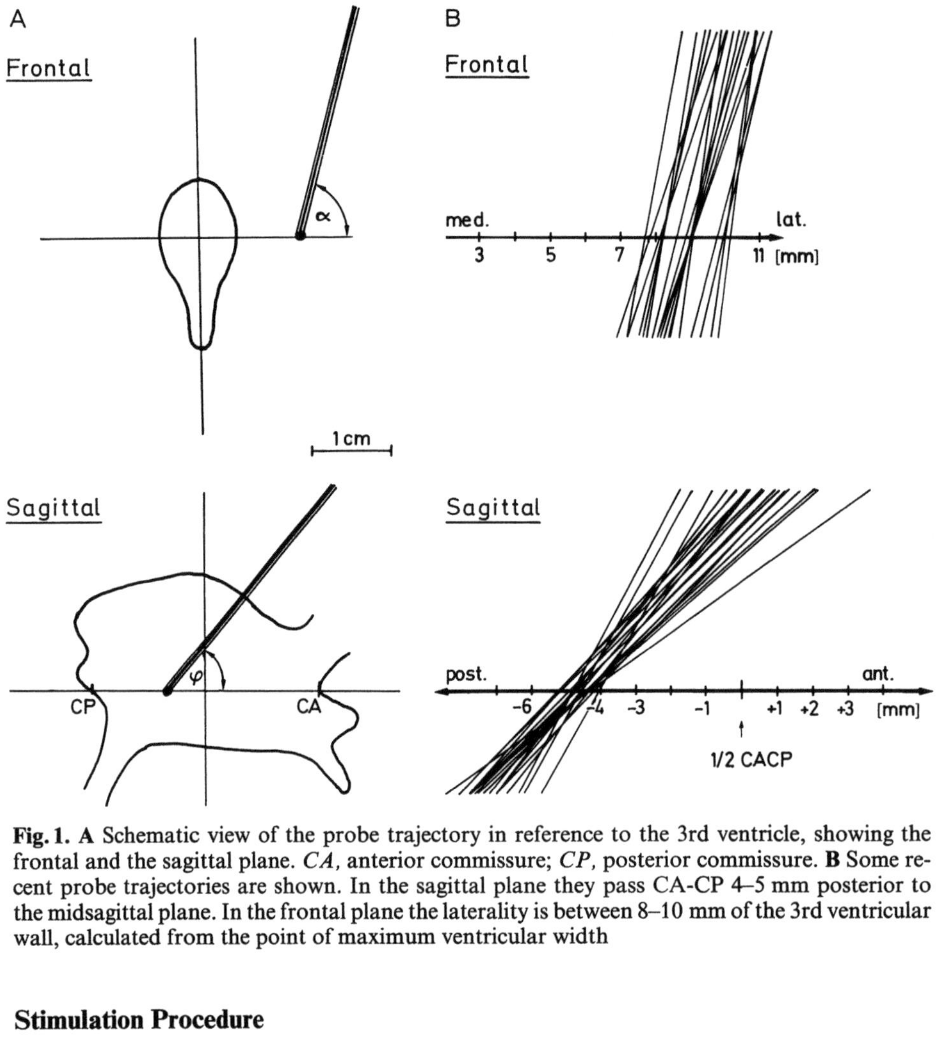

Fig. 1. A Schematic view of the probe trajectory in reference to the 3rd ventricle, showing the frontal and the sagittal plane. *CA*, anterior commissure; *CP*, posterior commissure. **B** Some recent probe trajectories are shown. In the sagittal plane they pass CA-CP 4–5 mm posterior to the midsagittal plane. In the frontal plane the laterality is between 8–10 mm of the 3rd ventricular wall, calculated from the point of maximum ventricular width

Stimulation Procedure

The correct position of the probe and the target point is controlled by stimulation in the correlative thalamic and subthalamic area. For the stimulation macro- or semimicroelectrodes, and, most recently, micro-electrodes have been used. The stimulus frequency is 50 Hz, square pulses of 0.2 ms duration. The applied stimulus trains last about 0.5 s. The threshold amplitude for the motoric effects in the periphery varies up to 1 mA or about 2–4 V.

Following the stimulation within the target area some facilitatory effects concerning tremor and muscle tone most frequently are observed. The tremor is facilitated, seen as a stronger burst synchronization of motor units in the EMG and an increase in amplitude. Regularly tonic and slow contractions of skeletal muscle groups in the extremities are observed. Often these visible effects are accompanied by sensations described by the patients as tension in the depth of the tissue or the muscles.

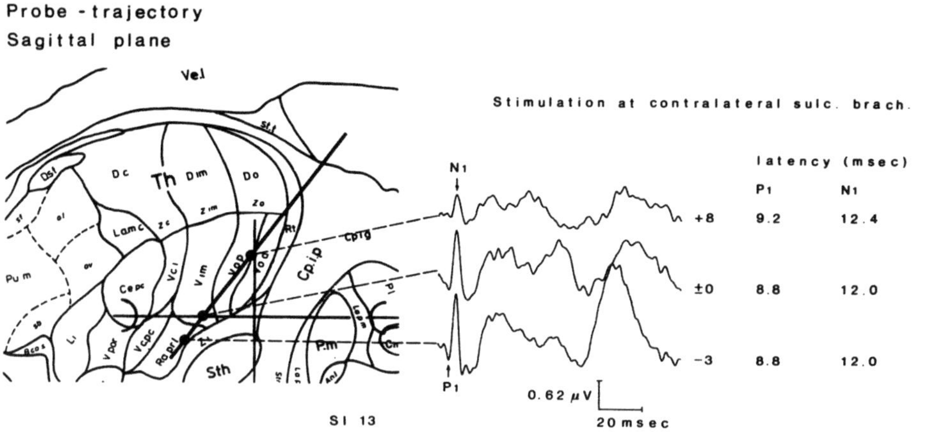

Fig. 2. Evoked potentials following mixed nerve stimulation in reference to the stereotaxic atlas of Schaltenbrand-Wahren [9] (plane sagittal lateral 13). The evoked potentials initially are triphasic. The latencies of the first positivity (*P1*) and the first negativity (*N1*) show a slight decrease with depth (9.2 ms to 8.8 ms, 12.4 ms to 12.0 ms, respectively), whereas the amplitudes show a slight increase. The probe passes ventralis oralis posterior (*V.o.p.*), ventralis intermedius (*V.i.m.*) on the way to the subthalamus (radiatio praelemniscalis/*Ra.prl.*). The target point is located in the V.i.m. and Ra.Prl. border region (depth 0 to −3)

In Fig. 3 those movement effects are plotted according to the stereotaxic coordinates of the stimulation sites and the affected extremity. A pattern of somatotopic distribution may be recognized, e.g., the leg is represented as more lateral, proximal structures as neck and shoulder as more medial, respectively.

Recording Procedure

Evoked potentials are recorded in the target area following electrical stimulation of peripheral nerves. Monopolar recording was performed by using a semimicroelectrode (reference to the stereotaxic frame). The stimulation of mixed peripheral nerves was carried out with bipolar surface electrodes; skin nerves were stimulated by two pairs of ring electrodes placed on fingers II–V.

The stimulus strength was in the range of 2–3 times above motor or sensory threshold; the stimuli consisted of square pulses of 0.1–0.3 ms duration. 400–500 sweeps were averaged "on line" by a Nicolet pathfinder, with the filter ranging from 0.5 Hz to 3 kHz.

Following stimulation of the median and ulnar nerve, in most cases triphasic potentials are recorded initially within the target area. During stimulation at the brachial sulcus the latencies of the first positivity (P1) ranged from 9.2 ms to 8.8 ms (decreasing with depth), and those of the first negativity (N1), from 12.3 to 12.0 ms. A slight increase of the amplitude is observed with increasing depth (Fig. 2).

In contrast to these findings, no significant evoked potentials could be recorded following stimulation of skin afferents (Fig. 4).

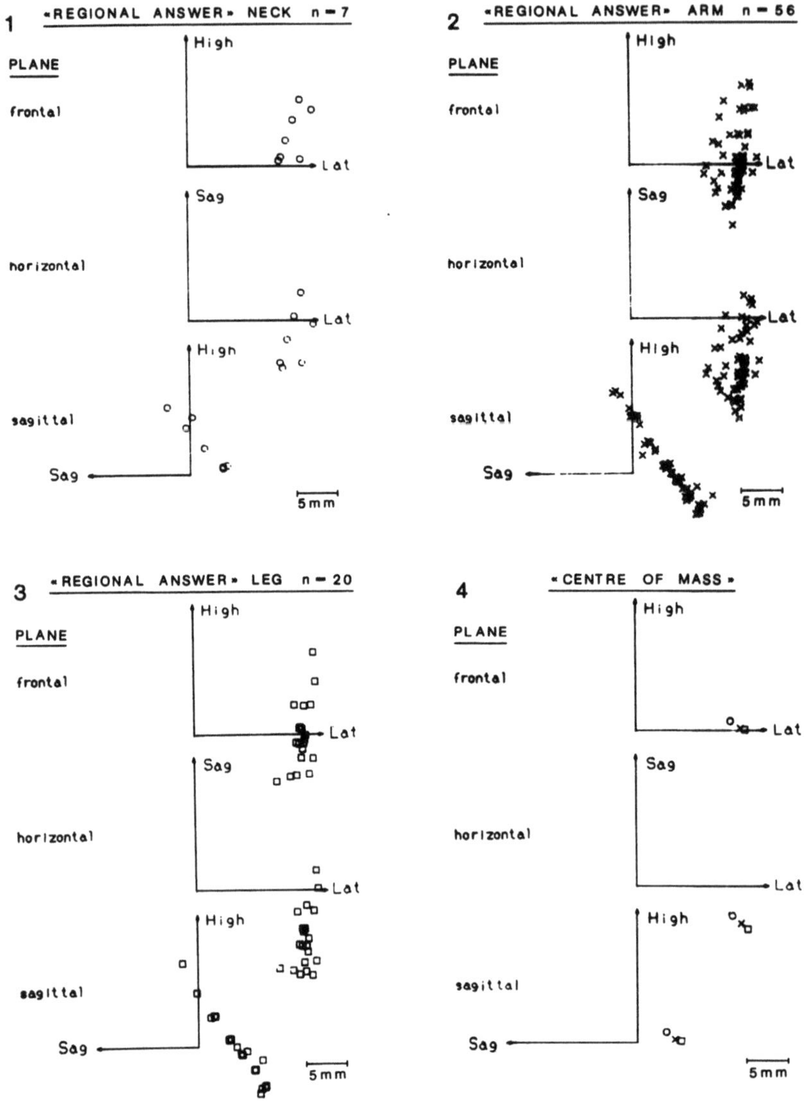

Fig. 3. Slow tonic movement effects are plotted, according to the stereotaxic coordinates of their stimulation sites and their respective localization in the body. Open circles (○) represent movements of neck muscles, crosses (×) those of muscle groups in the arm, and squares (□) those of the leg, respectively. The coordinate system is adapted to CA-CP and midventricular plane. For a better spatial orientation, the coordinate system is shown in the frontal, horizontal and sagittal plane. In *4d* the centers of mass of the respective movement effects are plotted and a medio-lateral somatotopic distribution is to be seen

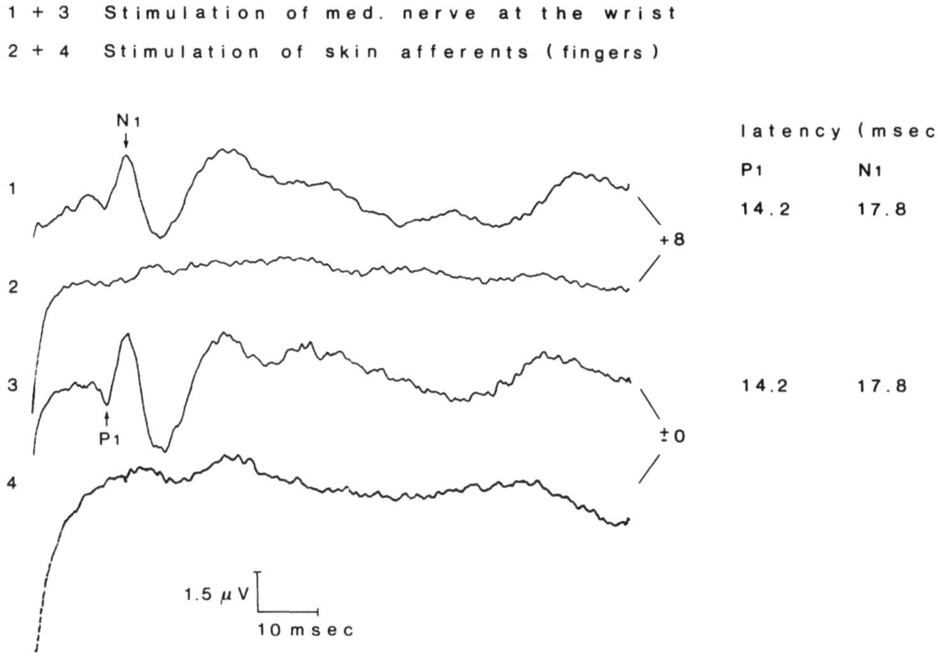

Fig. 4. In contrast to stimulation of mixed nerves, stimulation of skin afferents evokes no significant potential within the target region. For the anatomical localizations see Fig. 2. ± 0 and -3 symbolize the depth according to the stereotaxic system

Conclusions

Our observations suggest that the target point is in an area which may serve as a relay station involved in the control of muscle tone.

1. The target area appears to receive direct spinal input which does not originate from skin afferents and therefore may represent proprioceptive signals. We assume that this input originates in the skeletal muscles.

2. During stimulation in the target area muscle tone and concomitant tremor are modified. Those stimulation effects show some somatotopic organization. The modifying effect may be mediated via thalamocortico-spinal pathways. According to the normally low stimulus thresholds and the possibility of eliciting similar stimulation effects with semimicro- or microelectrodes within the thalamus, we believe that these effects are not mediated by directly descending pathways. This is in accordance with our recording data. Furthermore, animal studies show a strong afferent projection to the ventrolateral nucleus complex. The input in the termination area can be divided in pallido-thalamic, cerebello-thalamic and spino-thalamic pathways [1–3, 5–8).

Whether the somatosensory evoked potentials (SEP) recorded are mediated through the spino thalamic or lemniscal pathways is not clear. According to the transmission time, it should be a fast-conducting system, perhaps of lemniscal origin. However, the triphasic shape of the potential may reflect a mixed origin. This needs further investigation.

Summary

Within the target area (VL) used for the stereotaxic treatment of parkinsonian tremor and spasmodic torticollis electrical stimulation as well as recording of SEP was performed.

The effects of stimulation in the target area are facilitation of muscle tone showing some degree of somatotopic distribution. The recorded SEPs indicate a projection of an afferent system (probably of muscle afferents) to the target area.

We assume that the target area is a relay station involved in the control of muscle tone. The interruption of muscle afferents in combination with the correct somatotopic localization of the lesion is important for the therapeutic efficacy in parkinsonian tremor and spasmodic torticollis.

References

1. Asanuma H (1981) Functional role of sensory inputs to the motor cortex. Prog Neurobiol 16:241–262
2. Asanuma H, Thach WF, Jones EG (1983) Distribution of cerebellar terminations and their relation to other afferent terminations in the ventral lateral thalamic region of the monkey. Brain Res Rev 5:237–265
3. Berkley KJ (1983) Spatial relationships between the terminations of somatic sensory and motor pathways in the rostral brain stem of cats and monkeys. II. Cerebellar projections compared with those of the ascending somatic sensory pathways in lateral diencephalon. J Comp Neurol 220:229–251
4. Birk P, Struppler A, Riescher H, Keidel M (1986) Somatosensory evoked potentials in the human ventrolateral thalamus. Symposion on Evoked Potentials, Tübingen 1984 (to be published)
5. Glees P (1945) The interrelation of the striopallidum and the thalamus in the macaque monkey. Brain 68:331–345
6. Hassler R, Mundinger F, Riechert T (1979) Stereotaxis in Parkinson syndrome. With an atlas of the basal ganglia in Parkinsonism by R. Hassler. Springer, Berlin Heidelberg New York
7. Ilinsky JA, Kultas-Ilinsky K (1982) Comparative study of pallido-thalamic projections in the cat and monkey. Neurosci Lett 10:248–249
8. Jones EG, Friedman DP (1982) Projection pattern of functional components of thalamic ventrobasal complex on monkey somatosensory cortex. J Neurophys 48:521–544
9. Schaltenbrand G, Wahren W (1971) Atlas for stereotaxy of the human brain, 2nd edn. Thieme, Stuttgart
10. Struppler A (1982) Contribution of electromyography to stereotaxy. In: Schaltenbrand G, Walker AE (eds) Stereotaxy of the human brain. Thieme, Stuttgart New York, pp 436–448
11. Struppler A (1985) Stereoencephalotomy and motor control. In: Struppler A, Weindl A (eds) Electromyography and evoked potentials. Springer, Berlin Heidelberg New York Tokyo, pp 37–44

Electrical Stimulation in Human of the Sensory Thalamic Nuclei and Effects on Dyskinesias and Spasticity

J. Siegfried [1] and M. N. Pamir [1]

Introduction

Electrical stimulation of the diencephalon for localization purposes has been performed since the earliest use of stereotactic procedures to treat neurological diseases, and it has been particularly useful in the extrapyramidal movement disorders. However, the first reported deep brain stimulation used therapeutically in movement disorders seems to be from Bechtereva in Leningrad in 1972–1975 [2–4]. The number of cases operated on, the technique, and the cerebral targets are unfortunately not mentioned. The authors report a series of cases of Parkinson's disease, Wilson's disease, and dystonia musculorum deformans, all with deep brain stimulation having a favorable effect, especially for tremor. The real impetus to therapeutic brain stimulation, however, was given by Cooper [6], who placed surface electrodes directly on the anterior cerebellar cortex. Mundinger [15] was the first to report the implantation of chronic electrodes in the thalamus, pulvinar, and dentate nucleus for movement disorders, particularly spasmodic torticollis. Good results in (a) facial dyskinesias in a series of torsion dystonia, (b) double athetosis, and (c) arm dystonia in a case of traumatic spasticity are reported in his last paper, apparently with stimulation of the ventrolateral part of the thalamus and/or the zona incerta. The improvement in movement disorders, which occurred a few minutes after stimulation began, usually outlasted the stimulation by 20 min to 2 h. Over the course of weeks to months however, daily stimulation brought even further improvements. Mundinger and Neumüller [16] emphasized the advantage of using programmed stimulation individually adapted to each patient. The same deep brain structure, namely the subthalamic region, was also successfully used by Brice and McLellan [5], who reported suppression or decrease of severe intention tremor in 3 female multiple sclerosis patients. They used continuous stimulation with 75–150 Hz and square-wave pulses of 0.5–1 mA of 200–400 ms duration. To reduce the total amount of stimulation, the transmitter was controlled by a switching device triggered by electromyographic signals from the deltoid muscle of the appropriate arm. More recently, Andy [1] reported good results in 9 cases, particularly for tremor control, with stimulation of the center median, nucleus parafascicularis, and intralaminary system.

Sensory thalamic nuclei as the target point in therapeutic stimulation for dyskinetic movements has been emphasized by Mazars et al. [14], who reported marked improvement of involuntary movements in 17 cases of deafferentation pain of different origins, all associated with dyskinesias. The largest experience of deep brain stimulation for involuntary movements, however, is probably the

[1] Neurosurgical Department, University Hospital, CH-8091 Zürich, Switzerland.

Clinical Aspects of Sensory Motor Integration
Ed. by A. Struppler and A. Weindl
© Springer-Verlag Berlin Heidelberg 1987

one collected by Cooper [8]. He also mentions the favorable effect of electrodes placed in the posterior part of the thalamic ventrolateral nucleus adjacent to the internal capsule, in a series of 49 patients. In these patients, the placement of electrodes was confirmed by recording the thalamic somatosensory responses following stimulation of the median nerve. After a trial of stimulation, electrodes were internalized only in cases demonstrating clinical improvement, and 27 good results were reported out of 49 patients.

This review of the literature confirms that thalamic and subthalamic stimulation can control movement disorders, but favorable effects have been obtained by stimulation of the subthalamic region, the ventrolateral group of thalamic nuclei, as well as the specific and nonspecific thalamic sensory nuclei. Thus, because the rationale for deep brain stimulation in the treatment of movement disorders is still lacking, we were quite reluctant to use this technique, in spite of the fact that we have used it for pain control since 1977. However, observations made in patients over the years brought our attention to the decisive role played by the activation of primary sensory nuclei in motor control.

Material and Technique

Four cases of thalamic pain syndrome presenting with marked abnormal dyskinetic movements and two cases of very strong spasticity due to traumatic paraplegia have been selected for analysis. All cases have received thalamic stimulation for deafferentation pain. Our target for deafferentation pain is the nucleus ventroposteromedialis thalami when the pain is localized in the face and the nucleus ventroposterolateralis if it is in the hand or foot. For the 4 cases of postapoplectic thalamic syndrome, with pain more marked in the hand and also showing choreoathetotic movements, the electrodes were introduced stereotactically 3 mm in front of the commissura posterior on the baseline anterior commissure-posterior commissure (AC-PC), 1–2 mm below the line, and 13 mm laterally. For the 2 cases of paraplegic pain, electrodes on each side of the thalamus were implanted 3 mm in front of the commissura posterior on the baseline AC-PC, 1–2 mm below this line and 16 mm laterally. In order to minimize brain tissue trauma during electrode insertion, a very smooth, flexible monopolar electrode (Avery) with a central rigid stylet has been used. This electrode has the great advantage of eliminating the extra channel created by the guiding canula in other electrodes. The stimulating tip is a 5-mm long cylinder made of pure platinum, a well-established biocompatible material for tissue stimulation, and high resistance to fatigue fracture is provided by the electrode's helical design. This makes the electrode highly flexible and allows it to move with the brain and also gives it virtually no resiliance, thus allowing the electrode to remain bent upon stylet removal. Also a pure titanium screw has been designed to secure the electrode in a burr-hole of 2.5 mm before the stylet is removed, providing a water-tight seal over the stimulating electrode [21]. The electrodes were connected to a percutaneous extension and a trial of stimulation was used, and if satisfactory pain control was obtained, the devices were internalized with implantation of a programmable total implant system (ITREL-Medtronic) in the upper thorax region.

Results

Dyskinesias

In all 4 cases, stimulation has given satisfactory paresthesias in the painful area with impulses of 1–2 mA, 200 ms, and 33 Hz, and suppressed almost instantly the involuntary movements associated with the thalamic pain syndrome. As long as stimulation continued, the involuntary movements were controlled, but at its completion, the dyskinesias reappeared slowly over 1–2 min. Using this knowledge, the total implants were programmed with an on period of stimulation of 1 min every 2 min. Follow-up from 1 to 2 years confirms the continued effectiveness of the stimulation.

Spasticity

In the 2 cases of paraplegic pain associated with significant lower-extremity spasticity, the stimulation of the sensory thalamic nucleus suppressed almost instantly the clonus of the contralateral leg. At the end of stimulation, however, the clonus reappeared slowly over 2–3 min. The bilateral deep brain stimulation systems were in these 2 cases programmed with 1 min of stimulation every 2 min.

Discussion

Some mechanisms by which the electrical stimulation of sensory thalamic nuclei can improve or suppress involuntary movements will be reviewed.

Specific Role of Sensory Thalamic Nuclei

Activation of an Inhibitory Sensory System

It has been demonstrated by DeLong and Georgopoulos [9] that a significant proportion of neurons in the putamen, globus pallidus, and subthalamic nucleus that are related to active movements of specific body parts also respond to somatosensory stimulation of the same body part; thus, abnormal modulation of pallidal output may lead to the uncontrolled movements of patients with involuntary movements [10]. Consequently, somatosensory inputs of a specific and restricted nature may be used to control or monitor movements. A clinical example of cutaneous input modifying motor activity is "geste antagoniste" (putting the index finger on the chin), which in some cases is sufficient to control involuntary torticollis.

The favorable effects of intermittent sensory thalamic nuclei stimulation on movement disorders have been seen in our series of patients suffering from deafferentation pain due to lesions of the sensory pathways or thalamic relay systems. Therefore, the electrical stimulation of the sensory thalamic nuclei may restore or reinforce the requisite activity of this nucleus with activation of a deficient inhibitory system.

Inhibition of a Facilitatory System

Cooper [7], in an attempt to justify a general theory on causation and potential reversibility of involuntary movement disorders, suggests the concept of involuntary movement disorders of sensory communication. Consequently, restoration of the inhibitory-facilitatory balance may reverse syndromes which clinically appear to be long-lasting and irreversible, and Cooper feels that deep brain stimulation in posterior ventrolateral nuclei inhibits or suppresses the posterior ventrolateral sensory funnel.

Nonspecific Sensory Thalamic Nuclei (Stimulation of Other Structures)

Inhibition of a Facilitatory System

Improvement in movement disorders obtained by electrical stimulation of the ventrolateral thalamus, the pulvinar, the zona incerta, and other subthalamic regions, and of the dentate nucleus by Mundinger and Neumüller [15, 16], Brice and McLellan [5], and Schvarcz et al. [20], may suggest that the results are not related to a specific nucleus. Since all these structures, except the dentate nucleus, which was actually stimulated for spasticity, are near the sensory thalamic nuclei, it can be suggested that the sensory inputs were facilitated by the stimulation. It can also be argued that the spread of stimulating current can excite the reticular nucleus and the zona incerta, which belongs to the reticular formation; this latter has a well-recognized inhibitory function [12, 17, 19].

Activation of Cerebral Cortex

The role of the cerebral cortex in neural mechanisms of dyskinesias is unquestionable. The sensory thalamic nuclei project extensively not only to the sensory, but also to the motor cortex [17]. Kornhuber [13] stated that the motor cortex, from the sensory point of view, is a somatic and proprioceptive-vestibular association area, and the inability of athetotic patients to control their involuntary movements without the help of tactile stimuli indicates that the motor cortex is not just a motor function generator. The role of sensory inputs from sensory thalamic nuclei to cerebral cortex on motor activity has been emphasized by Evarts and Wise [11] and Porter [18]. It can therefore be suggested that the activation of sensory inputs via the sensory thalamic nuclei to the cerebral cortex is necessary to modulate normal motor control.

Summary

Intermittent electrical stimulation of sensory thalamic nuclei has been able to control the involuntary movements of 4 cases of thalamic pain syndrome and the spasticity in 2 cases of posttraumatic paraplegia. These 6 cases have been followed on a long-term basis after implantation of a programmable total implant stimulation system. Some mechanisms by which the electrical stimulation of sensory thalamic nuclei may improve or suppress movement disorders are also reviewed.

References

1. Andy DJ (1983) Thalamic stimulation for control of movement disorders. Appl Neurophysiol 46:107–111
2. Bechtereva NP, Bondartchuk AN, Smirnov VM, Meliutcheva LA (1972) Therapeutic electrostimulations of the brain deep structures. Vopr Neirokhnir 1:7–12
3. Bechtereva NP, Bondartchuk AN, Smirnov VM, Meliutcheva LA, Shandurina AN (1975a) Method of electrostimulation of the deep brain structures in treatment of some chronic diseases. Confin Neurol 37:136–140
4. Bechtereva NP, Kambarova DK, Smirnov VM, Shandurina AN (1975b) Using the brain's latent abilities for therapy: chronic intracerebral electrical stimulation. In: Sweet WH, Obrador S, Martin-Rodriguez JG (eds) Neurosurgical treatment in psychiatry, pain and epilepsy. University Park Press,Baltimore London Tokio, pp 581–613
5. Brice J, McLellan L (1980) Suppression of intention tremor by contingent deep brain stimulation. Lancet II:1221–1222
6. Cooper IS (1973) Effect of chronic stimulation of anterior cerebellum on neurological disease. Lancet I:1321
7. Cooper IS (1982) A general theory of causation and reversibility of involuntary movement disorders. Appl Neurophysiol 45:317–323
8. Cooper IS, Upton ARM, Amin I (1982) Chronic cerebellar stimulation and deep brain stimulation in involuntary movement disorders. Appl Neurophysiol 54:209–217
9. DeLong MR, Georgopoulos A (1979) Motor functions of the basal ganglia. In: Brookhart JM et al. (eds) The nervous system. American Physiological Society, Bethesda, pp 1017–1061 (Handbook physiology)
10. DeLong MR, Georgopoulos AP, Crutcher MD, Mitchell SS, Richardson RT, Alexander GE (1984) Functional organisation of the basal ganglia: contributions of single cell recording studies. Functions of the basal ganglia. Ciba Foundation Symposium 107. Pitman, London, pp 64–82
11. Evarts EV, Wise SP (1984) Basal ganglia outputs and motor control. Functions of the basal ganglia. Ciba Foundation Symposium 107. Pitman, London, pp 83–102
12. Jasper HH (1961) Thalamic reticular system. Electrical stimulation of the brain. In: Sheer DE (ed) Hogg Foundation research series. University of Texas Press, Austin, pp 277–287
13. Kornhuber HH (1978) Cortex, basal ganglia and cerebellum in motor control. Contemporary clinical neurophysiology. In: Cobb WA, Van Duijn H (eds) EEG and clinical neurophysiology, suppl 34. Elsevier, Amsterdam New York Oxford, pp 449–455
14. Mazars G, Merienne L, Cioloca G (1980) Control of dyskinesias due to sensory deafferentation by means of thalamic stimulation. Acta Neurochir [Suppl] (Wien) 30:239–243
15. Mundinger F (1977) Neue stereotaktisch-funktionelle Behandlungsmethode des Torticollis spasmodicus mit Hirnstimulatoren. Med Klinik 72:1982–1986
16. Mundinger F, Neumüller H (1982) Programmed stimulation for control of chronic pain and motor disease. Appl Neurophysiol 45:102–111

17. Nauta WSH, Domesick VB (1984) Afferent and efferent relationships of the basal ganglia. Functions of the basal ganglia. Ciba Foundation Symposium 107. Pitman, London, pp 3–29
18. Porter R (1984) Basal ganglia links for movement mood and memory (general discussion). Functions of the basal ganglia. Ciba Foundation Symposium 107. Pitman, London, pp 103–107
19. Russell GV (1961) Interrelationship within the limbic and centrencephalic systems. Electrical stimulation of the brain. In: Sheer DE (ed) Hogg Foundation research series. University of Texas Press, Austin, pp 167–181
20. Schvarcz JR, Sica R, Morita E (1980) Chronic self-stimulation of the dentate nucleus for the relief of spasticity. Acta Neurochir [Suppl] (Wien) 30:351–359
21. Siegfried J, Comte P, Meier RR (1983) Intracerebral electrode implantation system. J Neurosurg 59:356–359
ND

VIII. Posture and Movement: Interactions and Disturbances

Multi-Joint Arm Posture – New Perspectives on the Control of Arm Posture

E. Bizzi [1], F. A. Mussa-Ivaldi [2] and N. Hogan [2]

Introduction

Studies of the mechanical properties of the musculoskeletal apparatus have been found to be useful in gaining insights into the "rules" of the neural controller. This approach is based on the assumption that the neural control system must not only control but also take advantage of the musculoskeletal apparatus.

A case for this approach was made years ago by Feldman [8, 9], who investigated the spring-like properties of the human arm. Muscles do indeed behave like tunable springs in the sense that the force generated by them is a function of length and level of neural activation [16]. In addition, muscles are arranged about the joints in an agonist-antagonist configuration. If we attribute spring-like properties to muscles, then a limb's posture is maintained when the forces exerted by the agonist and antagonist muscle groups are equal and opposite. This implies that when a force is applied, the limb is displaced by an amount proportional to both the external force and the stiffness of the muscles. When the external force is removed, the limb should return to the original position. Evidence supporting the idea that muscles in vivo have indeed spring-like properties is briefly discussed in the first part of this contribution. In the second part, the implication of these findings for trajectory control are discussed.

Muscle Spring-Like Behavior

The idea that muscles display spring-like properties is not new; several recent studies have emphasized this point and have stressed that we can understand the organization of voluntary movements by studying the way in which the mechanical properties of the neuromuscular system set constraints upon the motor controller [4, 7, 10, 11].

Recently we provided a unique demonstration of the degree to which the neuromuscular system is spring-like [14]. We showed that when the hand is displaced from a maintained posture and held at zero velocity, arm muscles generate a restoring force which, when the arm is released, returns the hand to the initial position in a manner which is entirely compatible with what would be expected of a spring system. By recording the response to displacements of variable size and direction, we succeeded in describing the field of elastic forces associated with

[1] Department of Brain and Cognitive Sciences and [2] Department of Mechanical Engineering, E25-526 Massachusetts Institute of Technology, Cambridge, MA 02139, USA.

Clinical Aspects of Sensory Motor Integration
Ed. by A. Struppler and A. Weindl
© Springer-Verlag Berlin Heidelberg 1987

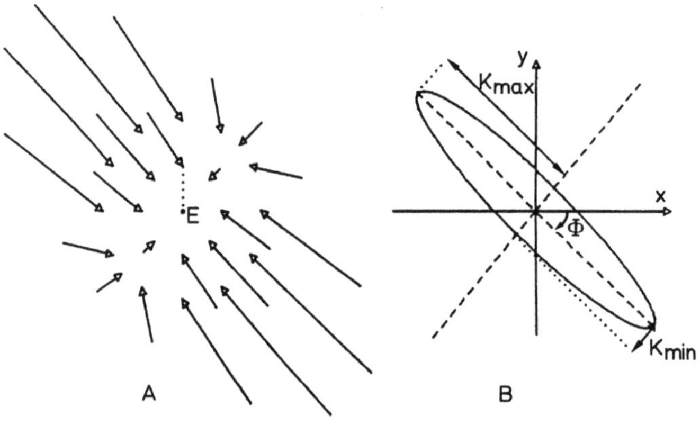

Fig. 1. Two representations of the postural force field. **A** Pin-cushion diagram: force vectors *F* (*arrows*) are shown in correspondence to each displacement from the equilibrium point, *E*. **B** Stiffness ellipse: the stiffness matrix which generates the force field is plotted with respect to a frame centered at the equilibrium point. The ellipse has a size (its area), an orientation (the angle Ø), and a shape (the ratio K_{max}/K_{min}). (Bizzi et al. [5])

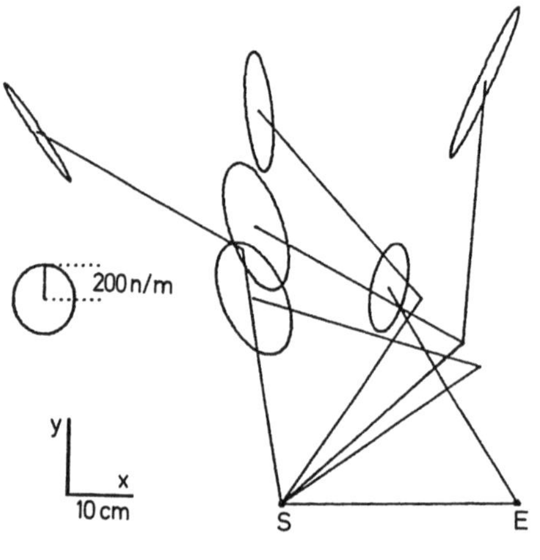

Fig. 2. Stiffness ellipses obtained from a subject during posture in six work-space locations. *S*, shoulder joint; *E*, elbow joint. The calibration for the stiffness is given by the *circle*, which represents an isotropic stiffness of 200 N/m. (Bizzi et al. [5])

the hand. At any given hand position, this field was characterized by a shape, a magnitude and an orientation (Fig. 1). We also found that the stiffness field, associated with hand posture varies substantially at different positions of the hand in the work space.

Specifically, the shape and the orientation display a common pattern, which can be summarized as follows: the stiffness is more isotropic (circular) in proximal positions and more anisotropic (elongated) in distal positions; the direction of maximum stiffness is approximately oriented along a radial line joining the hand to the shoulder (Fig. 2).

Experimentally, we determined the muscle stiffness properties by imposing a displacement to the hand and measuring the static restoring forces as a function of hand displacement. We then computed the elements of the stiffness tensor and checked for symmetric and anti-symmetric terms. By comparing the magnitude of the symmetric and anti-symmetric components of the stiffness, we were able to quantify the extent to which the muscles are spring-like. Our results have indicated that the stiffness is essentially symmetric, and this rigorously proves, for the first time, the spring-like nature of hand posture. In addition, these data allowed us to derive some important conclusions about the reflex regulation of the posture. In fact, the symmetry of the hand stiffness matrix implies that the matrix relation joint torques to joint displacement is also symmetric. The symmetry of the joint stiffness, in turn, indicates that a displacement of the elbow induces a torque at the shoulder which is equal to the elbow torque induced by a shoulder displacement. In other words, the spring-like behavior experimentally observed in hand posture results in the constraint that the mechanical coupling between the joints should be symmetric. If this is so, then heteronomous proprioceptive reflexes must have equal gain; a conclusion that could not have been reached on the basis of single-joint experiments [14].

From Posture to Movement

The studies on multi-joint arm posture and previous experimental work on visually triggered head and arm posture in trained monkeys have shown that final position of the limb is an equilibrium point between sets of opposing forces [2, 15]. If the control signal were to specify a new length-tension relationship for the springs, movement would occur until a new equilibrium point was reached. In accordance with this hypothesis, movements are, at the simplest level, transitions in posture. Our results show that in the case of simple forearm movements the control signal induces a gradual shift of the equilibrium point.

The movements which we have examined involve pointing with the forearm to a visible target. To accomplish this goal, the time course of the neural signals executing the transition from the initial to a final position was investigated [3]. Monkeys performing a pointing task were studied. Force and positional disturbances were applied to the forearm with a torque motor coupled to the shaft of the pivot arm on which the elbow rested. In some experiments, the animal was prevented from detecting disturbances of forearm position by surgically interrupting sensory roots conveying afferent activity from the arm, neck and upper torso (C1-T3). After deafferentation, the forearm pointing responses learned before the operation could easily be evoked by presentation of the targets [3]. The movement were similar to those observed before deafferentation.

In the first set of experiments both intact and deafferented animals were used and the following procedures adopted. In randomly selected trials evoked by the presentation of a target, the arm was clamped in its initial position before the onset of visually triggered movements and was released at various times after the onset of evoked agonist electromyographic (EMG) activity (no visual feedback was allowed). The duration of the holding period was varied randomly from 100 to 600 ms. The acceleration of the arm immediately after its release was plotted as a function of the holding period (i.e., the time elapsed since the onset of EMG activity in the agonists). In both intact and deafferented monkeys, the initial ac-

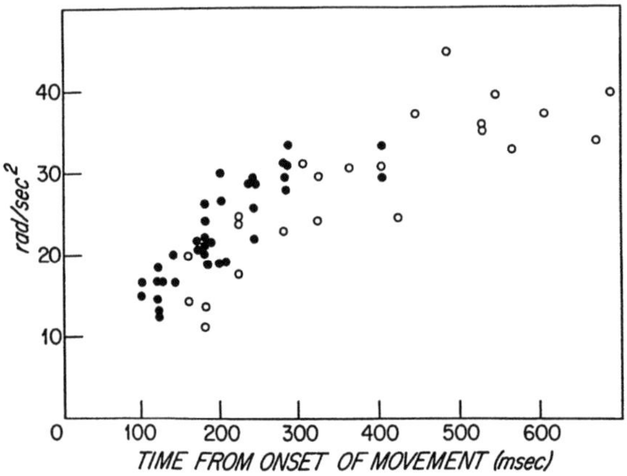

Fig. 3. The forearm of intact and deafferented animals was held in its initial position while the animal attempted to move toward a target light and released at various times. Plot of acceleration immediately following release versus holding time. Abscissa: time in milliseconds (msec); ordinate: radians/sec^2. *Solid dots,* intact animal; *circles,* deafferented animal. (Bizzi et al. [3])

celeration increased gradually with the duration of the holding period, for holding periods up to 400–600 ms, and the movements were, therefore, progressively faster after the release of the arm (Fig. 3).

Our experimental results indicate that for a movement of at least 600 ms duration, the mechanical expression of the alpha motoneuronal activity does not reach steady state until at least 400 ms have elapsed following the onset of action potentials in the muscle (Fig. 3). These findings are consistent with the notion of a centrally generated control signal specifying a gradual shift in the equilibrium point. As the equilibrium point moves further away from the position at which the limb was restrained, progressively larger torques are generated, resulting in progressively increasing values of acceleration after release and progressively faster movement trajectories.

Similar conclusions were drawn from another set of experiments, which involved quickly forcing the forearm to the target position. In this set of experiments, as the intact monkey began to move toward a new target position, a brief torque pulse (150 ms), whose onset was triggered by the initial increase in EMG activity in the agonist muscles, drove the elbow quickly to the intended final position. However, instead of remaining in this position, the forearm returned to a point intermediate between the initial and final positions before reversing direction again and moving back toward the target position. During the return movement to the intermediate position, which required extension, the flexor muscles showed EMG activity. Again, the findings suggest that these simple forearm movements cannot be ascribed to a rapid shift in the equilibrium point.

These findings suggest the existence of a gradually changing control signal during movement of the forearm from one equilibrium position to another; they are not consistent with the hypothesis of a single step-like shift to a final equilibrium point [6, 12]. Thus, in the transition from the initial to the final position, the alpha motoneuronal activity defines a series of equilibrium positions which constitute a trajectory whose endpoint is the desired final position. It should be emphasized that this hypothesis is based on analysis of forearm movements performed at moderate speeds, and that the character of the control signal may vary with the type of movement.

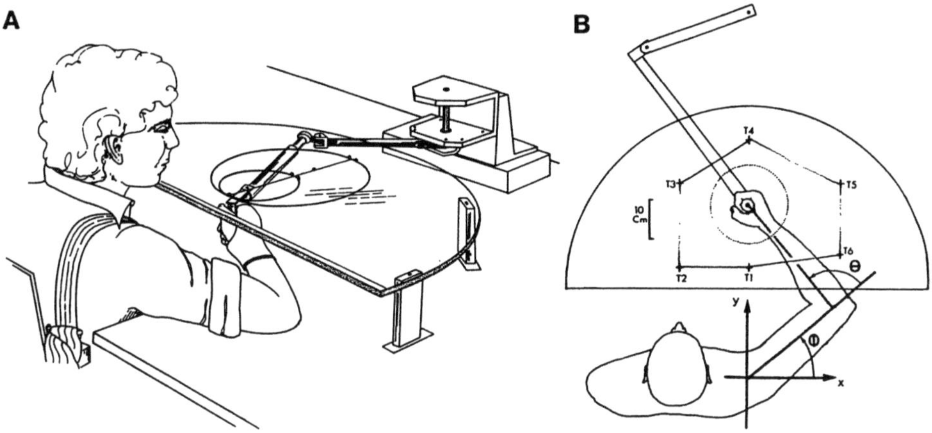

Fig. 4. **A** Oblique, and **B** plan views of a seated subject grasping the handle of the hand-position transducer. The right arm was raised to shoulder level and moved in a horizontal work space. Movement of the handle was measured by way of potentiometers located at the two mechanical joints of the apparatus (J1, J2). The subject was positioned so that J1 lay on the Y axis. The two links of the apparatus were made of aluminium tubes. A clutch was mounted on the joint J1 of the apparatus. When it was activated, the handle was constrained to move in a circular path around J2. (Abend et al. [1])

The hypothesis that movements are centrally represented as gradual shifts in the equilibrium position of the limbs has also been tested for planar multi-joint arm movements [13]. Subjects were asked to perform pointing movements between two targets while gripping the handle of a two-link planar manipulandum. Unexpected activation of the clutch at J1 at the beginning of a movement (see Fig. 4) caused subsequent hand movement to be along an arc with radius equal to the length of the more proximal (J2) link of the manipulandum. The target locations and the timing of the clutch engagement were arranged such that the constrained path took the handle away from the intended path. The clutch could then be released at different times, thereby removing the path constraint.

Following clutch release, a tendency for the path to return to the intended path rather than proceed directly to the final position was observed. In further experiments, handle force was measured. Movement duration was within the 600–1200 ms range. While the clutch was engaged, handle force was found to be always strongly oriented so as to restore the hand to the unconstrained path and not to the end-point of the path [13]. These findings allow us to extend to multi-joint arm movements the hypothesis that trajectories are executed as a sequence of equilibrium points of the hand.

Conclusions and Summary

Our experiments have been directed at understanding the way in which the motor system achieves the complex goal of controlling simultaneously the many degrees of the arm's freedom of movement. This contribution presents experimental findings which reveal some of the control solutions that have evolved for generating coordinated single- and multi-joint movements. Its main focus is on muscle elastic properties, i.e., the field of elastic forces associated with end-point of the limb. This field is described by the stiffness which relates forces to displacements and which is characterized by shape, size and orientation. The hand is at static equilibrium at the center of the field. In the second part of this contribution, we have presented experimental evidence supporting the hypothesis that arm movements are generated and controlled as a gradual shift of the end-point elastic field.

Acknowledgments. This research was supported by National Institute of Neurological Disease and Stroke Research Grant NS09343, National Institute of Arthritis, Metabolism, and Digestive Diseases Grant AM26710, and National Eye Institute Grant EY02621.

References

1. Abend W, Bizzi E, Morasso P (1982) Human arm trajectory formation. Brain 105:331–348
2. Bizzi E, Polit A, Morasso P (1976) Mechanisms underlying achievement of final head position. J Neurophysiol 39:435–444
3. Bizzi E, Accornero N, Chapple W, Hogan N (1982) Arm trajectory formation in monkeys. Exp Brain Res 46:139–143
4. Bizzi E, Accornero N, Chapple W, Hogan N (1984) Posture control and trajectory formation during arm movement. J Neurosci 4:2738–2744
5. Bizzi E, Mussa-Ivaldi FA, Hogan N (1986) Regulation of multi-joint arm posture and movement. Prog Brain Res 64:345–351
6. Cooke JD (1979) Dependence of human arm movements on limb mechanical properties. Brain Res 165:366–369
7. Feldman AG (1966) Functional tuning of the nervous system during control of movement or maintenance of a steady posture. III. Mechanographic analysis of the execution by man of the simplest motor tasks. Biophysics 11:766–775
8. Feldman AG (1974a) Change of muscle length due to shift in the equilibrium point of the muscle-load system. Biofizika 19:534–538
9. Feldman AG (1974b) Control of muscle length. Biofizika 19:749–751
10. Hogan N (1980) Mechanical impedance control in assistive devices and manipulators. In: Proceedings of the Joint Automatic Control Conference, San Francisco, vol 1
11. Hogan N (1984) Impedance control of industrial robots. Robots Comput Integrated Mfg 1:97–113
12. Kelso JAS, Holt KG (1980) Exploring a vibratory system analysis of human movement production. J Neurophysiol 43:1183–1196
13. McKeon B, Hogan N, Bizzi E (1984) Effect of temporary path constraint during planar arm movement. Abst Soc for Neurosci 10, part 1 102.4:337
14. Mussa-Ivaldi FA, Hogan N, Bizzi E (1985) Neural, mechanical, and geometric factors subserving arm posture in humans. J Neurosci 5:2732–2743
15. Polit A, Bizzi E (1978) Processes controlling arm movements in monkeys. Science 201:1235–1237
16. Rack PMH, Westbury DR (1974) The short-range stiffness of active mammalian muscle and its effect on mechanical properties. J Physiol (Lond) 240:331–350

Bimanual Load-Lifting Task. A Model for the Study of Coordination Between Posture and Movement

M. Dufossé [1], M. Hugon [2], J. Massion [1], and Y. Paulignan [1]

Introduction

The study of motor coordination has gained more and more interest over recent years. It is worthy of interest because in everyday life most of our movements are not restricted to a single joint but involve at the same time different segments of the same limb or several parts of the body. Various types of motor coordination have been considered.

First, a combined control over different segments is used for reaching a given goal. An example of this is a reaching movement with the hand toward a given point in space; here a multijoint control is organized by the central nervous system (see Bizzi, this volume); another example on eye and head muscles has been observed [6, 19].

Motor control coordination of the second category occurs when two or several motor actions are combined for the performance of the same motor behavior. Taking a glass with the hand is a combination of two coordinated actions, i.e., reaching for the glass with the arm and shaping the hand to grasp the glass. In this action, two separate control systems come into action with their own sensory cues, and higher level coordinating structures are responsible for the coordination [3, 18]. The coordination of changes in posture and equilibrium associated with the movement being performed belongs to the same category. This type of coordination, however, poses a question which has not yet been solved concerning the hierarchical organization involved. Posture and movement may either be organized by two separate control systems which are coordinated at a higher level, or the postural adjustments may be governed by the movement control system. This will be one of the topics of the present report.

Thirdly, two separate motor acts can be performed at the same time, such as walking and reading a paper. Here the two acts are controlled independently. Some interaction between the two acts does occur, however, and the coordination probably takes place at the highest hierarchical level.

The present report will center on the coordination between posture and movement. A short survey of the literature will illustrate the key problems involved in this topic.

Babinski [4] at the turn of the century was the first to investigate the changes in posture associated with movement in the standing man. He observed that upper trunk and neck movements are accompanied by associated movements of hip

[1] Laboratoire de Neurosciences fonctionelles, CNRS, B.P. 71, F-13402 Marseille Cedex 9, France.
[2] Laboratoire de Psychophysiologie, U.A. CNRS 372, Université de Provence, F-13397 Marseille Cedex 13, France.

Clinical Aspects of Sensory Motor Integration
Ed. by A. Struppler and A. Weindl
© Springer-Verlag Berlin Heidelberg 1987

and knee which result in maintaining the center of gravity at approximately the same place. He called these associated movements "synergies" and observed the loss of these synergies or "asynergy" in cerebellar patients. According to recent observations by Dichgans and Mauritz [11], the lesions were probably located in the vestibulo-cerebellum. These static or near static coordinations between posture and movement were also observed by Martin [20] with arm movement and by Gurfinkel et al. [13] with respiratory movements.

When the movement is fast, dynamic changes of posture are observed which precede the onset of the voluntary movement. When raising the arm, for example, leg muscle activity changes some 50–100 ms before the onset of activation of the prime mover of the arm [5, 7, 9, 10]. These anticipatory postural changes were interpreted as aiming to minimize the perturbation of posture and equilibrium due to movement performance. Hess [14] had already expressed a similar view in proposing that each movement consisted of two components: a teleokinctic one, i.e., the movement directed toward its goal, and an ereismatic one, i.e., the associated changes in posture and equilibrium.

Most of the studies concerning the coordination between posture and movement have been made in the standing man, with the moving segment causing a change in the position of the center of gravity and a dynamic perturbation of posture. One of the difficulties in a study of this type of coordination results from the continuity between the moving segment and the rest of the body, so that a change in myographic activity at a given time is not always related to a well-defined functional action involving equilibrium, posture or movement.

Coordination Between Posture and Movement in a Bimanual Task

An interesting model for the study of the coordination between posture and movement was explored in a previous study by Hugon et al. [15]. When both hands are utilized to manipulate an object, usually one hand is "postural" and supports the object and the other is the manipulatory hand [8]. In the present model, the subject is sitting, and one forearm maintained in a horizontal position is bearing a load. The subject is asked to lift the load with the other hand. The voluntary lifting of the load results in a perturbation of the load-bearing arm. However, notwithstanding the perturbation, the forearm position remains approximately at the same place. This is due to an "anticipatory" inhibition of the flexors of the postural forearm which precedes the onset of unloading (Fig. 1). The inhibition starts at the same time as the increase in EMG activity of the flexors of the lifting arm, indicating a simultaneous control of both sides. Thus in the situation of a bimanual task, the anticipatory adjustment of the forearm position minimizes the effect of the perturbation caused by the voluntary movement of the other arm. The aim of the local adjustments of posture in a bimanual task is thus comparable to the aim of the postural adjustments associated with voluntary movements in the standing man, except that in the bimanual task there is no need for a control of the whole body equilibrium.

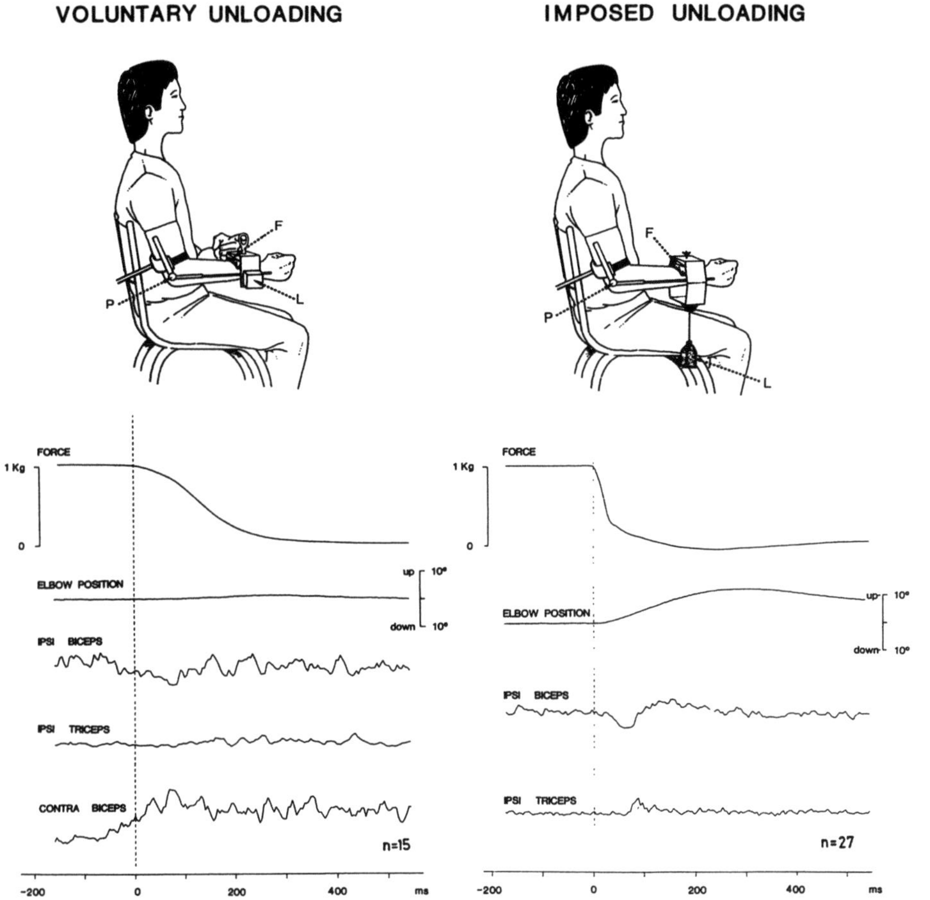

Fig. 1. Comparison between imposed and voluntary unloading in a bimanual task. The forearm is supporting a strain-gauge equipped platform (*F*) bearing the load. A potentiometer (*P*) is placed at elbow level. EMG of biceps is rectified and integrated with a leaky integrator (time constant: 10 ms). Note the reduced forearm rotation and the anticipatory biceps inhibition when the load (*L*) is released by a voluntary movement of the arm

Mechanisms Involved in the Coordination

The aim of the research that will be outlined here is to gain an insight into the central organization of this coordination.

Two main hypotheses concerning the coordination between posture and movement have been proposed (Fig. 2):

1. Two parallel and separate control systems could exist, the one for posture and the other for movement. The control of both could be coordinated at a higher level, by a timing signal (see start on Fig. 2; right). This model is in line with those proposed by Belenkiy et al. [5] and Cordo and Nashner [10].

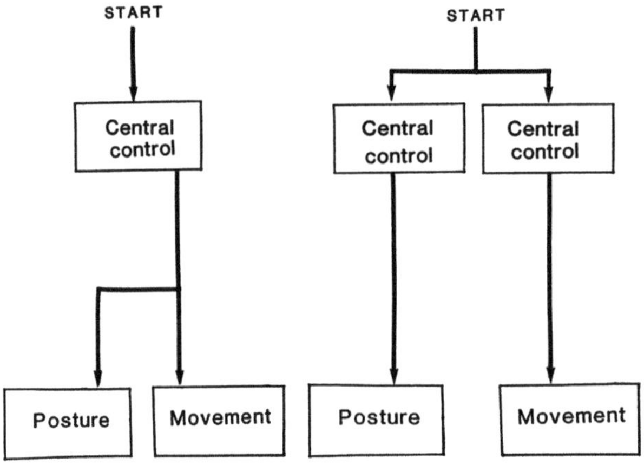

Fig. 2. Two models for the coordination between posture and movement

2. A single control system might exist which is also triggered by a timing signal (see start on Fig. 2; left), and which might control not only the timing of the coordination but also the force of the voluntary movement on one hand and the intensity of the postural adjustment on the other. This could be an example of a coordination through a corollary discharge or internal collateral. This model was proposed by Gahery and Massion [12] and by Alstermark [1] on the basis of experiments on cats.

In order to investigate this problem, new paradigms were tested on sitting subjects with one forearm maintained in a horizontal position, the instruction being to keep the forearm horizontal. A strain-gauge equipped platform was placed on that forearm and supported a 1 kg weight through an electromagnet. Unloading of the arm was obtained by the release of the load through disconnection of the electromagnet (Fig. 1). Elbow angle and integrated EMG from flexors of both arms were recorded.

The three following new conditions were explored. For each of them the load release was obtained by disconnecting the electromagnet.

1. An external cue on the timing of unloading (tone preceding unloading with a fixed interval) was given.
2. The subject was asked to press a key with the thumb of one hand in order to release the load supported by the other arm. This voluntary movement needs only a gentle force.
3. The subject was asked to lift a load of 1 kg with one hand. The very onset of unloading provokes the release of the load supported by the other arm. Here the voluntary lifting movement is comparable to that performed in bimanual coordination, but the link between lifting and unloading is made through an electronic circuit.

In regard to the single control system hypothesis, one might predict that the only effective paradigm for eliciting the anticipatory inhibition of the postural arm will be the situation where the subject performs a voluntary load-lifting movement (condition 3). With the two other conditions, either no movement con-

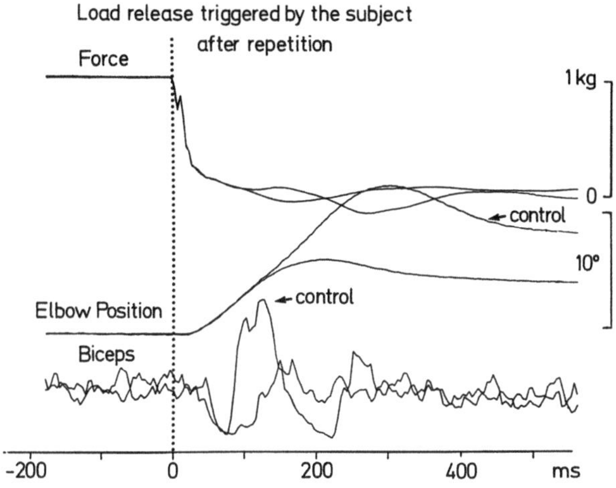

Fig. 3. Lack of anticipation when the load release is induced by pressing a key. Pressing a key with the thumb of the left hand releases the load supported by the right forearm. Averaged curves over 20 trials. Note the lack of anticipation, even after 5 series of 20 trials. Arm rotation is reduced with respect to the unpredictable situation (*control*), and the late biceps activation disappears from the biceps EMG trace

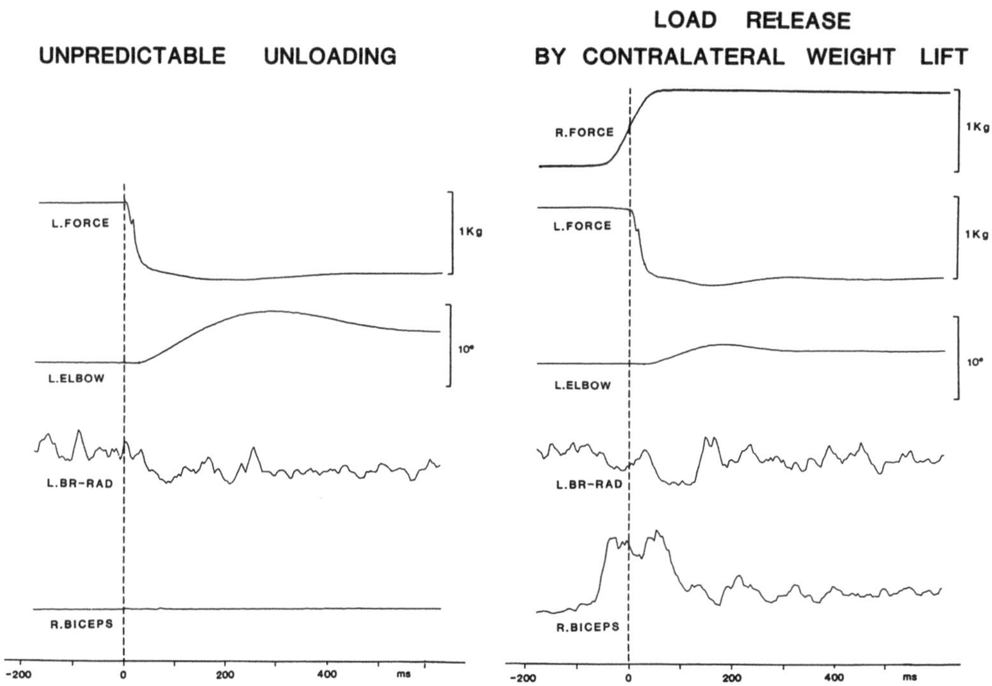

Fig. 4. Role of force control in coordination. A 1-kg load supported by a force platform is lifted by the left hand (*upper trace, right*). The load release on the right forearm is triggered when the load-lifting by the left arm has reached 200 g (*second trace, right*). Notice (a) the reduced arm rotation with respect to the unpredictable unloading shown on the left, (b) the anticipatory inhibition of the brachio-radialis, and (c) the later unloading reflex increasing the inhibition

trol exists (condition 1) or the movement control signal might not provide the force signal required for postural control (condition 2).

In the present experiments, the subjects were subjected to series of 20 trials which were repeated 4–6 times with one of the three new conditions. These results were compared to the reference paradigm, i.e., unloading imposed in an unpredictable way, and to the normal bimanual coordinated action, i.e., unloading caused by the other arm voluntary movement.

When an external cue on the unloading time was given or when unloading was elicited by pressing a key with the thumb, no anticipatory inhibition of the flexors of the postural arm was observed, even after up to six series of 20 trials. The amplitude of the forearm rotation was reduced with respect to the unpredictable unloading by 20–40%. However, the slope of the curve remained the same during the first 120 ms after the onset of unloading, indicating that the reduced elbow rotation did not result from an anticipatory inhibition of the forearm flexors but from a more efficient reflex action. An increase in the amplitude and duration of the unloading reflex was seen as well as a decrease in the late excitatory response described by Angel et al. [2] (Fig. 3).

When the third condition, i.e., lifting a 1-kg load with one hand, triggered the release of the load supported by the other forearm, then an anticipatory inhibition of the forearm flexors did appear, at a time when the EMG activation associated with the voluntary lifting movement started (Fig. 4). The anticipatory inhibition appeared after a few trials and improved markedly during the first two series of 20 trials. The forearm rotation could be reduced as much as during the normal bimanual lifting task. The reduced forearm flexor activity usually increased 20–30 ms after the onset of unloading. This was interpreted as indicating that an unloading reflex is added to the anticipatory inhibition of the flexors.

Functional Implication

Bimanual tasks are part of everyday experience from childhood. The anticipatory changes in the forearm position found in the type of coordination under study can be explained by the fact that the manipulation of heavy objects causes a perturbation of the forearm position so that the anticipatory change of position compensates in advance for these perturbations. As a result, the manipulation is performed with increasing accuracy in terms of force and displacement. The anticipatory postural changes are not due to a special long-term training for each new task. In the present series, when a subject lifts a weight supported by the other forearm, one or two trials usually suffice to obtain the optimal level of coordination. There is no need for a specific reinforcement such as information about the coordination obtained. The improvement of the coordination is unconscious as in the other examples of coordination between posture and movement. We are dealing, not with a voluntary bimanual control, but with an automatic adjustment of the position of one forearm as a function of the voluntary movement executed by the other arm.

The experimental series reported here was designed to search for a specific factor responsible for the anticipatory forearm inhibition. The lack of anticipation when the subject was provided with an external cue informing him on timing of the unloading was at first sight surprising. An inaccurate evaluation of the time was a possible explanation for the lack of anticipation. More surprising was the lack of anticipation with a load release induced by pressing a key. Here, an internal control signal, associated with the thumb movement, was available and was in fact not "utilized" for the control of the anticipatory positional adjustment of the other forearm. These negative results argue against the hypothesis that two separate control systems might exist in the present type of coordination, the one for the voluntary movement and the other for the postural adjustment. Finally, the positive result obtained when the voluntary movement consists of lifting a 1-kg load indicated that there is a relationship between the force control on one side and the adjustment of forearm position on the other. This result is in agreement with the hypothesis that a single control system for the movement control may exist, the postural adjustment depending on an efferent copy of the movement control; however, the fact that a short learning occurs before the optimization of the anticipatory forearm position adjustment indicated that the coordination depends on a gate and gain control. The cerebellum could be a good candidate for this function [17]. Do these results indicate that the only way to control posture in coordination with voluntary movements is by collaterals from the control pathway for movement and that no separate control of posture and movement occurs in a coordinated action? In agreement with Alstermark [1] and Ioffe et al. [16], we would like to propose that the two modes of coordination coexist and depend on the type of motor activity. When, for example, the adjustment of posture accompanying a voluntary movement results in a change of the whole body position in order to prevent a disequilibrium, as in the experiments by Nashner [21] and Gurfinkel [5], two separate control systems are probably operating. This mode of control would allow a control of the whole posture before the onset of the movement. A simple action on posture by internal collaterals of the movement control pathways would be inappropriate because it would not permit a change of the whole body posture prior to the movement. In fact, it might be proposed that the two modes are not mutually exclusive and that they may even coexist during the same motor act, depending on the postural function being regulated during the movement.

Summary

A model was developed for the study of the coordination between posture and movement. Human subjects sitting in a chair performed a load-lifting task, one forearm bearing the load (postural forearm), the other lifting the load (moving forearm). In this task, an "anticipatory" inhibition of the flexors of the postural forearm precedes the onset of unloading.

Two main hypotheses concerning the coordination are discussed. The first predicts two parallel and separate control systems, one for posture and the other for movement. The second suggests that postural control can be exercised by internal collaterals of the command signal to the moving segment.

Results from different experimental conditions suggest that for this load lifting task, the control of posture is exercised by internal collaterals of the command pathways for the lifting movement.

Acknowledgment. This work was supported by a grant from INSERM (PRC 133025).

References

1. Alstermark B (1984) Postural adjustments and propriospinal systems. Neurosci Lett [Suppl 18]:S15
2. Angel RW, Eppler W, Iannone A (1965) Silent period produced by unloading of muscle during voluntary contraction. J Physiol (Lond) 180:864–870
3. Arbib MA (1981) Perceptual structures and distributed motor control. In: Brooks VB (ed) Handbook of physiology. Section 1, The nervous system. vol II, Motor control, part 2. Amer Physiol Soc, Bethesda, pp 1449–1480
4. Babinski J (1899) De l'asynergie cérébelleuse. Rev Neurol 7:806–816
5. Belenkiy VE, Gurfinkel VS, Paltsev EI (1967) On elements of control of voluntary movements. (in Russian) Biofizica 12:135–141
6. Bizzi E, Kalil RE, Tagliasco V (1971) Eye-coordination in monkeys: evidence for centrally patterned organization. Science 1973:452–454
7. Bouisset S, Zattara M (1981) A sequence of postural movements precedes voluntary movement. Neurosci Lett 22:263–270
8. Brinkman C (1981) Lesions in supplementary motor area interfere with a monkey's performance of a bimanual coordination task. Neurosci Lett 27:267–270
9. Clement G, Gurfinkel VS, Lestienne F, Lipshits MI, Popov KE (1984) Adaptation of postural control to weightlessness. Exp Brain Res 57:61–72
10. Cordo PJ, Nashner LM (1982) Properties of postural adjustments associated with rapid arm movements. J Neurophysiol 47:287–302
11. Dichgans J, Mauritz KH (1983) Patterns and mechanisms of postural instability in patients with cerebellar lesions. In: Desmedt JE (ed) Motor control mechanisms in health and disease. Raven, New York, pp 633–659
12. Gahery Y, Massion J (1981) Coordination between posture and movement. TINS 4:199–202
13. Gurfinkel VS, Kots JM, Paltsev FI, Feldman AG (1971) In: Gelfand IM, Gurfinkel VS, Fomin SSV, Tsetlin ML (eds) Models of the structural functional organization of certain biological systems. MIT Press, Cambridge, MA, pp 382–395
14. Hess WR (1943) Teleokinetisches und ereismatisches Kräftesystem in der Biomotorik. Helv Physiol Pharmacol Acta 1:C62–C63
15. Hugon M, Massion J, Wiesendanger M (1982) Anticipatory postural changes induced by active unloading and comparison of passive unloading in man. Pflügers Arch 393:292–296
16. Ioffe ME, Frolov AA, Gahery Y, Frolov AG, Coulmance M, Davydov VI (1982) Biomechanical study of the mechanisms of postural adjustment accompanying learned and induced limb movements in cats and dogs. Acta Neurobiol Exp 42:469–482
17. Ito M (1984) The cerebellum and neural control. Raven, New York
18. Jeannerod M (1984) The contribution of open-loop and closed-loop control modes in prehension movements. In: Kornblum S, Requin J (eds) Preparatory states and processes. Lawrence Erlbaum, Hillsdale, pp 323–337
19. Lestienne F, Vidal PP, Berthoz A (1984) Gaze changing behaviour in head restrained monkey. Exp Brain Res 53:349–356
20. Martin JP (1967) The basal ganglia and posture. Pitman, London
21. Nashner LM (1983) Analysis of movement control in man using the movable platform. In: Desmedt JE (ed) Motor control mechanisms in health and disease. Raven, New York, pp 607–619

Neuromotor Psychophysical Aspects of Central Programming and Peripheral Regulation of Movement in Humans

J. N. SANES [1]

Introduction

A continuing issue in motor control research is the relative contributions of central motor commands and somesthetic peripheral inputs in regulating movement. There is experimental evidence that supports the seemingly contradictory viewpoints that movements can be controlled entirely by central commands (e.g. [6, 9]) or that peripheral inputs provide significant information for accurate motor performance [11, 12]. In part, the differences between the two sets of results may be resolved by recognizing that the dependence of movements on sensory inputs is related to the intended speed and accuracy of a movement. With this formulation in mind it may be argued that rapid and relatively uncontrolled movements can be programmed centrally [4, 10, 13], but that slow tracking or accurately intended movements make substantial use of somesthetic inputs [11, 16].

Whereas there are controversial views about the experimental results on somesthetic influences on movement control, there is little disagreement concerning the clinical impairments in movement skills that are exhibited by patients with sensory peripheral neuropathy [1, 7, 10, 15]. Despite extensive description of the clinical motoric deficits in patients with sensory neuropathy there has been little experimental analysis of the deficits observed in such patients.

In what follows, some recent observations will be presented on patients with a large-fiber sensory neuropathy. These results indicate that the patient's deficit in motor control is accompanied by disruption of the perceptual processes of kinesthesis.

Patients

The patients which were studied had originally presented with subacute onset of a predominantly sensory neuropathy. Neurological examination revealed reduced sensation to light touch, temperature, pin prick, limb position, and vibration sense. Graphesthesia and stereognosis were invariably impaired. A striking ataxia of the hands and choreoathetoid movements of the fingers were present when the patients outstretched their arms. Muscle strength was nearly normal and deep tendon reflexes were absent. Electromyography showed mild denervation in only two patients while motor nerve conduction velocity was normal in all but 1 of the 9 patients. Sensory potentials were absent in all patients.

[1] Human Motor Control Section, Medical Neurology Branch, NINCDS, NIH, Building 10, Room 5N-226, Bethesda, MD 20892, USA.

Clinical Aspects of Sensory Motor Integration
Ed. by A. Struppler and A. Weindl
© Springer-Verlag Berlin Heidelberg 1987

Methods

Flexion-extension movements of the wrist were evaluated while the hand was coupled to a torque motor. The patients were unable to see their hand, but instead, viewed a video display on which a target cursor (experimenter-controlled) and position cursor (patient-controlled) appeared. The patients' task was to always maintain alignment between the target and position cursors. In these experiments, patients were required either to move a specific angular rotation or to maintain a steady position with or without visual guidance. Accuracy of movement or posture and EMG activities were evaluated.

Results

Although the patients with a large-fiber sensory neuropathy could hold a relatively steady position when posture was visually guided, they were greatly impaired in their postural abilities when vision could no longer be used (Fig. 1). When visual guidance was removed, the patients' hand often moved quickly away from the initial position. The drift could continue in the same direction or it could be reversed. The amount of average positional drift for normal controls was negligible ($<1°$), whereas for patients it ranged from $1°–17°$. In addition, the mechanical load opposing postural maintenance was not a major factor in determining the magnitude or direction of drift by the patients. Thus, although the average drift was lowest for 3 of the patients when no load opposed postural maintenance,

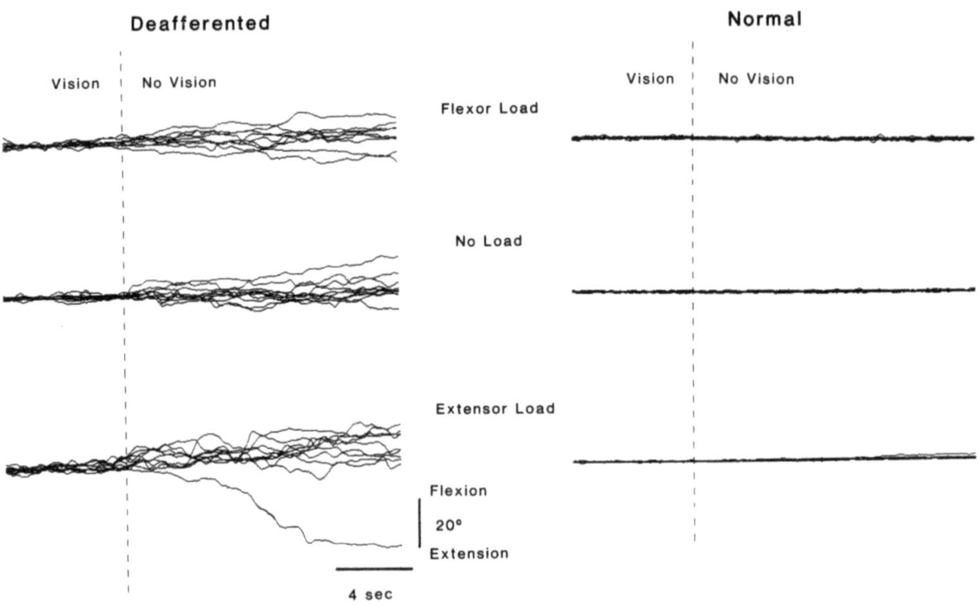

Fig. 1. Postural control of a patient with large-fiber sensory neuropathy and a normal control. Several trials are superimposed from each of three load conditions

there was no consistent relationship (except in one patient) between opposing load and direction or magnitude of drift. The deficits in maintaining steady postures were further exhibited by the impaired ability of patients to adequately compensate against loads that simulated spring tension. In these experiments, patients exhibited a variety of responses when they were required to maintain a steady posture against a spring load without visual guidance. The patients sometimes yielded to the spring, maintained stability at the required tension or increased tension against the spring load. These responses often occurred in a random fashion. Similar deficits in accuracy were observed when these patients performed intended movements of different sizes, especially when these movements were performed without visual guidance. In the initial phases of movement, the patients tended to overshoot the intended target and then often exhibited substantial instabilities and inaccuracy during the end-point positioning phase of movement. The initial overshoot and end-point error exhibited by patients was 2–3 times greater than that observed for normal controls.

Examination of the muscle activity of the wrist flexors and extensors revealed a possible basis for inaccuracies in postural and movement control. As Fig. 2 illustrates, the patients had significant difficulties in maintaining a constant EMG in either antagonist or agonist muscles. In the examples provided in Fig. 2, pa-

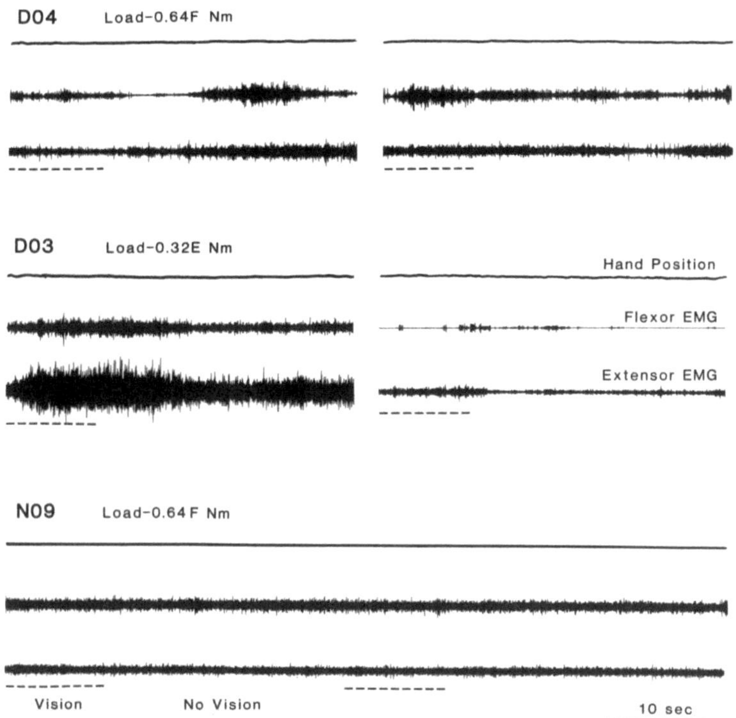

Fig. 2. Muscle activity patterns during postural maintenance. Muscle co-contraction and fluctuations in antagonist muscles were common while patients attempted to maintain a steady posture. These EMG fluctuations persisted even when postural maintenance was assisted by vision. Note the stable amounts of muscle activity exerted by the normal control

tients were required to maintain a constant posture against one of several constant loads. The common strategy for doing this is shown by the EMGs from a normal subject in the lower section; one simply sets a level of activity in agonist and antagonist muscles and maintains it. In contrast to setting a constant level of EMG, patients maintained a relatively stable posture in quite a different manner. The muscle activities showed considerable fluctuations from high to low levels of co-contraction, but the ratio of agonist to antagonist muscle activity remained relatively constant. This impairment was unrelated to the loading conditions or to the availability of visual guidance. Furthermore, the patients seemed unable to control their muscle activities since the EMGs fluctuated from moment to moment during postural maintenance and from trial to trial when active movements were performed. One implication of these results is that somesthetic inputs provide information for initial calibration and re-calibration (e.g., during postural control) for central specification of muscle activity. Without such proprioceptive signals the CNS must rely on alternative mechanisms to update motor brain regions. Seemingly, even these alternatives, such as memory, visual or auditory feedback, are not sufficient to compensate for the loss of proprioceptive afferent inputs.

Effort Sense and Deafferented Patients

The observation that patients with a large-fiber sensory neuropathy have difficulties in accurately setting and maintaining phasic or tonic levels of muscle activity suggests that they lack an awareness of such muscle activity. Thus, it would appear that these patients do not have an accurate sense of muscular effort. This notion, however, has not been supported by the observations of Gandevia and McCloskey [5] and Rothwell et al. [10]. Nevertheless, with the availability of such a large sample of deafferented patients we decided to reexamine the sense of muscular effort in patients with a selective peripheral neuropathy.

Methods

Unimanual and bimanual sense of effort was assessed. Two tasks were employed for the unimanual evaluation. In the first, patients were required to maintain a constant posture of 0° wrist rotation while torques were changed every 4–5 s to one of six levels opposing flexion (0–0.8 Nm). In the second task, the right hand of a patient was first pushed against a mechanical stop (45° extension) by a torque opposing flexion and the patient was then required to move the hand to 0° of flexion. In both of these tasks the patients had to discriminate whether the torque had changed in comparison to the torque of the previous trial. Patients responded "more", "less", or "same" on all trials, and the proportion of correct responses was tabulated for trials when torque actually changed and for trials when there was no change in torque. Typically, 240 trials were performed for each task. A bimanual matching task was also used in which a constant torque opposing flexion (0.16–0.8 Nm) was continuously applied to the left hand and a varying torque opposed flexion of the right hand. Subjects were required to judge which hand had the larger applied torque. A modified Method of Limits procedure was used to determine ascending and descending thresholds. With this procedure torque on the matching right hand started at 0 Nm and was raised in 0.5-Nm steps until the patient made four successive responses that torque was greater on the matching arm. Then torque was lowered in 0.5-Nm steps until four responses of torque greater on the reference arm occurred. Increases and decreases in torque were done for 5–10 approximations of the ascending and descending thresholds.

Effort Sense Results

The results from these experiments showed that patients with a large-fiber peripheral neuropathy have a severely impaired sense of muscular effort (Figs. 3 and 4). This was apparent in both the unimanual and the bimanual tasks. Closer inspection of the data revealed that these patients do have a crude sense of effort. In particular in the unimanual tasks, most errors that patients and normal controls

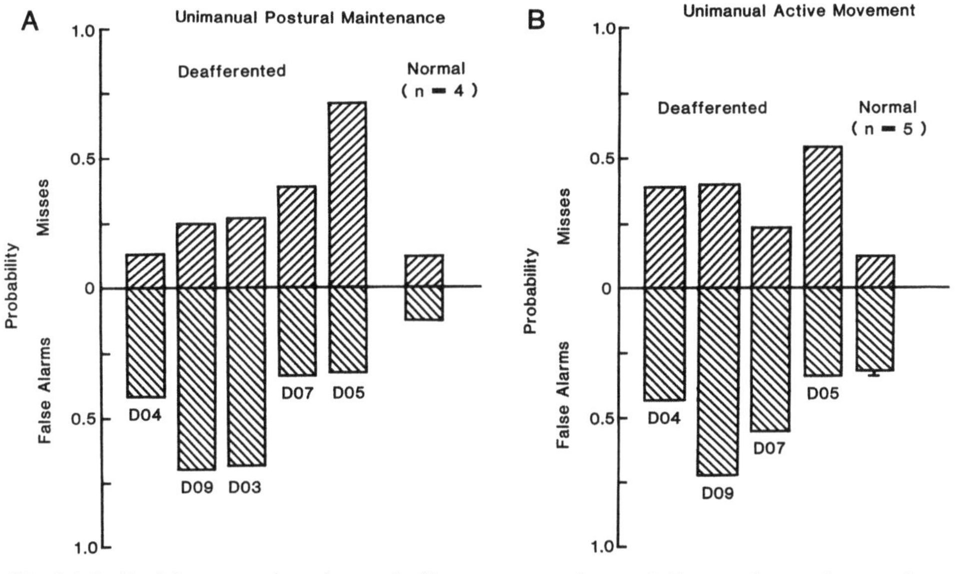

Fig. 3 A, B. Decision errors in unimanual effort sense experiment. **A** Errors when patients and a group of normals were required to detect differences or similarities between torques on successive trials while a posture of 0° of flexion was maintained. Misses refer to the cumulative probability of incorrect decisions when torque changed on two successive trials. The incorrect decision could take the form of responding that no change in torque had occurred or that torque increased when torque actually decreased (or the converse). False alarms refer to the cumulative probability of the incorrect statement that torque changed (either increased or decreased) when in fact torque remained constant on successive trials. **B** Errors when patients and a group of normals were required to detect torque shifts during active movement. Probability of misses and false alarms as defined in **A**. The absence of errors bars for normal subjects indicates extremely low between-subject variability

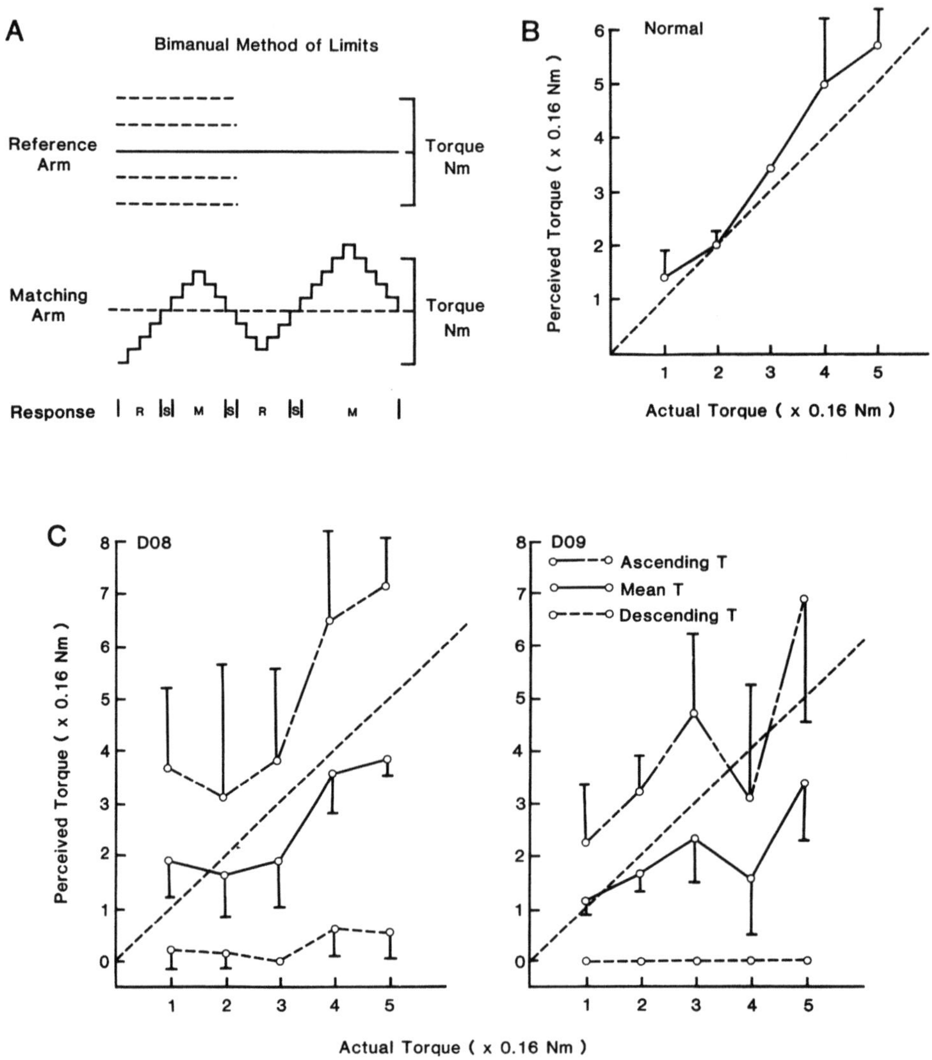

Fig. 4 A–C. Performance during bimanual effort detection. **A** Method used to assess weight perception using two arms simultaneously. The *upper section* (reference arm) depicts that one of five torques was chosen to oppose flexion continuously. The *lower section* illustrates a sample run to derive ascending and descending thresholds (see text for details). **B, C** Performance of normal subjects ($n=5$) and two patients with large-fiber sensory neuropathy. The mean threshold is shown for the group of normals (the ascending and descending thresholds were not different), whereas the ascending, descending and average thresholds are illustrated for patients. Note that both ascending and descending thresholds of both patients deviated from unity

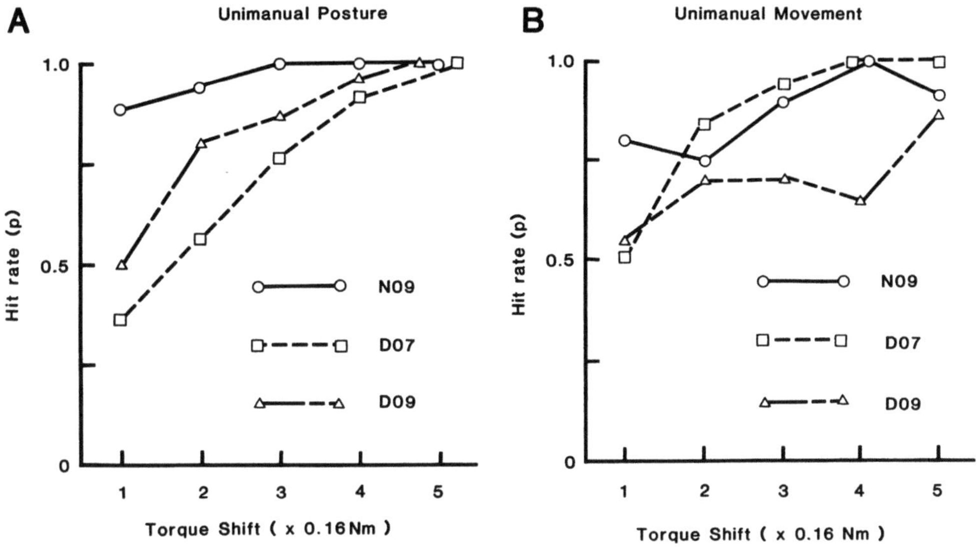

Fig. 5. A, B. Cumulative correct decisions versus amount of torque shift. The cumulative probability of successful decisions in both unimanual tasks plotted against the magnitude of torque shift on successive trials is shown for one normal subject and two patients. For both the patients and the normal subject the error rate was inversely proportional to the difference in torque on successive trials. Patient performance equaled or was close to that of the normal for torque shifts greater than 0.5 Nm but was significantly worse when the difference between torques was less than 0.5 Nm

made were when the change in torque was less than 0.5 Nm (Fig. 5), but of course patients made more of these errors.

Discussion

The composite results of these experiments illustrated that patients with a large-fiber sensory neuropathy were grossly impaired in their abilities to make voluntary movements or to maintain constant posture, especially when they were deprived of visual guidance. These results are compatible with those of Rothwell et al. [10]. The basis for the disordered motor performance in these patients appeared to be an inability to precisely control muscle activity for prolonged periods. Although it seems that the CNS programs levels of EMG to attain intended final positions [8], this strategy also appears to depend heavily on proprioceptive inputs. We have argued before that both the initial EMG bursts, accompanying rapid movements, and the steady-state levels of EMG at final position require somesthetic inputs [15]. The primary evidence for this is extreme variation in EMG activity at movement onset, even when a movement is performed repetitively, and while an end point is maintained (Fig. 2).

The observation of an unstable EMG during movement and posture is not an unexpected concomitant of disordered movements in patients with neuropathy.

However, the mechanism for this deficit has been unclear. Previously, we have noted that postural control deteriorates rapidly when deafferented patients lose feedback (visual) to correct inaccuracies in movement. It seemed as if the memory trace for a posture required nearly continuous updating. As for active movements, subjects must retrieve from memory the appropriate signals to move rapidly toward an intended location. This mechanism is apparently defective in deafferented patients since a normally configured triphasic EMG pattern is not always observed in patients with a large-fiber peripheral neuropathy [10, 15]. It is also possible that the diminished sense of muscular effort shown here can explain the fluctuations in muscle activity observed during final position control.

The impaired effort sense in the patients described here would seem to be at variance with the results of Gandevia and McCloskey [5] and those of Rothwell et al. [10], indicating that effort sense is primarily mediated by the mechanism of corollary discharge. In this scheme, it is assumed that motor commands, possibly from motor cortex, in addition to being directed to alpha motor neurons, are sent also to brain regions concerned with sensation (possibly the somatic sensory cortex). The current data are only partially inconsistent with such a scheme, insofar as only the deafferented patients exhibited impairment in effort sense for loads less than 0.5 Nm. Discrimination between larger loads was nearly equivalent in normal subjects and the deafferented patients. These results suggest a binary process for effort perception whereby recognition of small torques requires somatic sensory inputs while perception of large torques is independent of proprioceptive inputs. This conclusion is related to the mechanisms of central control for large and small movements, whereby performance of only small movements is disproportionately affected by disturbances in proprioception[11, 12, 14, 15]. These psychophysical results are paralleled by neurophysiological studies by Fromm and Evarts [2, 3] on pyramidal tract neurons (PTNs) in the monkey motor cortex. In these experiments, similar proprioceptive inputs occurring during performance of small or large movements had different effects depending on the movement size. The kinesthetic stimulus greatly increased PTN discharge when occurring during small movements but had relatively little effect during large movements. It is possible that the pathways for kinesthetic responsiveness are blocked during performance of large movements or appreciation of large changes in torques, thereby requiring central mechanisms to judge movement efficacy.

The nature of the residual sense of effort in deafferented patients could indeed be mediated by a mechanism of corollary discharge. However, it is possible that the remaining afferents (those with small diameter axons) signal crude information about muscle contractions. These afferents include those signaling chemical changes in muscle, which undoubtedly are greater for larger than for small muscle contractions.

Summary

The importance of signals from the central nervous system and those from the periphery for voluntary movement was evaluated in patients with a selective large-fiber sensory neuropathy. Motor performance of these patients was generally impaired, though performance deficits were greatest when movements were made without visual guidance. The patients were unable to maintain steady postures, move accurately against unpredictable loads or reproduce movements of various sizes. Examination of muscle activities accompanying the performance deficits revealed irregular electromyograms (EMG) during attempts to maintain constant postures and inconsistent bursting of muscle activity during voluntary movement. The basis for the difficulty in central specification of muscle activity appeared to be an inability to sense muscular effort.

References

1. Foerster O (1927) Schlaffe und spastische Lähmung. In: Bethe A, Bergman GV, Embden G, Ellinger A (eds) Handbuch der normalen und pathologischen Physiologie, vol 10. Springer, Berlin Heidelberg New York, pp 900–901
2. Fromm C, Evarts EV (1977) Relation of motor cortex neurons to precisely controlled and ballistic movements. Neurosci Lett 5:259–265
3. Fromm C, Evarts EV (1978) Motor cortex responses to kinesthetic inputs during postural stability, precise fine movement and ballistic movement in the conscious monkey. In: Gordon G (ed) Active touch: the mechanism of recognition of objects by manipulation. Pergamon, Oxford, pp 105–117
4. Hallett M, Shahani BT, Young RR (1975) EMG analysis of stereotyped voluntary movements in man. J Neurol Neurosurg Psychiatry 38:1154–1162
5. Gandevia SC, McCloskey DI (1977) Sensations of heaviness. Brain 100:345–354
6. Kelso JAS, Holt KG (1980) Exploring a vibratory system analysis of human movement production. J Neurophysiol 43:1183–1196
7. Landry O (1855) Mémoire sur la paralysie du sentiment d'activité musculaire. Gaz Hôpitaux Civils Militaires 28:269–271
8. Lestienne F, Polit A, Bizzi E (1981) Functional organization of the motor process underlying the transition from movement to posture. Brain Res 230:121–131
9. Polit A, Bizzi E (1979) Characteristics of motor programs underlying arm movements in monkeys. J Neurophysiol 42:183–194
10. Rothwell JC, Traub MM, Day BL, Obeso JA, Thomas PK, Marsden CD (1982) Manual motor performance in a deafferented man. Brain 105:515–542
11. Sanes JN, Evarts EV (1983) Effects of perturbations on accuracy of arm movements. J Neurosci 3:977–986
12. Sanes JN, Evarts EV (1983) Regulatory role of proprioceptive input in motor control of phasic or maintained voluntary contractions in man. In: Desmedt JE (ed) Motor control mechanisms in health and disease. Raven, New York, pp 47–59
13. Sanes JN, Jennings VA (1984) Centrally programmed patterns of muscle activity in voluntary motor behavior of humans. Exp Brain Res 54:23–32
14. Sanes JN, Mauritz K-H, Evarts EV, Dalakas MC, Chu A (1984) Motor deficits in patients with large-fiber sensory neuropathy. Proc Natl Acad Sci USA 81:979–982
15. Sanes JN, Mauritz K-H, Dalakas MC, Evarts EV (1985) Motor control in humans with large-fiber sensory neuropathy. Human Neurobiol 4:101–114
16. Woodworth RS (1899) The accuracy of voluntary movement. Psychol Rev 3:1–114

IX. Effects of Growth, Degeneration and Regeneration on the Sensory Motor System

Neurologically Effective Nerve Growths in the Mammalian Brain: Recent Work of Tsukahara and Kawaguchi

J. C. ECCLES [1]

Until recently it was generally believed that all the embryonic know-how had been lost in the mammalian post-natal period so that there was no regeneration of the mammalian brain and spinal cord after injury. But now we have the encouraging prospect arising from well-designed experiments that there is a considerable regenerative capacity in several regions of the mammalian brain, even in adults. Optimistically, one can predict that we are only at the beginning of an enterprise in which various surgical procedures plus rehabilitation therapy, with, for example, local administration of nerve growth factors, will be able to reduce some of the disabilities suffered by patients with lesions of the brain and spinal cord. Some experiments of Tsukahara and Kawaguchi have been selected for review because they have demonstrated not only the histological evidence of regeneration, but also that it is functionally effective.

The experiments of Tsukahara [10] are built around the spatial relations of two types of synapses on red nucleus neurons. The input from the interpositus nucleus is on the soma and adjacent proximal dendrites, whereas the cortico-rubral fibres make synapses on the more distal dendritic zones. Its unique value also derives from the fact that regenerated synapses are tested by intracellular recording of the excitatory postsynaptic potentials (EPSP) they generate.

Figure 1 illustrates the normal monosynaptic responses of a red nucleus neurone evoked by cerebral peduncle stimulation (Fig. 1 A) and by nucleus interpositus (IP) stimulation (Fig. 1 B). A comparison of the two gives an excellent example of the electrotonic slowing of EPSPs generated on distal dendrites (Fig. 1 A) (cf. Fig. 1 F).

When the interpositus nucleus in kittens or cats is destroyed, it is found that after 2–3 weeks, stimulation of the cerebral peduncle (CP) results in an EPSP (Fig. 1 C) with superposition of a newly developed fast "somatic" type of EPSP on the original slow dendritic type (Fig. 1 A). It was postulated that axonal sprouting of the CF fibers occupied the vacant sites left by the degenerating IP synapses, as is indicated in Fig. 1 F. The histograms of Fig. 1 D and E show the change in the summit times of the EPSPs for the whole series of experiments. This postulated new growth of synapses on the soma and proximal dendrites has been corroborated by rigorous histological studies by Murakami et al. [9], using horseradish peroxidase (HRP) techniques to demonstrate that normally the cortico-rubral synapses are located on small distal dendrites, but some weeks after interpositus destruction many synapses are on the larger dendrites and even on the soma. Destruction of the cortico-rubral synapses results in three types of new

[1] Max-Plank-Institut für Biologische Chemie, D-3400 Göttingen, FRG.

Clinical Aspects of Sensory Motor Integration
Ed. by A. Struppler and A. Weindl
© Springer-Verlag Berlin Heidelberg 1987

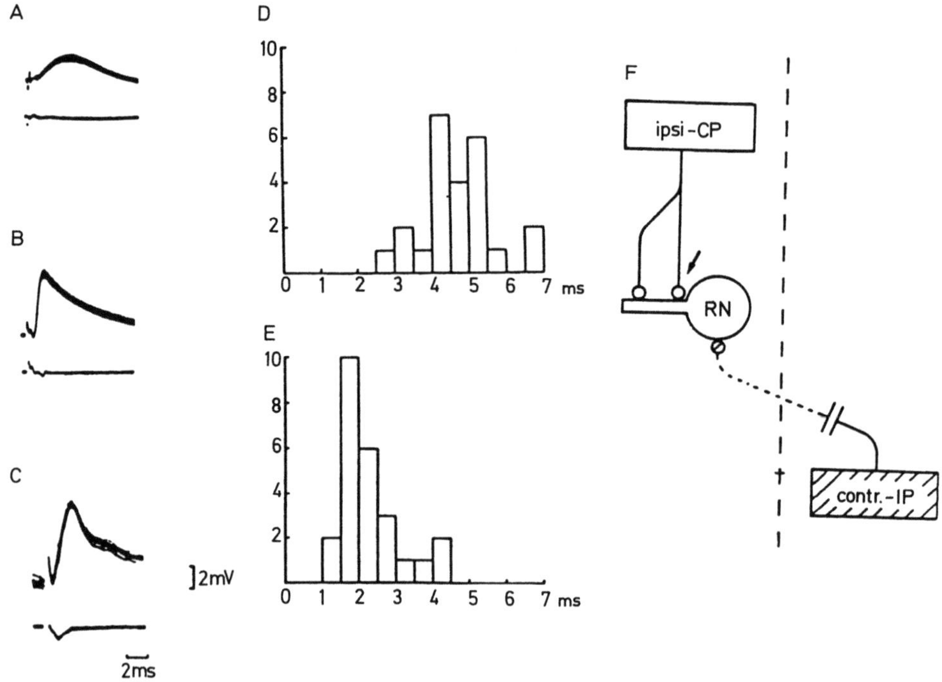

Fig. 1 A–F. Intracellular responses of the red nucleus neurons. *Upper traces* are intracellular responses in red nucleus (*RN*) neurons, while the *lower traces* show the corresponding field potentials recorded at a just extracellular position. **A** Cerebral peduncle (*CP*) EPSP. **B** Interpositus and an (*IP*) EPSP. **C** CP EPSP after IP destruction. Time and voltage calibrations for all intra- and extracellular responses are shown in **C. D, E** The histograms in **D** (normal cats) and **E** (after IP lesion) illustrate the frequency distribution (number of cells on the ordinate) of the "time-peak" of CP EPSP. **F** Illustrative diagram of pathways with *arrow* indicating a newly grown synapse on or close to the soma. (Tsukahara et al. [11, 12])

growth to occupy the vacated synaptic sites with growth distances of hundreds of microns [12].

Kawaguchi and associates have investigated the remarkable regeneration that occurs when the tract from the cerebellar nuclei to the thalamus is severed. This tract can be seen in the line drawing of Fig. 2 from the dentate nucleus (DE) to the thalamus with complete decussation. In the original experiments on kittens [3, 4] the tract was cut in a hemicerebellectomy removing the cerebellar nuclei on one side. It was shown that there developed a pathway from the intact cerebellar nucleus (the dentate) to the ipsilateral ventrolateral (VL) thalamic nucleus that is never observed normally. Both electrophysiological and HRP tract tracing methods demonstrated that the fibres of the intact cerebellar nuclei sprouted to innervate the denervated ipsilateral VL thalamus and that this innervation was powerful enough to evoke responses of the ipsilateral frontal motor cortex, which however were smaller than the contralateral. This sprouting probably occurred at the decussation, where the intact fibres would be closely admixed with the degenerating fibres, the sprouts then following the degenerated fibres to the VL thalamus.

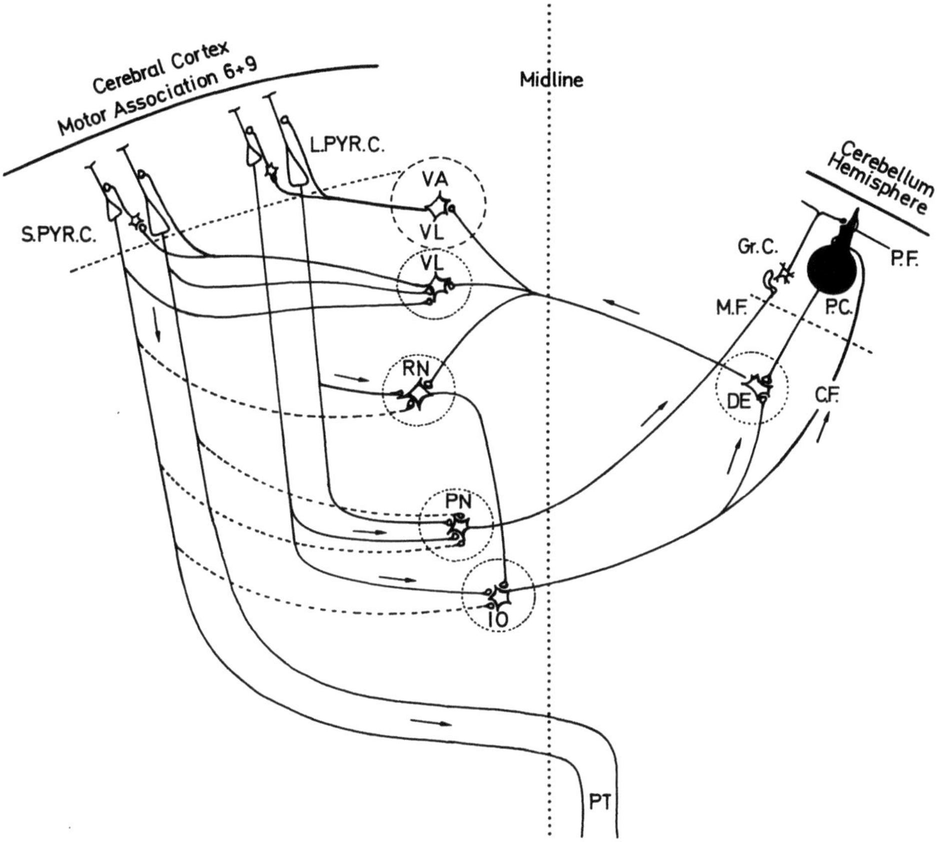

Fig. 2. Line drawings of the cerebrocerebellar pathways. *DE*, nucleus dentatus with axon crossing the midline in the complete decussation of the brachium conjunctivum on the way to the VA-VL thalamic nuclei. Pyramidal cells (*PYR.C.*) are shown in both the motor and association cortex together with the thalamic projections thereto. The Purkinje cell (*P.C.*) is shown with its axon inhibiting a DE cell. *S.PYR.C*, small pyramidal cell; *L.PYR.C*, large pyramidal cell; *RN*, red nucleus; *PN*, pontine nuclei; *IO*, inferior olive; *PT*, pyramidal tract; *P.F.*, parallel fibres; *M.F.*, mossy fibres; *Gr.C.*, granule cell; *C. F.*, climbing fibers

More recently, Kawaguchi et al. [5, 8], have reported in kittens an extraordinary regeneration following complete transection of the decussation of the brachium conjunctivum, which is the pathway from the cerebellar nuclei to the thalamus, the red nucleus (Fig. 2), and also the inferior olive and the pontine tegmental nucleus of the contralateral side. Extreme precautions were taken in the initial transection to ensure that it was complete, and this was verified by histology postmortem.

The cutting device (Fig. 3 A) was a structure of tungsten wire 200 μm in diameter, two parallel longitudinal sections of over 5 cm joined by a razor-sharp transverse of almost 1 cm. Under deep anaesthesia the device was aseptically inserted by a manipulator with carefully designed stereotaxis (Fig. 3 B) so as to sever

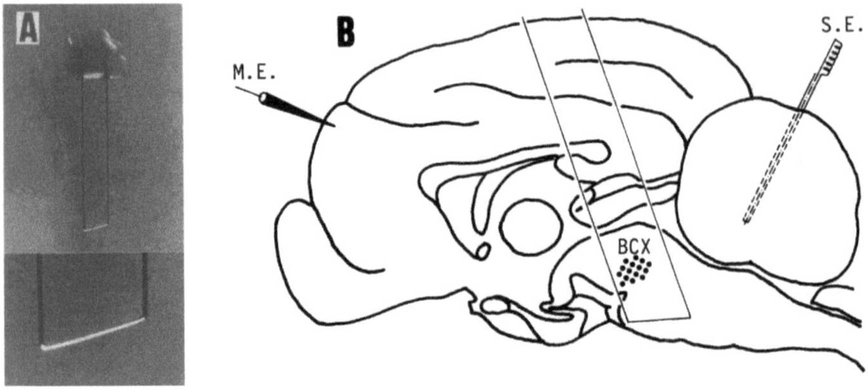

Fig. 3. A Cutting device consisting of a trilateral tungsten wire (200 μm in diameter) carrying a sharpened bottom side, the cross-bar enlarged below. B Experimental design. The decussation of the brachium conjunctivum was transected completely at the midline by the cutting device, then the animals were reared with the cutting device in situ. *M.E.*, microelectrode; *S.E.*, stimulating electrode; *BCX*, decussation of brachium conjunctivum

completeley the decussation of the brachium conjunctivum and to reach the base of the brain stem. The lateral wires were then cut above the dura mater so that the cutting device was left in situ for the whole duration of the experiment. In the final stage before histological examination it was removed from the excised brain from the ventral side, only the minute holes of the lateral wires being thereafter recognizable (Fig. 4 B, inset).

In 8 of these 82 transected kittens (ages 0 day to 3 months) there was regeneration across the lesion, the regenerating fibres mostly reinnervating the normal projection areas (Fig. 4 B).

By terminating the experiment at varying times after the section, the stages of the regenerative process were observed by the anterograde HRP technique. A few hours after the section there was swelling of the transected fibres that by 14 h had developed into growth cones very deeply stained blue with dark field illumination. The regenerating fibres were followed across the section, being intensely stained by HRP, and travelled along the correct path to their normal destination at 1–2 mm/day. This regeneration produced a dense mass of terminals in the thalamus by 19 days. The only serious deviation from normality was in the ipsilateral course of many fibres, resembling that after hemicerebellectomy. These fibres can be seen deviating at the thin arrow in Fig. 4 B. This ipsilateral projection was, however, to the correct locations and so is called paranormal by Kawaguchi et al. [8].

The regeneration from the cerebellar nuclei (interpositus and lateral) to the contralateral VA-VL thalamus becomes of even greater interest by the discovery that effective transmission from the cerebellum to the cerebral cortex is thereby restored. In Fig. 5 A there is the normal depth profile of the responses of the motor cortex to stimulation of the interpositus nucleus on the other side. Transection of the decussation of the brachium conjunctivum completeley abolishes this response (Fig. 5 B). However, in the kitten with the excellent regeneration (Fig. 4 B),

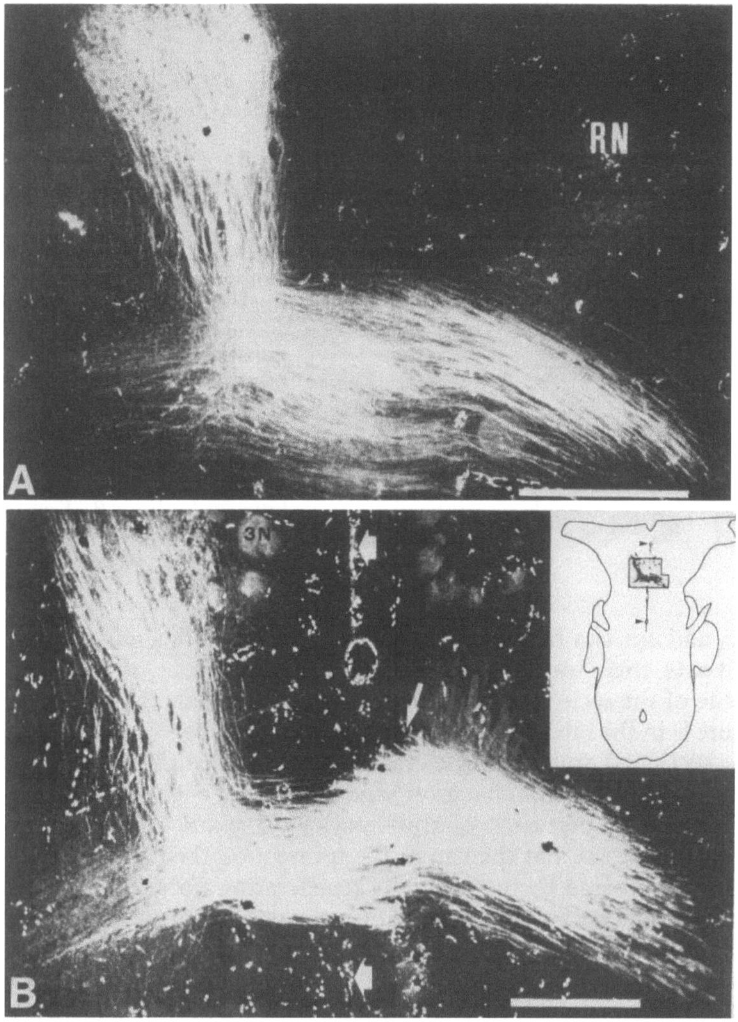

Fig. 4. A Dark field photomicrographs of the decussation of the superior cerebellar peduncle in a horizontal section of a 6-day-old kitten labelled with HRP injection into the cerebellar nuclei on the right side. **B** As in **A**, but showing regenerated fibres after complete transection of the decussation of the brachium conjunctivum in a kitten 6 days old and prepared 19 days later. Glia scars in the lesion rostral and caudal to the area of fibre crossing are shown by *thick arrows*. The *thin arrow* indicates some fibres deviating to take an ipsilateral course. In the *inset drawing* the two *arrows* show the two holes left by ventral extraction of the vertical arms of the cutting device before the histological preparation. The scale bar is 500 μm for **A** and **B**. (Kawaguchi et al. [8])

the depth profile (Fig. 5 C) showed that there had been a remarkable recovery to about half the response in the contralateral normal profile (Fig. 5 A). There was a successful functional recovery in all kittens displaying good histological regeneration. The great outstanding question is, why does effective regeneration occur in only about 10% of the transected kittens? In the large kitten series, success is

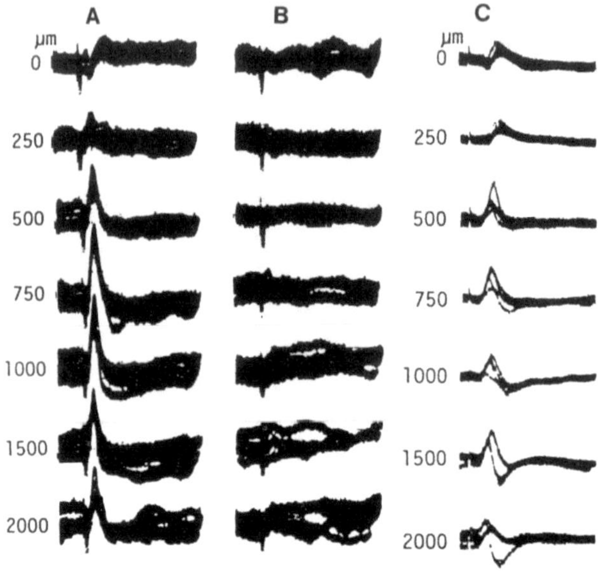

Fig. 5 A–C. Depth profiles of motor cortex in response to a single stimulus to the contralateral IP nucleus. **A** Normal kitten. **B** Immediately after transection of the decussation of the brachium conjunctivum. **C** In a kitten transected at 6 days and recorded 19 days later, being the same kitten as for Fig. 4 B. (Kawaguchi et al. [8])

not biased to age, and age can be excluded because, in a series of similar experiments on 19 adult cats, there was good regeneration in three [6].

Study of the site of the section in both kittens and cats reveals that the probable cause of failure is in the intense gliosis. Even in Fig. 4 B gliosis (thick arrows) can be seen in the sectioned area, the regeneration passing through a window. It is suggested that gliosis results from haemorrhage at the time of the section. Kawaguchi et al. [7] used local perfusion of arabinosylcytosine and 5-fluorodeoxyuridine to suppress the gliosis, but the gap at the transection then remained unfilled and could not be crossed by the normally regenerating fibres. Mechanical fibre guidance seems to be necessary; presumably for attachment of the NCAM or NgCAM of the growing fibres [2]. Once the gap is crossed it seems that the degenerating fibres provide the guidance for the regenerating fibres, just as with the bands of Büngner in peripheral nerve regeneration. This will account for the remarkable specificity of the regeneration to the "correct" sites.

Aguayo et al. [1] have demonstrated that nerve sprouts from neurones of the brain stem and spinal cord of adult rats can grow for several centimeters along an implanted peripheral nerve. The nerve degenerated, but the Schwann cells presumably formed the guide line for fibres that grew for over 30 mm. Unfortunately it was not possible in this way to bridge a transected spinal cord, because the fibres growing well along the nerve bridge were unable to re-enter the central nervous system for more than 2 mm of Schwann cell ensheathment and so could not establish functional connections.

From the experimental work of Tsukahara, Kawaguchi, and also of other investigators it can be concluded that regeneration of the mammalian CNS is possible even at the highest levels. Severed nerve fibres develop "regenerating" axonal sprouts with growth cones, which, if not blocked by gliosis, can travel quite long distances establishing synaptic connections, just as was done in the original neurogenesis. It is most encouraging that once the gap of the section is bridged the regenerating nerve fibres grow normally and selectively for millimeters, apparently following the degenerating fibre path. The block by gliosis and the necessity for guidance across the gap seem to be the principal impediments to effective regeneration in the CNS.

References

1. Aguayo AJ, Benfey M, David S (1983) A potential for axonal regeneration in neurons of the adult mammalian nervous system. In: Haber B, Perez-Polo JR, Hashim GA, Giuffrida Stella AM (eds) Nervous system regeneration. Alan R. Liss, New York, pp 327–340
2. Edelman GM (1984) Modulation of cell adhesion during induction, histogenesis and prenatal development of the nervous system. Ann Rev Neurosci 7:339–377
3. Kawaguchi S, Yamamoto T, Samejima A, Itoh K, Mizuno N (1979a) Morphological evidence for axonal sprouting of cerebello-thalamic neurons in kittens after neonatal hemicerebellectomy. Exp Brain Res 35:511–518
4. Kawaguchi S, Yamamoto T, Samejima A (1979b) Electrophysiological evidence for axonal sprouting of cerebellothalamic neurons in kittens after neonatal hemicerebellectomy. Exp Brain Res 36:21–39
5. Kawaguchi S, Miyata H, Kawamura M, Harada Y (1981) Morphological and electrophysiological evidence for axonal regeneration of axotomized cerebellothalamic neurons in kittens. Neurosci Lett 25:13–18
6. Kawaguchi S, Miyata H, Kato N (1982) Axonal regeneration of axotomized cerebellothalamic projection neurons in adult cats. J Physiol Soc Japan 44:383
7. Kawaguchi S, Miyata H, Kato N (1984) Mechanical guidance for axonal regeneration of cerebellothalamic neurons in cat. Neurosci Lett [Suppl 17] 520
8. Kawaguchi S, Miyata H, Kato N (1986) Regeneration of the cerebellofugal projection after transection of the superior cerebellar peduncle in kittens: morphological and electrophysiological studies. J Comp Neurol 245:258–273
9. Murakami F, Katsumaru H, Saito K, Tsukahara N (1982) A quantitative study of synaptic reorganization in red nucleus neurons after lesions of the nucleus interpositus of the cat: an electron microscope study involving intracellular injection of horseradish peroxidase. Brain Res 242:41–53
10. Tsukahara N (1981) Synaptic plasticity in the mammalian central nervous system. Ann Rev Neurosci 4:351–379
11. Tsukahara N, Hultborn H, Murakami F (1974) Sprouting of corticorubral synapses in red nucleus neurones after destruction of the nucleus interpositus of the cerebellum. Experientia 30:57–58
12. Tsukahara N, Fujito Y, Kubota M (1983) Specificity of the newly formed corticorubral synapses in the kitten red nucleus. Exp Brain Res 51:45–56

What can Microneurography Tell the Clinician About Nerve Regeneration or Disease?

R. MACKEL [1]

Introduction

The method of recording sensory nerve action potentials through the human skin was introduced by Dawson [3], and clinical use was soon made by Gilliatt and Sears [4]. The recording conditions were later improved by Buchthal and Rosenfalk [2], who inserted needle electrodes through the skin and used averaging techniques to improve the signal to noise ratio. Whether surface or near-nerve recording is used, the recorded signal is a compound action potential which cannot distinguish between its elementary components. Single unit analysis becomes mandatory to identify the latter. This type of analysis became possible with the introduction of the technique of percutaneous microneurography (MNG) [11]. For routine clinical application, recording the compound action potential (CAP) remains the method of choice, because it is less time consuming and provides basic information on nerve conduction in health and disease. Single unit analysis, used as a basic research tool, does, however, provide the detailed information that allows interpretation of the CAP and becomes crucial in the study of impulse generation and transmission.

The Compound Action Potential (CAP)

Electrical stimulation of a peripheral nerve in vitro [6] reveals three components, commonly referred to as A-alpha, A-delta, and C-fiber components. They reflect the conducting properties and caliber of differently sized axons and have been correlated with different types of sensation: touch-pressure, temperature, and pain. This is schematically illustrated in Fig. 1. All three components cannot, however, be detected with routine, in vivo, clinical procedures: surface recording typically displays only the first and largest A-alpha peak, while near-nerve or intrafascicular recording shows an additional A-delta peak when averaging techniques are used (see insets in Fig. 1).

Although conduction velocity and waveform measurements (amplitude and area) can be made on CAPs, their interpretations are limited. For example, accurate conduction velocity measurements can only be made for the earliest, fastest-conducting component of the CAP, without specifying the type of afferents

[1] Rockefeller University, 1230 York Avenue, and Department of Neurology, The New York Hospital-Cornell University Medical Center, 525 East 68th Street, New York, NY 10021, USA.

Clinical Aspects of Sensory Motor Integration
Ed. by A. Struppler and A. Weindl
© Springer-Verlag Berlin Heidelberg 1987

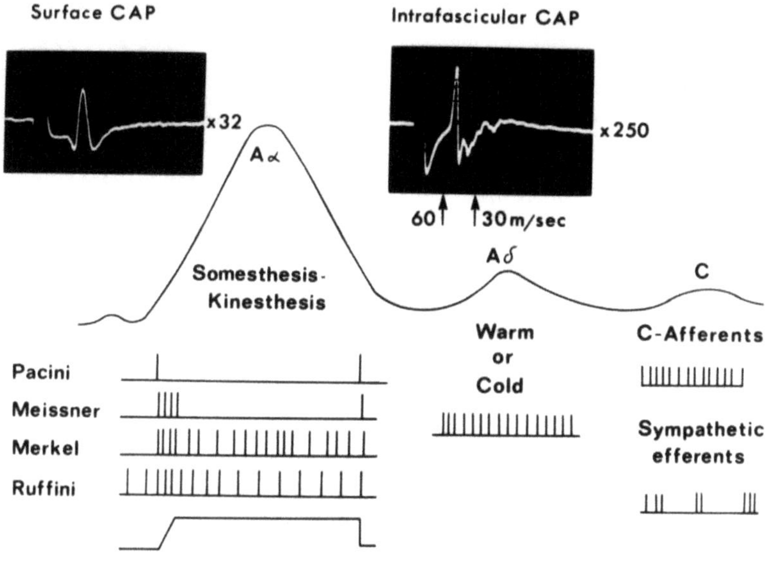

Fig. 1. Schematic illustration of the 3 components of the CAP and their sensory correlates. Underlying types of afferents with target structures (*A-α*) are indicated *below*. The *insets above* are surface (32 averages) and intrafascicular (250 averages) recordings from the median nerve at the wrist in response to electrical stimulation of digit II. Note that an A-δ component (30 m/s) can be detected with intrafascicular recording

(slowly adapting, rapidly adapting), their type of end organs (Ruffini endings, Pacinian corpuscles, Meissner end organs, or Merkel cells), and their destination in the periphery (skin, joints, muscles, ligaments, etc.), as shown in Fig. 1. Amplitude measurements are unreliable as an index of fibers in the nerve, firstly, because fluctuations in amplitude occur normally [2], and secondly, because amplitude measurements must necessarily be made from triphasic potentials (see insets in Fig. 1) rather than more accurately from the monophasic potentials that are recorded when the nerve is crushed between the recording electrodes (as done in in-vitro studies [6]). The CAP recorded from diseased or injured nerves typically displays alterations in shape and conduction when compared to normal. These distortions are reflected in the morphometry of the nerve and are typically associated with clinical impairments [1, 6]. As in the normal CAP, the type of injured or diseased fibers remains unidentified and their physiological properties undetected. In addition, distinction cannot be made between axons with or without peripheral connections or with peripheral connections but abnormal function, nor between degenerating and regenerating axons. Only single unit analysis can provide answers to these questions. A more detailed discussion on the physiological interpretation of the CAP and single unit analysis is found elsewhere [7].

Single Unit Analysis

Microneurographic analysis has provided a wealth of information on the physiology of normal afferents from muscle, skin, and sympathetic efferents (for review cf. [12]), but surprisingly few pathophysiological studies have been forwarded (for review cf. [5]). In the following a series of recent single unit studies [8–10] will be reviewed and their implications discussed in the light of their clinical relevance. These studies were performed on regenerated mechanosensitive afferents reinnervating the glabrous skin of the hand, following transsection with subsequent repair of the median or ulnar nerves. The aim of these studies was to see whether abnormal sensations, which typically occur following peripheral nerve injury, could result from abnormalities in the encoding properties of the afferents and alterations in the reinnervation pattern of their target structures.

It was found that it takes at least 6 months before functional connections between regenerated cutaneous afferents and peripheral end organs become established. But not all types of mechanoreceptors, known to exist in the normally innervated skin, become functionally reinnervated. Only three of the four types of mechanoreceptive units identified in the normal skin were encountered: RA (rapidly adapting, connecting to Meissner end organs), SAI (slowly adapting type I, connecting to Merkel cells), and SAII (slowly adapting type II, connecting to Ruffini endings). No Pacinian units were identified. However, an unusually large number of deep units were found, which are considered to innervate deeper tissues of the hand. These deep units respond only to strong mechanical stimulation and are unlikely to contribute to the return of tactile acuity. Deep units may be free nerve endings trapped in connective tissue or end organs which are morphologically connected but function abnormally. The low-threshold RA- and SAI-type afferents were most frequently found to reinnervate targets in the proximal fingers and palm. This is quite different from normal, where their concentration density is highest distally [12], in accordance with the good resolution capacity of the finger tips. This is illustrated in Fig. 2.

The clinical recovery of sensation was found to be associated with the type of regenerated afferents and their projections in the hand. The better the recovery, the more RA and SAI units were identified distally, as in normal innervation. The poorer the recovery, the more high-threshold deep units were encountered and all types of units were concentrated in the palm.

Some types of afferents do also display abnormalities in their discharge behavior to mechanical skin indentation and alterations in their receptive field configurations. These abnormalities are most prominent during early regeneration, when sensation is also poorest. The abnormalities are transitional, as they subside with the ongoing process of axonal maturation and clinical recovery. It takes, however, at least 2 years until the transitional properties have disappeared. The most distinct abnormalities were decreased responsiveness of SAI and deep units to repetitive mechanical stimulation (fatigue), low discharge rates of SAI units to sustained skin indentation, and weaker response of RA and SAI units to the velocity of skin indentations. SAII units behaved as normal in all respect. This is illustrated in Fig. 3.

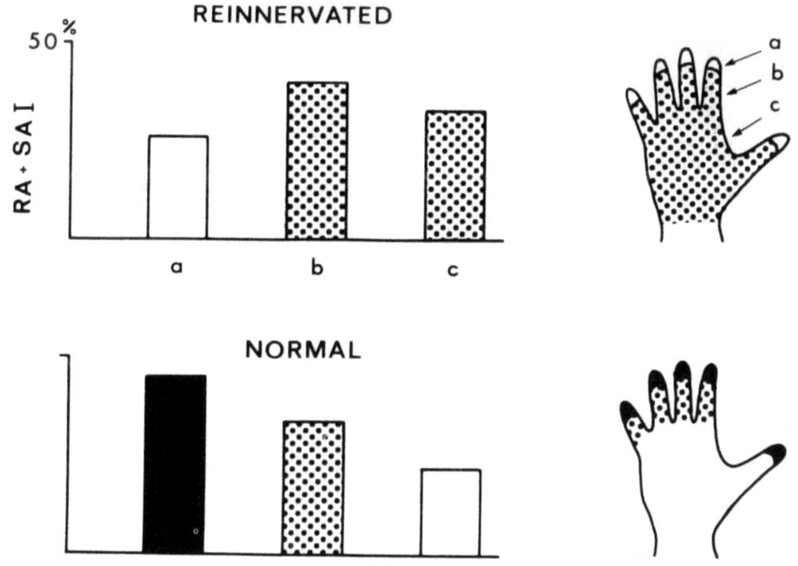

Fig. 2. Frequency of occurence of reinnervated and normally innervated RA and SAI units in the finger tips (*a*), proximal finger (*b*), and palm (*c*). *Lighter shadings* are areas of lower innervation density

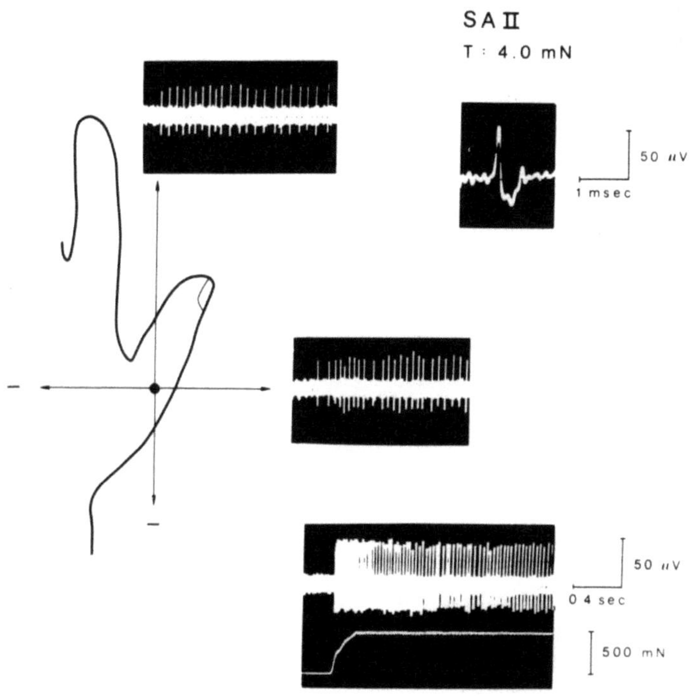

Fig. 3. Response behavior of a regenerated SAII unit to skin indentation (*bottom record*) and lateral skin stretch. One action potential (*upper right*) is generated at von Frey threshold (4.0 mN)

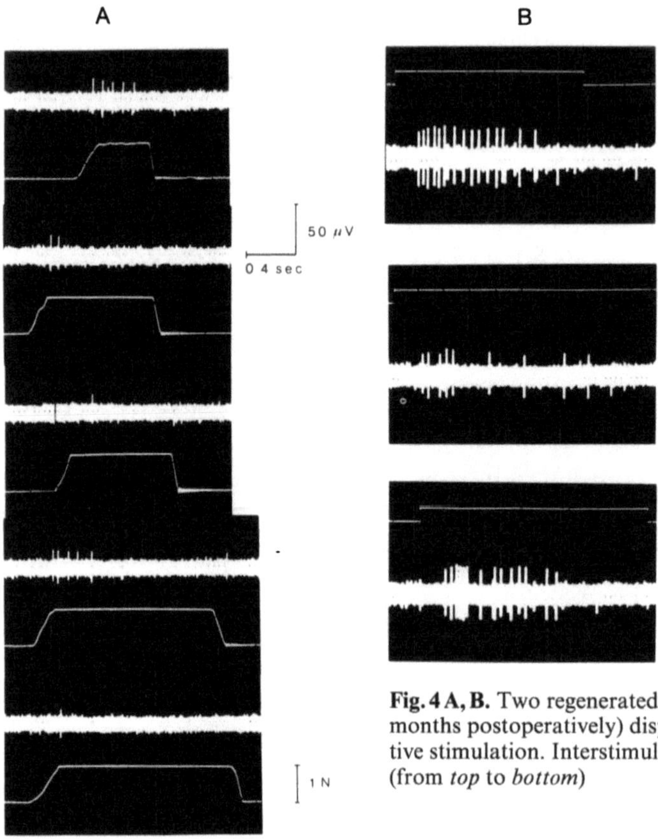

Fig. 4 A, B. Two regenerated deep units (**A, B** 12 months postoperatively) displaying fatigue to repetitive stimulation. Interstimulus interval was 3–4 s (from *top* to *bottom*)

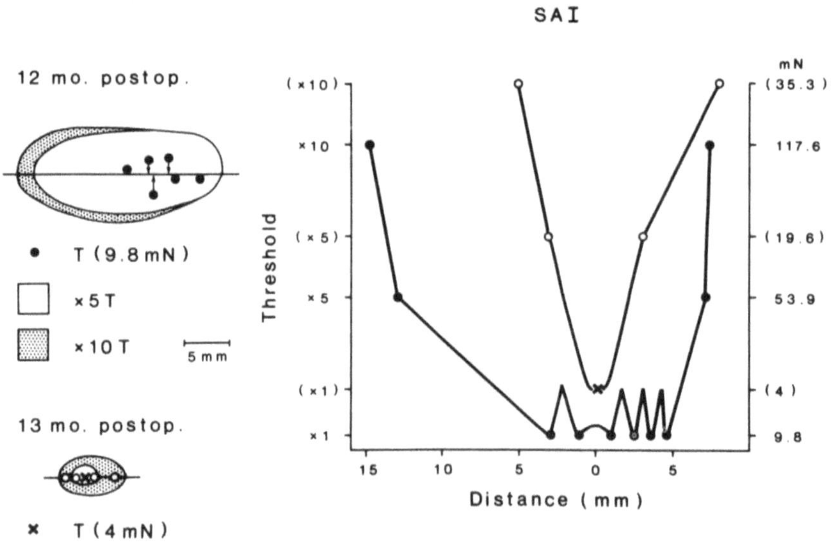

Fig. 5. Receptive field boundaries (*left*) and sensitivity profiles (*right*) of two reinnervated SAI units plotted in two patients 12 and 13 months postoperatively. The receptive fields were mapped at 5 and 10 times threshold (× 5, × 10). Absolute thresholds (*T*) are as indicated. Sensitivity profiles were plotted from left to right. Moving outward from the sensitive areas, the boundaries were determined where successively higher stimuli (× 5, × 10) were necessary to evoke a discharge in the afferent. These amplitudes of threshold are shown (*left ordinate*) with their absolute values in mN (*right ordinate*)

An example of two deep units displaying fatiguability to repeated stimulation is illustrated in Fig. 4. One unit stops responding, while the other afferent displays a decreased discharge when repetitively activated.

A second prominent abnormality was seen in the receptive field configuration of RA and SAI units. Firstly, in contrast to normal, individual RA and SAI afferents were found to reinnervate more than one receptive field (multiple receptive field innervation) [9]. Secondly, field sizes showed more variability, with more abnormally small fields encountered in RA units and both pin-point and abnormally large fields encountered in SAI units. The receptive fields and the sensitivity profiles of two regenerated SAI afferents are plotted in Fig. 5. The large receptive field is abnormal, the other within normal range. Both receptive fields have distinct boundaries and they were mapped at 5 and 10 times von Frey threshold. The receptive field sizes are reflected in the sensitivity profiles, which reflect the afferents' increasing response threshold to stimulation with increasing distance from the most sensitive points in the receptive field.

Conclusions

What have the findings from single unit analysis revealed that the CAP cannot tell us, and what provides at the same time information that is of use to the clinician? Firstly, it was found that it takes at least 6 months for regenerating axons to grow past the repair site at the wrist into the periphery and establish connections with target structures. This time span corresponds to the earliest reported from CAP studies (1). Compound action potential studies cannot, however, determine whether regenerated axons have established functional connections: This requires natural stimulation of the receptive end structures. The early sensory potentials evoked by electrical stimulation were probably due to stimulation of still outgrowing axons or axons in search of target structures. The absolute response threshold of identified cutaneous mechanosensitive afferents to von Frey stimulation were comparable to normal, once they became reconnected. This indicates that the commonly held belief that reduced sensitivity following nerve injury results from increased thresholds of mechanosensitive afferents has to be reconsidered. Single unit analysis has also shown that it takes at least 2 years until all connections have matured, most abnormalities in the response behavior of the afferents have disappeared, and the axons have regained their original caliber. This indicates that no prognosis can be made on the outcome of sensory recovery within 2 years postoperatively. Maximal recovery of the CAP was found to occur around 18 months postoperatively [1]. This is in good agreement with unitary analysis. Correlation between CAP measurements and tactile recovery [1] should, however, be treated cautiously. The size of the CAP which is typically used to make inferences on the number of underlying conducting axons and the functional state of clinical recovery can be misleading because the sensory CAP cannot differentiate between the types of afferents: single unit analysis has clearly shown that high-threshold deep units which contribute little to tactile sensibility are over-represented in regenerated peripheral nerves and their frequency of occurrence is typically associated with poor clinical recovery [8]. Single unit analysis has also clearly shown that functional recovery of the axons of regenerated high-threshold deep units is not distinguishable from low-threshold afferents [10]. As the CAP cannot distinguish between deep and cutaneous afferents, correlations between recovered conduction velocity and amplitude of the sensory potential and the degree of functional recovery could be based on the wrong type of afferents, which in turn can lead to false assumptions.

Unitary analysis has further shown that not all types of afferents known to exist in the normal glabrous skin become reinnervated and that the reinnervation density in the hand is quite different from normal. These findings, together with those which revealed abnormal impulse generation in regenerated afferents and distortion of the receptive field configuration, remain hidden to the CAP. The process of regeneration is certainly not uniform between those afferents which become reconnected with target structures, as some afferents display normal response behavior once reconnected, while others behave abnormally for a long time before resuming normal function.

In conclusion, single unit analysis of regenerated afferents gave access to a new type of peripheral nerve pathophysiology which is unavailable to routinely used electrodiagnostic procedures. It allowed reinterpretation of the CAP and clinically occurring phenomena because it provided new insights into the functional state of injured axons.

Summary

The advantages and disadvantages of unitary analysis of identified primary afferents are discussed in the light of routine electrodiagnostic procedures. Natural stimulation of the receptive fields of mechanosensitive cutaneous afferents has provided information which is inaccessible to the sensory compound action potential, elicited in response to electrical nerve stimulation. It was found that the response properties and receptive field configurations of particular types of regenerated cutaneous afferents undergo transitional changes which can explain the slow course of clinical recovery. These findings gave insight into a new type of peripheral nerve pathophysiology and led to a reinterpretation of clinical phenomena. Although unitary analysis can only be used on a research basis, it does provide the detailed information that allows interpretation of the sensory compound potential and thus becomes crucial in the study of the functional state of injured axons.

Acknowledgments. Part of this work was conducted while the author was recipient of a Teacher-Scientist Award from the Andrew W. Mellon Foundation.

References

1. Buchthal F, Kuhl V (1979) Nerve conduction, tactile sensibility, and the electromyogram after suture or compression of peripheral nerve: a longitudinal study in man. J Neurol Neurosurg Psychiatry 42:436–451
2. Buchthal F, Rosenfalk A (1966) Evoked action potentials and conduction velocity in human sensory nerves. Brain Res 3:1–122
3. Dawson GD (1956) The relative excitability and conduction velocity of sensory and motor nerve fibers in man. J Physiol 131:436–451
4. Gilliatt RW, Sears TA (1958) Sensory nerve action potentials in patients with peripheral nerve lesions. J Neurol Neurosurg Psychiat 21:109–118
5. Hagbarth K-E, Törebjörk HE, Wallin BG (1984) Microelectrode recordings from human skin and muscle nerves. In: Dyck PJ, Thomas PK, Lambert EH, Bunge R (eds) Peripheral neuropathy, 2nd edn. WB Saunders, Philadelphia, p 1016
6. Lambert EH, Dyck PJ (1984) Compound action potentials of sural nerve in vitro in peripheral neuropathy. In: Dyck PJ, Thomas PK, Lambert EH, Bunge R (eds) Peripheral neuropathy, 2nd edn. WB Saunders, Philadelphia, p 1030
7. Mackel R (1986) Human cutaneous mechanoreceptors during regeneration: physiology and interpretation. Ann Neurol 18:165–172
8. Mackel R, Kunesch E, Waldhör F, Struppler A (1983 a) Reinnervation of mechanoreceptors in the human glabrous skin following peripheral nerve repair. Brain Res 268:49–65
9. Mackel R, Brink EE, Wittkowsky G (1983 b)Transitional properties of afferents reinnervating mechanoreceptors in the human glabrous skin. Brain Res 276:339–343
10. Mackel R, Brink EE, Wittkowsky G (1985) Properties of cutaneous mechanosensitive afferents during the early stages of regeneration in man. Brain Res 329:49–69
11. Vallbo AB, Hagbarth K-E (1968) Impulses recorded with microelectrodes in human muscle nerves during stimulation of mechanoreceptors and voluntary contractions. Electroencephalogr Clin Neurophysiol 23:393
12. Vallbo AB, Hagbarth K-E, Torebjork HE, Wallin BG (1979) Somatosensory, proprioceptive and sympathetic activity in human peripheral nerve. Physiol Rev 59:919–957

Effects of Dopamine-Rich Grafts on Sensorimotor Impairments in Dopamine-Depleted Rats

S. B. DUNNETT [1] and A. BJÖRKLUND [2]

Introduction

Following the description, in the early 1970s, of reliable techniques for the transplantation of neural tissues to the brains of experimental mammals [10, 34, 41], research effort focused on the use of grafts, primarily at the anatomical level, to study developmental processes and regenerative potential in the CNS [2, 27, 28]. However, once it became apparent that neural grafts can survive and grow in the brain, the question of the functional contribution of such grafts was raised. In particular, could it be possible to reconstruct or replace damaged neurones by transplantation in a manner sufficient to ameliorate impairments or restore lost functions disrupted by neurological lesions? In 1979, two reports appeared [3, 35] which demonstrated that grafts of embryonic dopamine (DA)-rich tissue to the forebrains of rats could reverse motor impairments associated with DA-depleting lesions.

In the six years since then, graft-derived functional recovery has been reported following neural, neuroendocrine and genetic damage in diverse parts of the CNS. However, the original model system, involving DA denervation and replacement, has several distinct experimental advantages and remains the single most widely studied functional model. These advantages include the following:

1. Sensitive histofluorescent, immunohistochemical and biochemical procedures have enabled the detailed mapping of the anatomical distribution of forebrain DA systems in the embryonic as well as the adult brain.
2. Selective lesions can be made of catecholamine, including DA, neurones by central injection of the neurotoxin 6-hydroxydopamine (6-OHDA).
3. The behavioural consequences of such lesions have been well characterised, are dramatic in their effect, and are simple to measure.
4. Drugs are available for the selective pharmacological manipulation of pre- and postsynaptic elements of the DA innervation of the forebrain.
5. The reversal by grafts of impairments induced by central DA denervation may ultimately lead to direct clinical application in neurodegenerative diseases such as parkinsonism. Indeed, the first clinical report has recently appeared of autotransplantation of catecholamine-rich adrenal medulla tissue into the caudate nuclei of two patients with Parkinson's disease [1].

The following chapter will give a brief review of the sensorimotor consequences of central DA denervation in rats and its potential recovery following grafts of DA-rich tissue to the denervated forebrain.

[1] Department of Experimental Psychology, Cambridge CB2 3EB, UK.
[2] Department of Histology, S-223 62 Lund, Sweden.

Clinical Aspects of Sensory Motor Integration
Ed. by A. Struppler and A. Weindl
© Springer-Verlag Berlin Heidelberg 1987

Sensorimotor Deficits in 6-OHDA Lesioned Rats

Bilateral Lesions

Bilateral 6-OHDA lesions of the nigrostriatal DA pathway produce a debilitating behavioural syndrome in rats, including profound aphagia, adipsia and akinesia [39, 48, 52]. Indeed, it is probably the destruction of this ascending fibre system, passing through the lateral hypothalamus (LH) in the medial forebrain bundle, that underlies the classic LH syndrome [32, 33, 48]. Since several studies had suggested that the fundamental deficit following LH lesions may involve a failure of sensory control over motivated behaviours [29, 45], Marshall and colleagues developed a neurological test battery for the assessment of sensorimotor responsiveness of rats, which was then used to show very similar deficits in LH [31] and 6-OHDA [33] lesioned rats. These animals failed to orient to lateralized visual, olfactory and tactile stimuli, and showed deficits in a variety of reflexive placing, grasping and withdrawal responses [33]. Moreover, whereas neurotoxic lesions of the LH (which spare the ascending DA neurones) do produce residual regulatory impairments [23, 51], these animals show normal sensorimotor responsiveness [18].

In the subsequent decade, more detailed analyses of 6-OHDA rats' sensorimotor deficits have extended this original description. The aphagia and adipsia are initially accompanied by an active rejection of food but may subsequently recover through the same series of stages as had been first described for LH rats [44]. The akinesia is accompanied not just by catalepsy, but also by an active defence of posture and bracing to displacement [36]. Surprisingly, when studied over long time periods, the apparent akinesia is seen to involve continual small readjustments of posture such that the rat is in one sense continually active [50].

Unilateral Lesions

A practical disadvantage of the bilateral 6-OHDA lesion as a model system for the study of transplant function is that the profound debility of the rats necessitates their maintenance by tube-feeding several times daily over a period of months. By contrast, unilateral lesions leave one nigrostriatal system intact for the maintenance of essential regulatory function. Nevertheless, rats with unilateral lesions still show sensory neglect, restricted to the contralateral half of the body [12, 26, 30]. Moreover, as first described by Ungerstedt [46], they manifest a postural bias and, when activated, turn in an ipsilateral direction. This bias can be markedly enhanced by injection of the presynaptic DA stimulant amphetamine, and under this drug the rats turn in circles at a rate of 10–20 full rotations per minute over several hours [46]. Conversely, the postsynaptic agonist apomorphine induces similar rates of turning (but of shorter duration) in the contralateral direction, presumbably by activation of supersensitive receptors on the lesioned side [47]. This unilateral syndrome of behavioural deficits provides an easily quantified baseline for the assessment both of lesion completeness and of any functional consequences following transplantation of DA neurones.

Nature of the Deficit

In Marshall's tests of sensorimotor function following LH or 6-OHDA lesions, the rats' impairment did not appear to be primarily of sensory origin (sensitivity to the stimuli was evidenced by, e.g., pupillary constriction) nor primarily of motor origin (the rats would make the same muscle movements in performing automatic sequences, e.g. grooming). Rather, it was concluded that the impairment was one of a failure of sensorimotor integration [31, 32]. An experiment by Turner [45] attempted to resolve this issue in a more systematic way by training rats to make an ipsilateral or contralateral head-turn response to escape or avoid aversive stimulation to the ipsilateral and contralateral hind paws. He found that unilateral lesions only disrupted performance when the experimental contingency was to make a contralateral response to avoid a contralateral stimulus, but not in making contralateral responses to ipsilateral stimuli or vice versa, thereby excluding simple sensory or motor interpretations in favour of a deficit, like Marshall, in sensorimotor integration. However, this study only investigated the effects of unilateral LH or amygdala lesions, and not of nigrostriatal bundle damage per se.

Two studies have addressed this issue with 6-OHDA lesions. Carli et al. [9] employed an appetitive version of the Turner paradigm, and in contrast to his results found that the rats were deficient in making movements in the direction contralateral to a unilateral intrastriatal lesion, irrespective of the side of the eliciting stimulus. However, the contralateral responses were executed with normal speed and accuracy, whereas the time to *initiate* contralateral movement was increased. This suggests an impairment in the decision or selection of the appropriate response, rather than in the execution of the motor sequence itself. In a quite different paradigm, Dunnett and Björklund [11] trained rats to turn in circles for water reinforcement. Unilateral nigrostriatal lesions only disrupted rats trained to turn in the contralateral direction. Since these rats would still make turns of normal appearance and speed in the contralateral direction, but with much reduced frequency in favour of unreinforced ipsilateral turns, a conclusion similar to that of Carli et al. [9] was reached, i.e. that the lesions induced a deficit in the initiation of contralateral movements rather than in the ability to execute the appropriate motor programs.

One key issue arising from these observations is whether the full 6-OHDA syndrome can be fully accounted for in terms of a sensory neglect of internal and external stimuli which motivate, regulate or direct all voluntary behaviour. Thus, it has been suggested that the aphagia and adipsia following nigrostriatal damage reflects an inattention to both internal and external stimuli controlling food and water intake [26, 32], and that rotation reflects a combination of locomotor activation and a bias in attention to the external world on the ipsilateral side [25, 49]. Although there is considerable evidence of topographic and functional heterogeneity within the neostriatum it has not proved possible, using standard lesion techniques, to determine whether separate components of the syndrome can all be characterized as sensorimotor in origin. This issue is further clouded by the relative imprecision of the term "sensorimotor", frequently being defined, by exclusion, as neither primarily sensory nor primarily motor but some integrative function between these extremes.

Grafting DA Neurones in 6-OHDA Lesioned Rats

Techniques

The essential conditions to achieve viable survival of intracerebral grafts of neuronal tissues are the use of embryonic donor tissue of a particular gestational age (days E13–E15 in the case of mesencephalic DA neurones), and implantation to a host site that can provide adequate vascularization for the rapid incorporation of the graft into the host blood and cerebrospinal fluid circulations. Three alternative procedures have been used to achieve this latter condition in the rat neostriatum.

1. Lateral Ventricle Grafts. Freed and coworkers [19, 35] have implanted graft tissue, via a modified lumbar puncture needle, directly into the lateral ventricles of rats. The choroid plexus provides a rich vascular supply to nourish the graft tissue, although the ependymal lining of the ventricle may provide some restriction to the free growth of axonal connections between graft and host brain.

2. Two-Stage Cavity Grafts. Graft tissue placed into a newly made cavity overlying the neostriatum shows poor survival [41], which has been attributed to inadequate vascularization. Björklund and Stenevi [3, 42] therefore introduced a two-stage procedure in which the cortical cavity was made by aspiration 3–6 weeks prior to implantation surgery. In the intervening period, a new, highly vascularized pia forms over the floor and walls of the wound and provides a suitable nutrient bed for the transplanted tissue.

3. Suspension Grafts. Using modified tissue culture procedures, the dissected embryonic donor tissue is incubated in trypsin, washed and dissociated into a dense cell suspension that can be injected directly into host parenchyma [6, 7, 37]. The advantages of the suspension procedure lie in the rapid incorporation of the dispersed cells into the host circulation, with no special provision required as in the other techniques. Moreover, suspension injections can be stereotaxically placed in any site in the host brain, using single or multiple deposits. The main disadvantage lies in the inaccessability of the graft tissue for subsequent experimental manipulation (e.g. implantation of electrodes or injection of tracer dyes), which can be readily achieved under visual control for grafts made by the other two procedures.

Histology

The viability of grafts of DA-rich mesencephalic tissues to the neostriatum has in all cases been initially studied by catecholamine histofluorescence [4, 8, 19]. All three procedures yield good rates of graft survival (85%–95%). The grafts contain clusters of fluorescent DA neurons within a background of non-fluorescent neurones and glia, and occasional rearrangement of the cells is seen in a band with perpendicularly extending dendrites, reminiscent of the organization in the intact substantia nigra [4, 24]. Swirls of fluorescent fibres within the grafts can be seen to cross the graft-host border and ramify into a dense terminal plexus in the denervated host striatum proximal to the graft, which declines in density at greater distances up to a maximum of 3–4 mm [4, 8]. These observations have been confirmed both immunohistochemically [24, 40] and using computer-assisted densitometry [21].

Ultrastructure

Although at the light-microscopic level the graft-derived DA fibre ingrowth appears to provide a reinnervation of the denervated host striatum of normal appearance, it might be that this involves an ingrowth of free nerve endings that do not make direct connections with host neurones. However, it has now been confirmed ultrastructurally that graft-derived tyrosine hydroxylase immunoreactive (THir) terminals do establish normal symmetric synaptic contact with the dendritic shafts and spine necks of striatal medium spiny neurones [22]. However, the innervation pattern is not entirely normal: Freund et al. [22] found that, in addition to the medium spiny innervation, their fibres encapsulated the giant cholinergic cells of the striatum, making many synaptic contacts *en passant*. The cause of this additional, abnormal input is unknown but may reflect an additional compensatory mechanism in restoring the normal balance between cholinergic and DA influences on medium spiny neurons in the situation where graft-derived DA reinnervation is below the level of the intact intrinsic innervation.

Biochemistry

Post-mortem measurements of the neostriatum in our rats with unilateral 6-OHDA lesions have indicated a reduction of DA content on the ipsilateral side by >97%. A single graft can restore DA concentrations to 5%–15% of normal levels in the whole neostriatum [38], and multiple (suspension) grafts can increase this up to 50% or more in individual animals [39]. Moreover, DA synthesis, turnover and receptor sensitivity, which are markedly increased by the compensatory response to the lesions, are all reduced to close-to-normal levels in the vicinity of the graft [20, 38, 39]. Moreover, metabolic activity (assessed by 2-deoxyglucose autoradiography), which is reduced in the lesioned striatum, is restored to normal levels in the graft-reinnervated striatum [38].

Effects of DA-Rich Grafts on Sensorimotor Function

The anatomical and biochemical studies indicate that grafted DA-rich tissues can provide a DA reinnervation of the 6-OHDA lesioned brain that may reconstruct the damaged circuits in a functionally advantageous way.

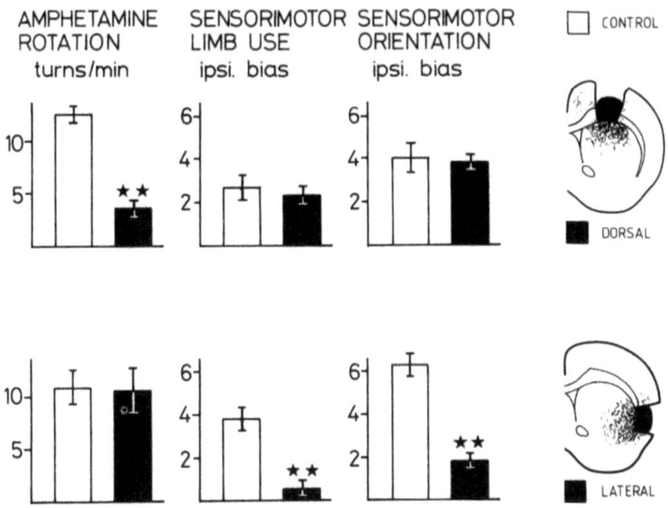

Fig. 1. Functional effects of "solid" DA-tissue grafts. All animals initially received a unilateral 6-OHDA lesion of the nigrostriatal DA pathway. Grafts were placed in either a dorsal or a lateral cortical cavity so as to provide a DA reinnervation of dorsal or lateral segments of the denervated striatum (see sketches on the *right*). Grafted animals were then compared with control rats (with 6-OHDA lesions alone) on three behavioural tests. The dorsal grafts provided a significant recovery of the amphetamine-induced rotation bias, whereas the lateral grafts provided a significant recovery of the ipsilateral sensorimotor bias (on two different measures), but not vice versa. (Redrawn from Dunnett et al. [13, 14])

Solid Grafts

The first studies of DA-rich graft function demonstrated recovery of the motor biases induced in unilateral lesioned rats by apomorphine [13, 19, 35] or amphetamine [3, 4, 13]. However, when the same animals were assessed on a modification of Marshall's sensorimotor battery, they did not differ at all from the rats with lesions alone (see Fig. 1) [4, 13].

One possible interpretation for this failure may be that in spite of providing a new DA innervation of the denervated striatum, some other critical connections to or from the graft fail to become established, in particular since the graft is located far from its "natural" site in the host ventral mesencephalon. Alternatively, the failure may have been attributable to the fact that, from its site in a dorsal parietal cavity, the graft only reinnervates dorsal portions of the neostriatum, whereas lesion studies have indicated that more ventral or lateral striatal sites are critical in normal sensorimotor performance [12].

In support of the second interpretation, Dunnett et al. [14] used lateral parietal cavities to provide a graft location closer to ventrolateral segments of the striatum. In this situation, the grafts provided a substantial recovery in the rats' sensorimotor performance, whereas their rotation biases remained as great as in the lesioned controls (see Fig. 1).

Suspension Grafts

One limitation of the solid graft procedure is that grafts can only be placed in superficial cortical sites or in the ventricles. What might be the effect of graft placement in deep sites in the ventral neostriatum, in the substantia nigra itself, or in the LH where lesions have their greatest effect? Using the suspension technique we have placed single grafts in each of these areas, or used multiple injections to achieve more extensive reinnervation of forebrain targets [16].

Rats with unilateral lesions and grafts into the head of the neostriatum show parallel patterns of recovery to animals with solid grafts (see "Solid Grafts"). Thus, a graft injected in the dorsal neostriatum reduces both amphetamine- and apomorphine-induced rotation, whereas grafts in ventrolateral striatum leave both these measures of motor asymmetry unameliorated. In contrast to the double dissociation obtained with the solid grafts, however, both dorsal and ventrolateral striatal placements significantly ameliorate sensorimotor asymmetry (see Fig. 2), although this effect is more complete for the lateral placements. One rea-

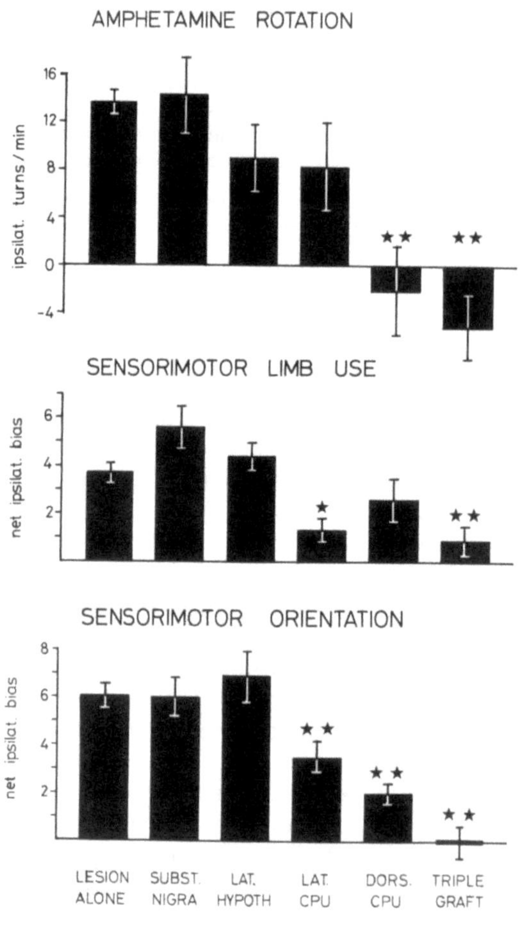

Fig. 2. Functional effects of "suspension" DA-tissue grafts. As in Fig. 1, all animals initially received a unilateral 6-OHDA lesion, followed by injections of DA-rich cell suspensions into the substantia nigra, the lateral hypothalamus, the lateral caudate-putamen (*LAT.CPU*), the dorsal CPU (*DORS.CPU*), or a triple graft involving injections into lateral CPU, dorsal CPU and nucleus accumbens. Control rats remained ungrafted. Rats were tested on the same three tests as shown for solid grafts in Fig. 1. Grafts outside the striatal target areas were without detectable behavioural effect. Grafts in the striatum revealed both a topographic organization of the target and additive effects following multiple placement. (Redrawn from Dunnett et al. [16])

son for this difference between the different graft procedures may lie in the observation that the "dorsal" suspension placement extends deep into the head of the neostriatum, and fibre outgrowth reaches ventral portions of the nucleus to a much greater extent than is obtained with solid grafts in a dorsal parietal cavity. Rats with multiple suspension grafts show a more extensive recovery on all tests, which is equivalent to an additive combination of the effects of each individual graft.

In contrast to striatal graft placements, rats with grafts injected into the host substantia nigra or LH show no functional differences from rats with lesions alone on any test so far investigated (see Fig. 2) [16]. Histologically, the grafts are seen to survive well, but they give rise to very limited outgrowth into host mesencephalon or diencephalon, in contrast to the extensive outgrowth seen from striatal placements. Thus it appears that reinnervation of the appropriate target rather than cell replacement in the region isotopic to lesion damage is the critical determinant of functional recovery in these tests.

Limitations of Recovery

Whereas the rotational and sensorimotor tests have revealed substantial or complete recovery following reconstruction of damaged forebrain DA systems by intracerebral DA-rich grafts, this has not been obtained for all components of the 6-OHDA syndrome. Whereas the locomotor hypokinesias of bilaterally lesioned rats can be ameliorated by ventral striatal grafts, in particular in the nucleus accumbens, the aphagia and adipsia that result from total forebrain DA depletions have remained resistant to any graft procedure so far attempted [4, 15, 17]. No recovery is seen even following placement of 15 injections into discrete forebrain sites that reinstate DA content to levels far higher than that necessary to induce any detectable deficits in partially lesioned animals. This suggests that for some components of the behavioural syndrome it is not simply necessary to restore DA concentrations above some critical level. Moreover, the failure is unlikely to be attributable to absence of reinnervation of some particular terminal area, as most potential sites (including all parts of the neostriatum, nucleus accumbens, amygdala, bed nucleus of the stria terminalis, and neocortex) have been investigated without effect. Rather, these observations suggest that, for certain functions, the nigrostriatal DA projection may transmit specific information rather than simply providing a minimal level of terminal activation for the neostriatum to function normally.

Implications for Striatal Function

The investigation of the conditions and patterns of recovery obtained by intrastriatal DA-rich grafts have several implications for the understanding of normal striatal function. The degree of recovery obtained from ectopic graft placements indicates that for many functional measures striatal DA input functions in a relatively "permissive" manner, providing a gating mechanism which enables an otherwise intact neostriatum to process information normally [6]. A similar conclusion has been reached by Stricker and Zigmond [43] from observations of the spontaneous recovery that can take place following partial damage to the nigrostriatal system. Nevertheless, the failure of recovery in other behavioural measures suggests that this does not provide a full description.

The dissociation between compensation on the motor bias and the sensorimotor tests following grafting to different striatal sites is important on two counts. First, it emphasizes the need to select carefully functional tests appropriate not only to the grafted tissue but also to its precise location. It was serendipity that the rotation tests first adopted to assess graft function were appropriate to the areas of reinnervation achieved in dorsal and medial striatum, as without this stimulation subsequent studies might never have been undertaken. Second, the observation that rotational bias can recover while the sensorimotor deficits remain intact, and vice versa, challenges the hypothesis that rotational asymmetry is a consequence of activation in a rat with unilateral sensorimotor neglect. A similar logic applies to the conclusion that the aphagia and adipsia following bilateral lesions cannot be attributed solely to a neglect of stimuli controlling appropriate feeding and drinking responses.

However, it remains the case that more sensitive measures of sensorimotor function, as described in "Nature of the Deficit", await investigation with the intracerebral transplantation paradigm. These procedures might disentangle some of the factors determining situations in which grafts can, and cannot, provide beneficial consequences for the brain-damaged recipient.

Summary

Lesion of ascending dopamine systems on one side of the rat brain results in a behavioural syndrome that includes sensorimotor impairments on the side of the body contralateral to the side of the lesion, and bilateral lesions produce bilateral impairments. Although the behavioural consequences of such lesions have been well described, the functional nature of the deficits remains unclear. Intracerebral transplantation of dopamine neurones has been used (a) to investigate the possibility of ameliorating the behavioural deficits by reconstruction of the damaged neural circuitry, and (b) to further investigate the functional organization of the ascending dopamine systems. The grafts can indeed restore many, but not all, components of the dopamine-denervation syndrome. The pattern of results suggests that intrinsic dopamine neurones exert a permissive influence within a topographically heterogeneous organization.

References

1. Backlund E-O, Granberg P-O, Hamberger B, Sedvall G, Seiger Å, Olson L (1985) Transplantation of adrenal medullary tissue to striatum in Parkinsonism. In: Bjorklund A, Stenevi U (eds) Neural grafting in the mammalian CNS. Elsevier, Amsterdam, pp 551–556
2. Björklund A, Stenevi U (1979a) Regeneration of monoaminergic and cholinergic neurons in the mammalian central nervous system. Physiol Rev 59:62–100
3. Björklund A, Stenevi U (1979b) Reconstruction of the nigrostriatal dopamine pathway by intracerebral nigral transplants. Brain Res 177:555–560
4. Björklund A, Dunnett SB, Stenevi U, Lewis ME, Iversen SD (1980a) Reinnervation of the denervated striatum by substantia nigra transplants: functional consequences as revealed by pharmacological and sensorimotor testing. Brain Res 199:307–333
5. Björklund A, Schmidt RH, Stenevi U (1980b) Functional reinnervation of the neostriatum in the adult rat by use of intraparenchymal grafting of dissociated cell suspensions from the substantia nigra. Cell Tiss Res 212:39–45
6. Björklund A, Stenevi U, Dunnett SB, Iversen SD (1981) Functional reactivation of the deafferented neostriatum by nigral transplants. Nature 289:497–499
7. Björklund A, Stenevi U, Schmidt RH, Dunnett SB, Gage FH (1983a) Intracerebral grafting of neuronal cell suspensions. I. Introduction and general methods of preparation. Acta Physiol Scand [Suppl] 522:1–7
8. Björklund A, Stenevi U, Schmidt RH, Dunnett SB, Gage FH (1983b) Intracerebral grafting of neuronal cell suspensions. II. Survival and growth of nigral cell suspensions implanted in different brain sites. Acta Physiol Scand [Suppl] 522:9–18
9. Carli M, Evenden JL, Robbins TW (1985) Depletion of unilateral striatal dopamine impairs initiation of contralateral actions and not sensory attention. Nature 313:679–682
10. Das GD, Altman J (1971) Transplanted precursors of nerve cells: their fate in the cerebellums of young rats. Science 173:637–638
11. Dunnett SB, Björklund A (1983) Conditioned turning in rats: dopaminergic involvement in the initiation of movement rather than the movement itself. Neurosci Lett 41:173–178
12. Dunnett SB, Iversen SD (1982) Sensorimotor impairments following localized kainic acid and 6-hydroxydopamine lesions of the neostriatum. Brain Res 248:121–127
13. Dunnett SB, Björklund A, Stenevi U, Iversen SD (1981a) Behavioural recovery following transplantation of substantia nigra in rats subjected to 6-OHDA lesions of the nigrostriatal pathway. I. Unilateral lesions. Brain Res 215:147–161
14. Dunnett SB, Björklund A, Stenevi U, Iversen SD (1981b) Grafts of embryonic substantia nigra reinnervating the ventrolateral striatum ameliorate sensorimotor impairments and akinesia in rats with 6-OHDA lesions of the nigrostriatal pathway. Brain Res 229:209–217
15. Dunnett SB, Björklund A, Stenevi U, Iversen SD (1981c) Behavioural recovery following transplantation of substantia nigra in rats subjected to 6-OHDA lesions of the nigrostriatal pathway. II. Bilateral lesions. Brain Res 229:457–470
16. Dunnett SB, Björklund A, Schmidt RH, Stenevi U, Iversen SD (1983a) Intracerebral grafting of neuronal cell suspensions. IV. Behavioural recovery in rats with unilateral implants of nigral cell suspensions in different forebrain sites. Acta Physiol Scand [Suppl] 522:29–37
17. Dunnett SB, Björklund A, Schmidt RH, Stenevi U, Iversen SD (1983b) Intracerebral grafting of neuronal cell suspensions. V. Behavioural recovery in rats with bilateral 6-OHDA lesions following implantation of nigral cell suspensions. Acta Physiol Scand [Suppl] 522:39–47
18. Dunnett SB, Lane DM, Winn P (1985) Ibotenic acid lesions of the lateral hypothalamus: comparison with 6-hydroxydopamine-induced sensorimotor deficits. Neurosci 14:509–518
19. Freed WJ, Perlow MJ, Karoum F, Seiger Å, Olson L, Hoffer BJ, Wyatt RJ (1980) Restoration of dopaminergic function by grafting of fetal rat substantia nigra to the caudate nucleus: long-term behavioral, biochemical, and histochemical studies. Ann Neurol 8:510–519
20. Freed WJ, Ko GN, Niehoff DL, Kuhar MJ, Hoffer BJ, Olson L, Cannon-Spoor ME, Morihisa JM, Wyatt RJ (1983) Normalization of spiroperidol binding in the denervated rat striatum by homologous grafts of substantia nigra. Science 222:937–939

21. Freed WJ, Hoffer BJ, Olson L, Wyatt RJ (1984) Transplantation of catecholamine containing tissues to restore the functional capacity of the damaged nigrostriatal system. In: Sladek JR, Gash DM (eds) Neural transplants: development and function. Plenum, New York, pp 373–406
22. Freund T, Bolam JP, Björklund A, Stenevi U, Dunnett SB, Powell JF, Smith AD (1985) Efferent synaptic connections of grafted dopaminergic neurons reinnervating the host neostriatum: a tyrosine hydroxylase immunocytochemical study. J Neurosci 5:603–616
23. Grossman SP, Dacey D, Halaris AE, Collier T, Routtenberg A (1978) Aphagia and adipsia after preferential destruction of nerve cell bodies in hypothalamus. Science 202:537–539
24. Jaeger CB (1985) Cytoarchitectonics of substantia nigra grafts: a light and electron microscopic study of immunocytochemically identified dopaminergic neurons and fibrous astrocytes. J Comp Neurol 231:121–135
25. Kelly PH, Moore KE (1976) Mesolimbic dopaminergic neurones in the rotational model of nigrostriatal function. Nature 263:695–696
26. Ljungberg T, Ungerstedt U (1976) Sensory inattention produced by 6-hydroxydopamine-induced degeneration of ascending dopamine neurons in the brain. Exp Neurol 53:585–600
27. Lund RD, Hauschka SD (1976) Transplanted neural tissue develops nerve connections with host rat brain. Science 193:582–584
28. Lund RD, Harvey AR, Jaeger CB, McLoon SC (1982) Transplantation of embryonic neural tissue to the tectal region of newborn rats. In: Morrison AB, Strick PL (eds) Changing concepts of the nervous system. Academic Press, New York, pp 361–375
29. MacDonnell MF, Flynn JP (1966) Control of sensory fields by stimulation of the hypothalamus. Science 152:1406–1408
30. Marshall JF, Gotthelf T (1979) Sensory inattention in rats with 6-hydroxydopamine-induced lesions of ascending dopaminergic neurons: apomorphine-induced reversal of deficits. Exp Neurol 65:389–411
31. Marshall JF, Teitelbaum P (1974) Further analysis of sensory inattention following lateral hypothalamic damage in rats. J Comp Physiol Psychol 86:375–395
32. Marshall JF, Teitelbaum P (1977) New considerations in the neuropsychology of motivated behaviors. In: Iversen LL, Iversen SD, Snyder SH (eds) Handbook of psychopharmacology, vol 7. Plenum, New York, pp 201–229
33. Marshall JF, Richardson JS, Teitelbaum P (1974) Nigrostriatal bundle damage and the lateral hypothalamic syndrome. J Comp Physiol Psychol 87:808–880
34. Olson L, Seiger Å (1972) Brain tissue transplanted to the anterior chamber of the eye: I, Fluorescent histochemistry of immature catecholamine and 5-hydroxytryptamine neurons innervating the rat iris. Z Zellforsch 195:175–194
35. Perlow MJ, Freed WJ, Hoffer BJ, Seiger Å, Olson L, Wyatt RJ (1979) Brain grafts reduce motor abnormalities produced by destruction of nigrostriatal dopamine system. Science 204:643–647
36. Schallert T, De Ryck M, Whishaw IQ, Ramirez VD, Teitelbaum P (1979) Excessive bracing reactions and their control by atropine and l-dopa in an animal analog of Parkinsonism. Exp Neurol 64:33–43
37. Schmidt RH, Björklund A, Stenevi U (1981) Intracerebral grafting of dissociated CNS tissue suspensions: a new approach for neuronal transplantation to deep brain sites. Brain Res 218:347–356
38. Schmidt RH, Ingvar M, Lindvall O, Stenevi U, Björklund A (1982) Functional activity of substantia nigra grafts reinnervating the striatum: neurotransmitter metabolism and [14C]-2-deoxy-D-glucose autoradiography. J Neurochem 38:737–748
39. Schmidt RH, Björklund A, Stenevi U, Dunnett SB, Gage FH (1983) Intracerebral grafting of neuronal cell suspensions. III. Activity of intrastriatal nigral suspension implants as assessed by measurements of dopamine synthesis and metabolism. Acta Physiol Scand [Suppl] 522:19–28
40. Schultzberg M, Dunnett SB, Björklund A, Stenevi U, Hökfelt T, Dockray GJ, Goldstein M (1984) Dopamine and cholecystokinin immunoreactive neurones in mesencephalic grafts reinnervating the neostriatum: evidence for selective growth regulation. Neuroscience 12:17–32

41. Stenevi U, Björklund A, Svendgaard N-Aa (1976) Transplantation of central and peripheral monoamine neurons to the adult brain: techniques and conditions for survival. Brain Res 114:1–20
42. Stenevi U, Björklund A, Dunnett SB (1980) Functional reinnervation of the denervated neostriatum by nigral transplants. Peptides 1 [Suppl] 1:111–116
43. Stricker EM, Zigmond MJ (1976) Recovery of function following damage to central catecholamine-containing neurons: a neurochemical model for the lateral hypothalamic syndrome. In: Sprague JM, Epstein AN (eds) Progress in physiological psychology. Academic Press, New York, pp 121–189
44. Teitelbaum P, Epstein AN (1962) The lateral hypothalamic syndrome: recovery of feeding and drinking after lateral hypothalamic lesions. Psych Rev 69:74–90
45. Turner BH (1973) Sensorimotor syndrome produced by lesions of the amygdala and lateral hypothalamus. J Comp Physiol Psychol 82:37–47
46. Ungerstedt U (1971 a) Striatal dopamine release after amphetamine or nerve degeneration revealed by rotational behavior. Acta Physiol Scand [Suppl] 367:49–68
47. Ungerstedt U (1971 b) Post-synaptic supersensitivity after 6-hydroxydopamine induced degeneration of the nigro-striatal dopamine system. Acta Physiol Scand [Suppl] 367:69–93
48. Ungerstedt U (1971 c) Adipsia and aphagia after 6-hydroxy-dopamine induced degeneration of the nigro-striatal dopamine system. Acta Physiol Scand [Suppl] 367:95–122
49. Ungerstedt U, Ljungberg T, Ranje C (1977) Dopamine neurotransmission and the control of behavior. In: Cools AR, Lohman AHM, van den Bercken JHL (eds) Psychobiology of the striatum. Elsevier/North-Holland, Amsterdam, pp 85–97
50. Whishaw IQ, Robinson RE, Schallert T, DeRyck M, Ramirez VD (1978) Electrical activity of the hippocampus and neocortex in rats depleted of brain dopamine and norepinephrine: relations to behavior and effects of atropine. Exp Neurol 62:748–767
51. Winn P, Tarbuck A, Dunnett SB (1984) Ibotenic acid lesions of the lateral hypothalamus: comparison with the electrolytic lesion syndrome. Neuroscience 12:225–240
52. Zigmond MJ, Stricker EM (1972) Deficits in feeding behavior after intraventricular injection of 6-hydroxydopamine in rats. Science 177:1211–1213

Subject Index

MIX
Papier aus verantwortungsvollen Quellen
Paper from responsible sources
FSC® C105338

If you have any concerns about our products,
you can contact us on
ProductSafety@springernature.com

In case Publisher is established outside the EU,
the EU authorized representative is:
Springer Nature Customer Service Center GmbH
Europaplatz 3, 69115 Heidelberg, Germany

Printed by Libri Plureos GmbH
in Hamburg, Germany